AF864496

Selected Titles in This Series

624 **P. Deift and K. T-R McLaughlin,** A continuum limit of the Toda lattice, 1998

623 **S. A. Adeleke and Peter M. Neumann,** Relations related to betweenness: Their structure and automorphisms, 1998

622 **Luigi Fontana, Steven G. Krantz, and Marco M. Peloso,** Hodge theory in the Sobolev topology for the de Rham complex, 1998

621 **Gregory L. Cherlin,** The classification of countable homogeneous directed graphs and countable homogeneous n-tournaments, 1998

620 **Victor Guba and Mark Sapir,** Diagram groups, 1997

619 **Kazuyoshi Kiyohara,** Two classes of Riemannian manifolds whose geodesic flows are integrable, 1997

618 **Karl H. Hofmann and Wolfgang A. F. Ruppert,** Lie groups and subsemigroups with surjective exponential function, 1997

617 **Robin Hartshorne,** Families of curves in $\mathbb{P}^3$ and Zeuthen's problem, 1997

616 **Serguei G. Bobkov and Christian Houdré,** Some connections between isoperimetric and Sobolev-type inequalities, 1997

615 **Michael A. Dritschel and Hugo J. Woerdeman,** Model theory and linear extreme points in the numerical radius unit ball, 1997

614 **Richard Warren,** The structure of k-CS-transitive cycle-free partial orders, 1997

613 **D. L. Flannery,** The finite irreducible linear 2-groups of degree 4, 1997

612 **Joan Porti,** Torsion de Reidemeister pour les variétés hyperboliques, 1997

611 **D. Ginzburg, I. Piatetski-Shapiro, and S. Rallis,** L functions for the orthogonal group, 1997

610 **Mark Hovey, John H. Palmieri, and Neil P. Strickland,** Axiomatic stable homotopy theory, 1997

609 **Liviu I. Nicolaescu,** Generalized symplectic geometries and the index of families of elliptic problems, 1997

608 **Christina Q. He and Michel L. Lapidus,** Generalized Minkowski content, spectrum of fractal drums, fractal strings, and the Riemann zeta-functions, 1997

607 **Adele Zucchi,** Operators of class C_0 with spectra in multiply connected regions, 1997

606 **Moshé Flato, Jacques C. H. Simon, and Erik Taflin,** Asymptotic completeness, global existence and the infrared problem for the Maxwell-Dirac equations, 1997

605 **Liangqing Li,** Classification of simple C^*-algebras: Inductive limits of matrix algebras over trees, 1997

604 **Hajnal Andréka, Steven Givant, and István Németi,** Decision problems for equational theories of relation algebras, 1997

603 **Bruce N. Allison, Saeid Azam, Stephen Berman, Yun Gao, and Arturo Pianzola,** Extended affine Lie algebras and their root systems, 1997

602 **Igor Fulman,** Crossed products of von Neumann algebras by equivalence relations and their subalgebras, 1997

601 **Jack E. Graver and Mark E. Watkins,** Locally finite, planar, edge-transitive graphs, 1997

600 **Ambar Sengupta,** Gauge theory on compact surfaces, 1997

599 **Tai-Ping Liu and Yanni Zeng,** Large time behavior of solutions for general quasilinear hyperbolic-parabolic systems of conservation laws, 1997

598 **Valentina Barucci, David E. Dobbs, and Marco Fontana,** Maximality properties in numerical semigroups and applications to one-dimensional analytically irreducible local domains, 1997

597 **Ragnar-Olaf Buchweitz and John J. Millson,** CR-geometry and deformations of isolated singularities, 1997

596 **Paul S. Bourdon and Joel H. Shapiro,** Cyclic phenomena for composition operators, 1997

(*Continued in the back of this publication*)

A Continuum Limit of the Toda Lattice

MEMOIRS
of the
American Mathematical Society

Number 624

A Continuum Limit of the Toda Lattice

P. Deift
K. T-R McLaughlin

January 1998 • Volume 131 • Number 624 (end of volume) • ISSN 0065-9266

American Mathematical Society
Providence, Rhode Island

1991 *Mathematics Subject Classification.*
Primary 58F07; Secondary 58E99.

Library of Congress Cataloging-in-Publication Data

Deift, Percy, 1945–
A continuum limit of the Toda lattice / P. Deift, K. T-R McLaughlin.
p. cm. — (Memoirs of the American Mathematical Society, ISSN 0065-9266 ; no. 624)
"January 1998, volume 131, number 624 (end of volume)."
Includes bibliographical references.
ISBN 0-8218-0691-2
1. Kortweg-de Vries equation. 2. Lattice theory. 3. Mathematical physics. I. McLaughlin, K. T-R (Kenneth T-R), 1969– II. Title. III. Series.
QC20.7.D5D45 1998
530.15′474—dc21 97-31680
CIP

Memoirs of the American Mathematical Society

This journal is devoted entirely to research in pure and applied mathematics.

Subscription information. The 1998 subscription begins with volume 131 and consists of six mailings, each containing one or more numbers. Subscription prices for 1998 are $435 list, $348 institutional member. A late charge of 10% of the subscription price will be imposed on orders received from nonmembers after January 1 of the subscription year. Subscribers outside the United States and India must pay a postage surcharge of $30; subscribers in India must pay a postage surcharge of $43. Expedited delivery to destinations in North America $35; elsewhere $110. Each number may be ordered separately; *please specify number* when ordering an individual number. For prices and titles of recently released numbers, see the New Publications sections of the *Notices of the American Mathematical Society.*

Back number information. For back issues see the *AMS Catalog of Publications.*

Subscriptions and orders should be addressed to the American Mathematical Society, P. O. Box 5904, Boston, MA 02206-5904. *All orders must be accompanied by payment.* Other correspondence should be addressed to Box 6248, Providence, RI 02940-6248.

Copying and reprinting. Individual readers of this publication, and nonprofit libraries acting for them, are permitted to make fair use of the material, such as to copy a chapter for use in teaching or research. Permission is granted to quote brief passages from this publication in reviews, provided the customary acknowledgment of the source is given.

Republication, systematic copying, or multiple reproduction of any material in this publication (including abstracts) is permitted only under license from the American Mathematical Society. Requests for such permission should be addressed to the Assistant to the Publisher, American Mathematical Society, P. O. Box 6248, Providence, Rhode Island 02940-6248. Requests can also be made by e-mail to reprint-permission@ams.org.

Memoirs of the American Mathematical Society is published bimonthly (each volume consisting usually of more than one number) by the American Mathematical Society at 201 Charles Street, Providence, RI 02904-2294. Periodicals postage paid at Providence, RI. Postmaster: Send address changes to Memoirs, American Mathematical Society, P. O. Box 6248, Providence, RI 02940-6248.

© 1998 by the American Mathematical Society. All rights reserved.

This publication is indexed in *Science Citation Index®*, *SciSearch®*, *Research Alert®*, *CompuMath Citation Index®*, *Current Contents®/Physical, Chemical & Earth Sciences.*
Printed in the United States of America.

∞ The paper used in this book is acid-free and falls within the guidelines established to ensure permanence and durability.
Visit the AMS home page at URL: http://www.ams.org/

10 9 8 7 6 5 4 3 2 1 03 02 01 00 99 98

Contents

ABSTRACT. We describe a continuum limit of the Toda lattice ODE system, obtained by taking as initial data for the finite lattice successively finer discretizations of two smooth functions.

Using the integrability of the finite Toda lattice, we adapt the method introduced by Lax and Levermore [**13**] for the study of the zero dispersion limit of the KdV equation, to the case of the Toda lattice.

In this Memoir, we establish rigorous leading asymptotics for the spectral problem associated to the Toda lattice, as the discretization parameter tends to zero. Using this information, the analysis of the continuum limit is then transformed via the Lax - Levermore method into a constrained maximization problem in function space, which we study in detail. By differentiating the Euler - Lagrange equations associated to this maximization problem, we reduce to a Riemann - Hilbert problem in the theory of analytic functions. By constructing solutions to this Riemann - Hilbert problem subject to the constraints of the maximization problem, we produce explicitly the maximizer, and derive detailed information concerning the continuum limit.

We obtain rigorous long time asymptotics for the Toda lattice, for a general class of initial data, which, for example, permit the formation of shocks. Near shock time, we show that the continuum limit is governed by one phase (and zero phase) oscillations for the entire spatial domain. Finally, we note that the methods developed in this Memoir imply, in particular, a new proof and generalization of the results of Tian [**19**] for the zero dispersion limit of the KdV equation.

P. Deift was supported in part by NSF Grant #9203771.
K. T-R McLaughlin was supported in part by AFOSR Grant # F49620-92-J-0272.

Acknowledgments

We would like to thank Nick Ercolani, Brian Hayes, Tom Hou, Thomas Kriecherbauer, Peter Lax, David Levermore, Henry McKean, David McLaughlin, F.R. Tian, and Stephanos Venakides for very helpful discussions.

In addition we would like to thank Tom Hou and Peter Lax for showing us their earlier calculations on related problems.

K. T-R M. would like to thank Percy Deift, for all of his guidance, support, and wisdom.

P. Deift
K. T-R McLaughlin

CHAPTER 1

Introduction

The Toda lattice is a mechanical system with Hamiltonian

$$H = \frac{1}{2}\sum_{k=1}^{n} y_k^2 + \sum_{k=1}^{n-1} e^{x_k - x_{k+1}} \quad . \tag{1}$$

The equations of motion are

$$\frac{d}{dt}x_j = y_j \ , \tag{2}$$

$$\frac{d}{dt}y_j = -\left[e^{x_j - x_{j+1}} - e^{x_{j-1} - x_j}\right] \ , \tag{3}$$

and we consider initial data

$$x_k(t=0) = x_k^0 \ , \tag{4}$$

$$y_k(t=0) = y_k^0 \ , \tag{5}$$

for $j = 1, \ldots, n$.

This system is completely integrable and is solved by the methods of Inverse Spectral Theory [**7**], [**6**], [**14**]. The idea is as follows. Following Flaschka [**7**], one makes the change of variables

$$a_j = -\frac{1}{2}y_j \ ,$$

$$b_j = \frac{1}{2}\exp\left(\frac{x_j - x_{j+1}}{2}\right) \ .$$

Under this change of variables, the Toda equations (2 - 3) become

$$\frac{d}{dt}a_k = 2\left(b_k^2 - b_{k-1}^2\right) \quad k = 1, \ldots, n \ , \tag{6}$$

$$\frac{d}{dt}b_k = b_k\left(a_{k+1} - a_k\right) \quad k = 1, \ldots, n-1 \ , \tag{7}$$

where we define $b_0 = b_n \equiv 0$, and the data becomes,

$$a_k(0) = a_k^0 \quad k = 1, \ldots, n \ , \tag{8}$$

$$b_k(0) = b_k^0 \quad k = 1, \ldots, n-1 \ . \tag{9}$$

One defines, out of the a's and b's, two $n \times n$ matrices,

[0]Received by the editors July 13, 1995

$$L = \begin{pmatrix} a_1 & b_1 & & 0 \\ b_1 & a_2 & b_2 & \\ & b_2 & a_3 & \ddots \\ 0 & & \ddots & \ddots \end{pmatrix}, \quad B = \begin{pmatrix} 0 & b_1 & & 0 \\ -b_1 & 0 & b_2 & \\ & -b_2 & 0 & \ddots \\ 0 & & \ddots & \ddots \end{pmatrix}.$$

Then some algebra shows that the Toda system can be re-written in Lax pair form in terms of these two matrices, $\frac{d}{dt}L = BL - LB$. It follows immediately from this that the eigenvalues of the matrix $L(t)$ are constant in time, and in addition, following Moser [**16**], one sees that the time dependence of the first component $f_j(1,t)$ of the jth eigenvector is simple and can be determined explicitly. As is well known, this is enough information to reconstruct the tri-diagonal matrix $L(t)$, and hence the original quantities $x_k(t)$ and $y_k(t) = \dot{x}_k$. This represents the core of the Inverse Spectral Method: one maps from the matrix $L(0)$ to the spectral data $\left\{\lambda_j^0, f_j(1,0)\right\}$. As $L(t)$ flows forward in time, the scattering data evolves in a simple and explicit way, and so one knows the spectral data of $L(t)$. Then one must reconstruct the matrix $L(t)$ from the spectral data.

The following formula reconstructs the b's from the knowledge of the scattering data of L. Set

$$c_k = \sum_{i=1}^{n} \lambda_i^k f_i(1,t)^2 \,, \tag{10}$$

$$D_k = det \begin{pmatrix} c_0 & c_1 & \cdots & c_k \\ c_1 & c_2 & \cdots & c_{k+1} \\ \vdots & \vdots & \ddots & \vdots \\ c_k & c_{k+1} & \cdots & c_{2k} \end{pmatrix}. \tag{11}$$

Then

$$b_k^2 = \frac{D_k D_{k-2}}{D_{k-1}^2}. \tag{12}$$

A similar formula exists for a_k (see chapter 2). Formulae such as (12) have been discovered and rediscovered by many authors over the years and have been known at least since the 1870's, in particular in the work of Heine (see [**18**]). More recently similar formulae occur in work on matrix models in field theory. Also, Deift, Venakides, and Oba have used a continuum version of (12) to analyze the Toda shock problem [**22**].

Continuum Limit of the Toda Lattice.

In this paper we study the Toda Lattice when the size of the system (i.e. the number of particles) tends to infinity, and with initial data arising from the discretization of two functions, as follows.

Given two smooth functions $a(x)$ and $b(x)$, ($b(0) = b(1) = 0, \quad b(x) > 0 \; for \; 0 < x < 1$), defined on $[0,1]$, consider the following initial conditions for the Toda Lattice in the Flaschka variables (6-7)

$$(13)\qquad a_k = a(k\epsilon)) \quad k = 1, \ldots, [\frac{1}{\epsilon}]\,,$$
$$b_k = b(k\epsilon) \quad k = 1, \ldots, [\frac{1}{\epsilon}] - 1\,.$$

For a fixed t, evolve the system forward to the (large) time $t' = \frac{t}{\epsilon}$, and for $x \in [0,1]$, set

$$(14)\qquad k(\epsilon, x) = \left[\frac{x}{\epsilon}\right]\,,$$

where the bracket denotes the integer part of the argument. Then define, for fixed t, two piecewise constant functions of x, $0 \le x \le 1$, indexed by ϵ,

$$(15)\qquad \hat{a}_\epsilon(x,t) = a_{k(\epsilon,x)}(t') \;=\; a_{k(\epsilon,x)}(\frac{t}{\epsilon})\,,$$

$$(16)\qquad \hat{b}_\epsilon(x,t) = b_{k(\epsilon,x)}(t') \;=\; b_{k(\epsilon,x)}(\frac{t}{\epsilon})\,.$$

The continuum limit is obtained by letting ϵ tend to zero. In this paper we analyze the limit of the quantities in (15-16). As we will see this problem has many elements in common with the zero dispersion limit for the KdV equation

$$\begin{aligned} u_t - 6uu_x + \epsilon^2 u_{xxx} &= 0\,, \\ u(x,0) &= u_0(x)\,, \end{aligned}$$

first analyzed as $\epsilon \downarrow 0$ by Lax-Levermore [**13**], and then by Venakides [**21**]. Formally, from (6-7), the continuum functions $\hat{a}_\epsilon(x,t)$ and $\hat{b}_\epsilon(x,t)$ should satisfy the partial differential equations

$$(17)\qquad \frac{\partial a}{\partial t} = 2\frac{\partial\,(b^2)}{\partial x}\,,$$
$$(18)\qquad \frac{\partial b}{\partial t} = b\frac{\partial a}{\partial x}\,,$$
$$(19)\qquad a(x,0) = a_0(x)\,,$$
$$(20)\qquad b(x,0) = b_0(x)\,,$$
$$(21)\qquad b(x=0,t) = 0 = b(x=1,t)\,.$$

Observe from (17) and (21) that as long as $b(x,t)$ is smooth , $a(x,t)$ is constant in time for $x = 0$ or 1, and so it is not necessary to specify boundary conditions explicitly for $a(x,t)$. The boundary condition (21) is natural in the following sense. If one solves the particle system numerically (see also chapter 3) with small ϵ and $b_0(0) \neq 0$, $b_1(1) \neq 0$, then one finds from (16) that $b(0,t)\left(\approx \hat{b}_\epsilon(0,t)\right)$ and $b(1,t)\left(\approx \hat{b}_\epsilon(1,t)\right)$ converge rapidly to zero for $t > 0$. Thus from all practical points of view, one might as well impose (21) from the start.

We note several things. First, (17-18) is a hyperbolic system, which may exhibit shocks. In fact, if we look for a solution of (17-18) satisfying $a = 2b$, the equations reduce to the familiar Burger's equation, for which the solution may develop shocks. In particular this means that (17 - 18) cannot describe the continuum limit of the Toda system for all t, and for times beyond the shock time we expect a more complicated description of the continuum limit in which rapid oscillations develop in analogy with the results in [**13**] and [**21**]. Second, the problem above is an initial - boundary value problem. This is to be contrasted with previous small dispersion results (in [**13**], [**21**], for example) for which the underlying dispersionless pde is considered on the whole line. The fact that the initial-boundary value problem is the continuum limit of an integrable system offers the rather unique opportunity of being able to analyze exactly the behavior of a hyperbolic system with boundary conditions. Third, although the Toda system is integrable for any finite number of particles, and although the commuting integrals have continuum limits, it is not clear how to use these integrals to describe the behavior of the continuum limit of the lattice and, in particular, how to linearize the flow (in this connection see, for example, the discussion following (28)).

We note finally that the system (17-18) was first considered in [**1**] and [**2**], where many of the basic properties of the equations were obtained. There the authors consider the PDE system itself, and their work is based on analysis of the continuum limits of the integrals of the finite dimensional system, and operates under the assumption that the initial data for the system is such that shocks do not develop. In [**11**], Kodama derives an explicit solution formula for the initial value problem, through a formal semiclassical limit.

This work involves the analysis of the continuum limit of the Toda lattice both before and after breaking. The problem divides naturally in two. First, in order to solve the Toda Lattice at time t using Inverse Spectral Theory, one needs to compute the spectral data for $L(0)$. As described above, the matrix $L(0)$ has entries which arise from the discretization of 2 functions, and its size is determined by the mesh size of the discretization. Thus we are interested in describing the spectral data as the mesh size of the discretization decreases to zero, while the size of the matrix tends to infinity. The WKB theory for the three term recurrence relation is analogous in many ways to the classical problem of describing the eigenfunctions of the Schrödinger operator as the coefficient of the second derivative tends to zero. Nevertheless the discrete problem introduces new analytical difficulties. Typically, discrete sums, $\sum e^{if(k)}$, which are rather subtle to estimate, appear in place of the familiar integrals $\int e^{if(x)}$, which are easy to estimate. We have established leading order asymptotics for the difference equation associated with the eigenvalue problem $L(0)f = \lambda f$. The second problem is the analysis of the continuum limit, using the knowledge of the spectral data, and the reconstruction formula (12). We use a variant of the method introduced by Lax and Levermore ([**13**]) to analyze this limit. The modifications to this method produce explicit results concerning the associated PDE system. These will be described below.

The Spectral Data.

In Appendix I we establish **leading order asymptotics** for the difference equation associated with the eigenvalue problem $L(0)f = \lambda f$,

$$b((j-1)\epsilon)f_{j-1} + a(j\epsilon)f_j + b(j\epsilon)f_{j+1} \;=\; \lambda f_j \;. \tag{22}$$

The asymptotics are valid for $j = 1, \dots, [\frac{1}{\epsilon}]$, as ϵ tends to zero. Asymptotics for this difference equation are obtained, for example, in [**8**], under the assumption that $\left|\frac{\lambda - a_j}{2b_j}\right| \geq \delta > 1$ for all j. The problem is that this assumption is not satisfied if λ is an eigenvalue of the matrix $L(0)$. We have obtained asymptotics in the more general situation, where $\left|\frac{\lambda - a_j}{2b_j}\right| \geq 1$ in some regions, and $\left|\frac{\lambda - a_j}{2b_j}\right| \leq 1$ in other regions. Since the completion of this work we have learned of the work of Costin [**4**] , who considers a more general family of difference equations, and obtains WKB asymptotics valid in regions of exponential decay and oscillation as well. His work provides a general WKB prescription for difference equations of any finite order, and is not restricted to symmetric difference equations.

As in the continuous WKB approximation, these asymptotics show that the solution exhibits exponential behavior (either exponentially growing or decreasing) in regions where $\left|\frac{\lambda - a_j}{2b_j}\right| > 1$, and that in regions where $\left|\frac{\lambda - a_j}{2b_j}\right| < 1$, the solution exhibits oscillatory behavior. A transition between these two regions occurs at a turning pont. The WKB asymptotics are not valid in a vicinity of a turning point, and asymptotics in such a vicinity are obtained from Bessel functions. Indeed, it turns out that the role played by Airy's equation in the neighborhood of a turning point in the continuum WKB theory is now played by a particular three - term recurrence relation for Bessel functions. These ingredients allow one to patch together solutions in the different regions, and construct the eigenvectors. The construction shows, for example, that for the general class of functions satisfying the boundary condition $b(0) = b(1) = 0$, and such that

$$\alpha(x) \;=\; \frac{a(x) - 2b(x)}{2}$$

is a monotone decreasing function of x, and

$$\beta(x) \;=\; \frac{a(x) + 2b(x)}{2}$$

possesses at most one critical point (a maximum), the eigenvectors of L exhibit oscillations in an interior region, surrounded by two regions of exponentially decreasing behavior. They satisfy an asymptotic eigenvalue condition

$$\Phi(\lambda_j) \equiv \int_{x_-(\lambda_j)}^{x_+(\lambda_j)} \cos^{-1}\left(\frac{\lambda_j - a}{2b}\right) dx \;=\; (j\pi + \pi/2)\epsilon \;, \tag{23}$$

and the first component of the eigenvector with this eigenvalue is approximated by

$$
\begin{aligned}
(24) \qquad f_\lambda(1) &= \exp\left[-\frac{1}{\epsilon}\int_0^{x_-(\lambda_j)} \cosh^{-1}\left(\frac{\lambda_j - a}{2b}\right)\, dx\right][1+O(\epsilon)] \\
&= e^{-\frac{1}{\epsilon}\theta(\lambda)}\,[1+O(\epsilon)]\ ,
\end{aligned}
$$

where

$$
(25) \qquad \theta(\lambda) = \int_0^{x_-(\lambda_j)} \cosh^{-1}\left(\frac{\lambda_j - a}{2b}\right)\, dx\ .
$$

These spectral functions play a central role in the calculation of the continuum limit of the Toda lattice.

The fact the functions $a(x)$ and $b(x)$ satisfy the boundary condition $b(0) = b(1) = 0$ implies that the eigenfunctions must exhibit exponential decay as x approaches 0 or 1. Although the initial data for which we evaluate the continuum limit of the Toda lattice satisfies this boundary condition, for completeness we also consider the difference equation for a class of functions which do not satisfy this boundary condition. Specifically, we show that for the case $b \equiv \frac{1}{2}$, and $a(x)$ symmetric about $x = 1/2$ and possessing exactly one minimum at $x = 1/2$, that a new group of eigenvectors exist, which exhibit oscillations in the wings, with exponential behavior in the interior. For this group, we see an **exponential splitting of eigenvalues**. Indeed, these eigenvectors occur in pairs. Two eigenvectors from a pair have eigenvalues which are exponentially close together. For each pair, the two eigenvalues satisfy the same asymptotic condition, which is similar to (23).

As mentioned above, this asymptotic description of the spectral data plays a crucial role in analyzing the continuum limit problem. We also present numerical studies to illustrate the WKB phenomena described above.

The Continuum Limit

From the reconstruction formulae (12) one sees that the crucial quantity to analyze is $D_{k(\epsilon,x)}$. Following [**13**], we obtain in section (2) the following asymptotic formula,

$$
\epsilon^2 \log D_{k(\epsilon,x)} \approx Q(x,t) = max \quad \left[\frac{2}{\pi}(t\lambda - \theta\ ,\psi) + \frac{1}{2\pi^2}(L\psi\ ,\psi)\right]\ .
$$

Here $(,)$ represents the standard real L^2 inner product, L is a convolution operator with kernel $\log(x-y)^2$, and the maximization is taken over the following set:

$$
\left\{\psi \in L^1(-1,1)\ ,\ 0 \le \psi \le -\Phi'\ ,\ \int \psi(\lambda)d\lambda = \pi x\ \right\},
$$

where $\Phi(\lambda)$ is the spectral function defined in (23), and $\theta(\lambda)$ is as defined in (25).

It is easy to show, following [**13**], that a solution to this problem exists, as the maximization is of a convex functional over a convex set. The maximizer will satisfy the associated Euler -Lagrange equation on a subset of the spectral interval, which for simplicity we take to be [-1,1], and will hit one of the constraints on the complement of this set. By differentiating with respect to λ, the Euler - Lagrange equation is turned into a Riemann - Hilbert problem, which takes the form

$$\begin{aligned} H\psi &= \theta'(\lambda) - t \ \ on \ \ I_1 \\ \psi &= 0 \ \ on \ \ I_2 \\ \psi &= -\Phi'(\lambda) \ \ on \ \ I_3 \end{aligned}$$

where H is the Hilbert Transform operator, and for the cases considered here the I_j's are finite unions of disjoint intervals such that $\bar{I_1} \cup \bar{I_2} \cup \bar{I_3} = \mathbf{R}$, the real axis.

Our work differs from the work of Lax and Levermore in the analysis of this Riemann - Hilbert problem. Rather than considering the problems for the differentiated functions ψ_x and ψ_t as in Lax-Levermore [**13**], we express the solution of this Riemann - Hilbert problem directly through standard techniques of complex analysis. For example, in the so called "zero gap" case, when

$$\begin{aligned} I_1 &= (\alpha \ , \ \beta) \ \subset \ [0,1] \ , \\ I_2 &= (-\infty \ , \ \alpha) \ \cup \ (1 \ , \ \infty) \ , \\ I_3 &= (\beta \ , \ 1) \ , \end{aligned}$$

the function ψ can be written as the real part of a function $f(z) \in H^p, 1 \leq p < 2$, where $f(z)$ has the representation

$$\begin{aligned} f(z) &= R(z)^{1/2}\frac{1}{\pi i}\left(\int_\alpha^\beta i\frac{\theta'(\lambda)-t}{R(\lambda)^{1/2}}\frac{d\lambda}{\lambda - z} + \int_\beta^1 \frac{-\Phi'(\lambda)}{R(\lambda)^{1/2}}\frac{d\lambda}{\lambda - z}\right) , \\ R(z) &= [(z-\alpha)(z-\beta)]^{1/2} \ . \end{aligned}$$

Clearly this function is in H^p for some $1 \leq p < 2$ only if the sum of the integrals behaves like $O(1/z^2)$ as z tends to infinity. This gives a nontrivial condition on the pair of functions $\alpha(x,t), \beta(x,t)$. Another condition is obtained by enforcing the constraint $\int \psi = \pi x$. Our conditions can be written in the form

$$\begin{aligned} T(\alpha,\beta) \equiv & \int_\alpha^\beta \frac{1}{[(\lambda-\alpha)(\beta-\lambda)]^{1/2}}\theta' d\lambda \\ & + \int_\beta^1 \frac{1}{[(\lambda-\alpha)(\lambda-\beta)]^{1/2}}(-\Phi')d\lambda = \pi t \end{aligned} \tag{26}$$

$$\begin{aligned} X(\alpha,\beta) \equiv & \int_\alpha^\beta \frac{\lambda - m}{[(\lambda-\alpha)(\beta-\lambda)]^{1/2}}\theta' d\lambda \\ & + \int_\beta^1 \frac{\lambda - m}{[(\lambda-\alpha)(\lambda-\beta)]^{1/2}}(-\Phi')d\lambda = \pi x \ , \end{aligned} \tag{27}$$

where $m = \frac{\alpha+\beta}{2}$. These conditions express a complete linearization of the initial - boundary value problem, and their analysis gives explicit results concerning the continuum limit. This technique generalizes to the case with more than one interval (α, β), and is the basis for our analysis of the continuum limit of the Toda equations. We note also that this technique is similar to that used by Deift, Venakides and Zhou [**5**].

We will consider three general classes of data. For the first class (see chapter 4), we show that the continuum limit of the Toda lattice is governed by a **zero gap ansatz for all** $t > 0$. For the second class of data, considered in chapter (5), we give explicit formulae both for the solution of the problem, as well as for the time of first shock formation. We show that for this class, the continuum limit is governed, even after shock formation, by a zero gap ansatz, indicating an absence of oscillations. The shock forms at the boundary $x = 0$: the flux $2\frac{\partial b^2}{\partial x}\big|_{x=0} = 4b\frac{\partial b}{\partial x}\big|_{x=0}$ is no longer zero, and from (17) we see that $a(0,t)$ is now free to move and its evolution can be computed explicitly. This situation is in contrast to the zero dispersion problem of Lax and Levermore [**13**], where shock formation always takes place together with the generation of oscillations.

Further, in chapters (6-8) we determine another class of data for which a shock forms, and **oscillations develop** in the continuum limit of the Toda lattice. We are able to prove the existence of two points of phase transition, defining a spatial interval in which the continuum limit is no longer governed by the continuous Toda equations. We prove that on this interval, a new gap opens up in the Riemann - Hilbert problem. The endpoints of the intervals defining $\{I_j\}_{j=1}^3$ evolve according to a Whitham averaged type PDE system in four unknowns, and we prove the existence of a solution to this PDE system which connects to the solutions of the continuous Toda PDE system describing the continuum limit on the complement of this interval.

For each class of data we present numerical studies of the Toda system with many particles, to illustrate the phenomena described.

In all three cases considered, we have explicit **long time asymptotics**. The continuum limit is governed, in the long time limit, by a zero gap solution of the Riemann - Hilbert problem, whose endpoints converge as follows,

$$b(x,t) = \frac{\beta(x,t) - \alpha(x,t)}{4} \longrightarrow 0 \quad as \quad t \to \infty$$

$$a(x,t) = \frac{\beta(x,t) + \alpha(x,t)}{2} \longrightarrow \Phi^{-1}(x) \quad as \quad t \to \infty \; .$$

These formulae give rise to the following interpretation for the long time behavior of $a(x,t)$. Just as the finite dimensional Toda lattice converges, as t tends to infinity, to a diagonal matrix whose entries are the eigenvalues of the matrix $L(t = 0)$, arranged in descending order [**16**], we see that the "off diagonal function" $b(x,t)$ converges to zero, while the "diagonal function" $a(x,t)$ converges to the asymptotic spectral measure $\Phi^{-1}(\pi x)$ arising from the difference equation $Lf = \lambda f$. Note that $\Phi^{-1}(\pi x)$ is a descending function of its argument. Some of these results were anticipated in [**2**].

In this work we take the viewpoint that the underlying PDE systems involved, either the continuous Toda equations before a shock, or the Whitham averaged equations after the shock, are determined by the Riemann - Hilbert problem. In previous work on the zero dispersion limit for the KdV equation ([**13**],[**21**]), the Riemann - Hilbert problem is differentiated with respect to x and t, yielding two problems whose solutions are ratios of the form $\frac{p(\lambda)}{R(\lambda)^{1/2}}$, where $p(\lambda)$ and $R(\lambda)$ are polynomials in λ, and $R(\lambda)$ is chosen to have a certain band-gap structure. From compatibility considerations regarding these two solutions, a PDE system is derived which governs the evolution of the endpoints of the bands and gaps. In order to construct the maximizer, i.e. integrate the differentiated Riemann - Hilbert problems, these equations must have appropriate solutions. The central task involved in describing the zero dispersion limit for the KdV equation becomes, in this setting, the problem of proving existence of a solution to these PDE systems. We show that for the continuum limit of the Toda lattice, the conditions produced by the Riemann - Hilbert problem, both (26-28) in the case of the zero gap ansatz, as well as their generalization to the one gap case, are equivalent to these underlying PDE systems (the continuous Toda equations, and the Whitham modulation equations, respectively). Further, we prove that these conditions indeed uniquely determine the endpoints of the bands and gaps, and as a consequence, we prove existence of solutions of the underlying PDE systems (both the continuous Toda equations and the Whitham modulation equations).

We remark that these ideas afford a **new proof** of the results of Tian [**19**], who establishes that under certain technical assumptions, the zero dispersion limit of the KdV equation is governed, beyond the shock time for the associated inviscid Burger's equation

$$u_t - 6uu_x = \lim_{\epsilon \to 0} u_t - 6uu_x + \epsilon^2 u_{xxx} = 0 ,$$

by a one gap Riemann - Hilbert problem. In a wonderful sequence of calculations, Tian proceeds first by "integrating" the Whitham modulation equations, using a hodograph transformation introduced by Tsarev [**20**], and then analyzing this transformation by means of an associated linear equation of Euler-Poisson-Darboux type. By contrast we work directly with the Riemann - Hilbert problem, and all considerations of the modulation equations and their integration are circumvented (see [**15**]).

The outline of this paper is as follows. In Appendix I we present the WKB theory and establish rigorous leading order asymptotics for the difference equation $Lf = \lambda f$. In chapter 2 we start with the reconstruction formulae as they appear in [**18**], derive the maximization problem that governs the continuum limit, and from this obtain the associated Riemann - Hilbert problem. In chapters 3 - 8 we analyze the continuum limit of the Toda lattice for different classes of data. We start in chapter 3 with very simple initial data, $a^0 \equiv 0$, $b^0 \equiv \frac{1}{2}$, for the purpose of illustration. Then in chapter 4 we consider a general class of initial data for which we prove that the continuum limit is governed by a "zero gap" ansatz for all $t > 0$. In chapter 5 we study a class of data for which a shock develops at the boundary, and oscillations *do not* form in the continuum limit of the lattice. In chapters 7 - 9 we determine another class of data for which we prove that a shock forms, and beyond shock time we prove that the continuum limit is described globally by a combination of one and zero gap ansatzes,indicating the existence of single phase oscillations in the continuum limit of the Toda lattice. Chapter 11 establishes rigorous long time asymptotics for all cases considered.

Finally we would like to state that this work is greatly influenced by the work of Hou and Lax [**10**] and Goodman and Lax [**9**] in discrete systems.

CHAPTER 2

Analysis of Log formula

We will follow Szego [**18**] in presenting the formulae for the reconstruction of the matrix elements $\{a_j, b_j\}$ from the spectral data $\{\lambda_j, f_j(1)\}$.

The procedure is to orthogonalize the polynomials $\{\lambda^k\}$, $k = 1, \dots, [\frac{1}{\epsilon}]$ with respect to the spectral measure

$$d\alpha(\lambda) = \sum_{j=1}^{[\frac{1}{\epsilon}]} f_j(1)^2 \delta(\lambda - \lambda_j) . \tag{28}$$

This produces polynomials $p_k(\lambda)$, given by the following explicit formula (see [**18**], page 26):

$$p_k(\lambda) = \frac{(D_{k-1}D_k)^{-\frac{1}{2}}}{k!} \int\int \cdots \int (\lambda - x_0)(\lambda - x_1)\dots(\lambda - x_{k-1}) \cdot \prod_{\nu<\mu} (x_\nu - x_\mu)^2 \prod_{j=0}^{k-1} d\alpha(x_j) , \tag{29}$$

where

$$D_k = \frac{1}{(k+1)!} \int\int \cdots \int \prod_{\nu<\mu} (x_\nu - x_\mu)^2 \prod_{j=0}^{k-1} d\alpha(x_j) ,$$

and using the definition of the spectral measure (28), we obtain

$$D_k = \sum_{j_0<j_1<\cdots<j_k} f_{j_0}(1)^2 \dots f_{j_k}(1)^2 \prod_{l\neq i} |\lambda_{j_i} - \lambda_{j_l}| . \tag{30}$$

The polynomials $p_k(\lambda)$ satisfy the three term recurrence relation

$$\lambda p_k(\lambda) = b_{k-1}p_{k-1}(\lambda) + a_k p_k(\lambda) + b_k p_{k+1}(\lambda) ,$$

which is, of course, precisely the eigenvalue equation for the matrix

$$L = \begin{pmatrix} a_1 & b_1 & & 0 \\ b_1 & a_2 & b_2 & \\ & b_2 & a_3 & \ddots \\ 0 & & \ddots & \ddots \end{pmatrix} .$$

On page 41 of [**18**], Szego presents the reconstruction formulae for the b_k's,

$$b_k^2 = \frac{D_{k+1}D_{k-1}}{D_k^2} , \tag{31}$$

and simple calculations yield a similar expression for a_k:

$$a_k = E_k - E_{k+1} ,$$

where

$$E_k = -\frac{1}{D_{k-1}k!}\int\int\cdots\int\left(\sum_{i=0}^{k-1} x_i\right)\prod_{\nu<\mu}(x_\nu - x_\mu)^2 \prod_{i=0}^{k-1} d\alpha(x_i) .$$

The matrix being reconstructed evolves according to the Toda ODE system [**6**]

$$\frac{d}{dt}L = BL - LB .$$

It follows immediately from this Lax - pair equation that the eigenvalues of the matrix $L(t)$ are constant in time, and in addition one sees that the time dependence of the first component $f_j(1,t)$ of the jth (normalized) eigenvector is simple and can be determined explicitly [**16**]:

$$f_j(1,t)^2 = \frac{f_j(1,0)^2 e^{2\lambda_j t}}{\sum_{m=1}^N f_m(1,0)^2 e^{2t\lambda_m}} .$$

Using this formula, $D_k = D_k(t)$ becomes

$$D_k(t) = F^{-(k+1)} \sum_{j_0<j_1<\cdots<j_k} f_{j_0}(1,0)^2 \ldots f_{j_k}(1,0)^2 \, e^{2t\sum_{i=0}^k \lambda_{j_i}} \prod_{l\neq i} |\lambda_{j_i} - \lambda_{j_l}| ,$$

where

$$F = \sum_{m=1}^N f_m(1,0)^2 e^{2\lambda_m t} .$$

Now as $b_k^2 = \frac{D_{k+1}D_{k-1}}{D_k^2}$, we can replace D_k by

$$\tilde{D}_k = \tilde{D}_k(t) = \sum_{j_0<j_1<\cdots<j_k} f_{j_0}(1,0)^2 \ \ldots \ f_{j_k}(1,0)^2 \, e^{2t\sum_{i=0}^k \lambda_{j_i}} \prod_{l\neq i} |\lambda_{j_i} - \lambda_{j_l}| .$$

One can easily show from this representation that a_k can be expressed solely in terms of the $\tilde{D}_k$'s,

$$a_k = \frac{\partial}{\partial t}\frac{1}{2}\log \tilde{D}_k - \frac{\partial}{\partial t}\frac{1}{2}\log \tilde{D}_{k-1} ,$$

and, moreover, using the Toda system, one can show that another formula can be obtained for b_k, which admits easier analysis in the continuum limit than the one presented in (31) above,

$$b_k^2 = \frac{\partial^2}{\partial t^2}\frac{1}{4}\log \tilde{D}_k . \tag{32}$$

As described in the Introduction, we are interested in the continuum limit of the ODE system,

$$\begin{aligned}\lim_{\epsilon\to 0} \hat{b}_\epsilon(x,t) &= \lim_{\epsilon\to 0} b_{k(\epsilon,x)}\left(\frac{t}{\epsilon}\right) \ , \\ \lim_{\epsilon\to 0} \hat{a}_\epsilon(x,t) &= \lim_{\epsilon\to 0} a_{k(\epsilon,x)}\left(\frac{t}{\epsilon}\right) \ ,\end{aligned}$$

where

$$k(\epsilon,x) = \left[\frac{x}{\epsilon}\right] \ ,$$

and $[z]$ denotes the integer part of z. Inserting these scalings into the formula for $\tilde{D}_k$, we obtain

$$\tilde{D}_{k(\epsilon,x)} = \sum_{j_0<j_1<\cdots<j_{k(\epsilon,x)}} f_{j_0}(1,0)^2 \ldots f_{j_{k(\epsilon,x)}}(1,0)^2 \exp\left(\frac{2t}{\epsilon}\sum_{i=0}^{k(\epsilon,x)} \lambda_{j_i}\right) \prod_{l\neq i} |\lambda_{j_i} - \lambda_{j_l}| \ ,$$

and in terms of these $\tilde{D}_{k(\epsilon,x)}$'s, we have

$$\hat{b}_\epsilon(x,t)^2 = \frac{\tilde{D}_{k(\epsilon,x)}\tilde{D}_{k(\epsilon,x)-2}}{\tilde{D}^2_{k(\epsilon,x)-1}} \ ,$$

along with the formulae

$$\begin{aligned}\hat{b}_\epsilon(x,t)^2 &= \frac{\partial^2}{\partial t^2}\frac{1}{4}\epsilon^2 \log \tilde{D}_{k(\epsilon,x)} \ , \\ \hat{a}_\epsilon(x,t) &= \frac{\partial}{\partial t}\frac{1}{2}\frac{\epsilon^2\log\tilde{D}_{k(\epsilon,x)} - \epsilon^2\log\tilde{D}_{k(\epsilon,x)-1}}{\epsilon} \ . \end{aligned} \tag{33}$$

In following the prescription set down in [**13**], the next step is to replace the spectral data $\left\{\lambda_j^0, f_j(1,0)\right\}$ by approximate spectral data, given by the discrete version of the WKB method presented in Appendix I.

We take initial data for a_j and b_j from the discretization of two smooth functions, and we define

$$\alpha^0(x) = a^0(x) - 2b^0(x) \ ,$$

and

$$\beta^0(x) = a^0(x) + 2b^0(x) \ .$$

Observe that if the continuous Toda equations (17 - 18) are written in terms of variables $\alpha(x,t) = a(x,t) - 2b(x,t)$, $\beta(x,t) = a(x,t) + 2b(x,t)$, we obtain

$$\begin{aligned}\alpha_t &= -\frac{\beta-\alpha}{2}\alpha_x \ , \\ \beta_t &= \frac{\beta-\alpha}{2}\beta_x \ ,\end{aligned}$$

which is in Riemann invariant form.

In the analysis of the behavior of the lattice as $\epsilon \to 0$, we will assume that $a_0(x)$ is a monotone decreasing function of x, and that $\beta_0(x)$ possesses at most one critical point, a maximum. As described in Appendix I, WKB theory says that the matrix L obtained from this type of data has eigenvalues determined (asymptotically) by the equation

$$\Phi(\lambda) = (j - 1/2)\pi\epsilon \ , \quad j = 1, \ldots, [1/\epsilon] \ , \tag{34}$$

where

$$\Phi \equiv \int_{x_-(\lambda)}^{x_+(\lambda)} \cos^{-1}\left(\frac{\lambda - a_0(x)}{2b_0(x)}\right) dx \ . \tag{35}$$

For these eigenvalues, we have the following asymptotic form for their eigenfunctions:

$$f_j(1,0) \approx \exp\left[-\frac{1}{\epsilon}\int_0^{x_-(\lambda)} \cosh^{-1}\left(\frac{\lambda - a_0(x)}{2b_0(x)}\right) dx\right] = \exp\left(-\frac{\theta(\lambda)}{\epsilon}\right) \ ,$$

where

$$\theta(\lambda) \equiv \int_0^{x_-(\lambda)} \cosh^{-1}\left[\frac{\lambda - a_0(x)}{2b_0(x)}\right] dx \ .$$

This gives an expression for $\tilde{D}_{k(\epsilon,x)}$:

$$\tilde{D}_{k(\epsilon,x)} \approx \sum_{j_0<j_1<\cdots<j_{k(\epsilon,x)}} \exp \sum_{i=0}^{k(\epsilon,x)} \left(\frac{-2\theta(\lambda_{j_i}) + 2t\lambda_{j_i}}{\epsilon}\right) \prod_{l\neq i} |\lambda_{j_i} - \lambda_{j_l}|$$

$$= \sum_{j_0<j_1<\cdots<j_{k(\epsilon,x)}} \exp\left[\frac{\epsilon \sum_{i=0}^{k(\epsilon,x)} \{2t\lambda_{j_i} - 2\theta(\lambda_{j_i})\} + \epsilon^2 \sum_{i\neq l} \log|\lambda_{j_i} - \lambda_{jl}|}{\epsilon^2}\right]$$

$$= \sum_{s\in \mathbf{S}_{k(\epsilon,x)}} \exp\left[\frac{Q_s}{\epsilon^2}\right] \ , \tag{36}$$

where $\mathbf{S}_k$ represents the collection of all subsets of the first $\left[\frac{1}{\epsilon}\right]$ integers of size $k+1$, the summation is over all such subsets, and for a particular subset s, we define

$$Q_s = \epsilon \sum_{j\in s} [2t\lambda_j - 2\theta(\lambda_j)] + \epsilon^2 \sum_{j\neq l,\ j,l\in s} \log|\lambda_j - \lambda_l| \ . \tag{37}$$

Again following [**13**], we rewrite this expression for Q_s in integral form,

$$Q_s = \frac{2}{\pi}\int (t\lambda - \theta(\lambda))\psi_s(\lambda)\, d\lambda + \frac{1}{2\pi^2}\int\int \log(\lambda - \mu)^2\, \psi_s(\lambda) \times' \psi_s(\mu)\, d\lambda d\mu \ ,$$

where

$$\psi_s = \epsilon\pi \sum_{j\in s} \delta(\lambda - \lambda_j) \ , \tag{38}$$

and

$$\psi_s(\lambda) \times' \psi_s(\mu) = \epsilon^2\pi^2 \sum_{j,l\in s \ ,j\neq l} \delta(\lambda - \lambda_j)\, \delta(\mu - \lambda_l) \ .$$

We define

$$\phi_\epsilon(\lambda) = \pi\epsilon \sum_{j=1}^{[1/\epsilon]} \delta(\lambda - \lambda_j) \ ,$$

$$\phi(\lambda) = -\Phi'(\lambda) \ . \tag{39}$$

The analysis in [**13**] for the zero dispersion limit of the KdV equation suggests that the problem of computing the behavior of $b_{k(\epsilon,x)}(t/\epsilon)$ is equivalent to finding the maximal element of the sum (36).

For each fixed value of ϵ, the measure of the functions ψ are fixed by the value of $k(\epsilon, x)$:

$$\int \psi_s(\lambda)\, d\lambda = k(\epsilon, x)\, \pi\, \epsilon \ ,$$

so for ϵ tending to zero, we see that this constraint is converging to the constraint

$$\int \psi(\lambda) = \pi x \ .$$

As in [**13**], one can prove that

$$\lim_{\epsilon\to 0} \epsilon^2 \log D_{k(\epsilon,x)}$$

exists, and is given by the solution of a maximization problem,

$$\lim_{\epsilon\to 0} \epsilon^2 \log D_{k(\epsilon,x)} = Q(x,t) \ , \tag{40}$$

where

$$Q(x,t) = \max_{\psi\in A} \left[\frac{2}{\pi} (t\lambda - \theta \ , \psi) + \frac{1}{2\pi^2} (L\psi \ , \psi) \right] \ , \tag{41}$$

A is the set of all Lebesgue measurable functions ψ defined on the spectral domain (which we arrange throughout this thesis to be the interval [-1,1]), satisfying

$$0 \leq \psi(\lambda) \leq \phi(\lambda) \ ,$$

$$\int \psi(\lambda)\, d\lambda = \pi x \ ,$$

the operator L is defined by

$$L\, \psi\, (\lambda) = \int \log (\lambda - \mu)^2\, \psi(\mu)\, d\mu \ ,$$

and (,) represents

$$(f,g) = \int f(\lambda)g(\lambda)\, d\lambda \ .$$

The proof of (40) is very similar to the proof of analogous formulae in [**13**], and we do not reproduce the details. There are minor modifications in the argument. For example, in place of uniform convergence on compact subsets of the x - axis, (see theorem 2.2, [**13**]), we have bounded convergence on compact subsets. That is, one can show that the limit in (40) exists pointwise for each x, and

$$\left|\epsilon^2 \log D_{k(\epsilon,x)}\right| \ < \ const. \ ,$$

for all small ϵ and for all x in any given compact subset of $[0,1]$. This is, of course, enough to prove the existence of limits of weak derivatives. For example, together with bounded convergence, (40) immediately implies that

$$w - \lim_{\epsilon\to 0} \log \hat{b}_\epsilon^2 \ = \ \frac{1}{2}Q_{xx} \ , \tag{42}$$

$$w - \lim_{\epsilon\to 0} \hat{b}_\epsilon^2 \ = \ \frac{1}{4}Q_{tt} \ , \tag{43}$$

and similarly,

$$w - \lim_{\epsilon\to 0} \hat{a}_\epsilon \ = \ \frac{1}{2}Q_{xt} \ . \tag{44}$$

Uniform bounds on the l^2 norms of $\{b_k^2\}$ and $\{a_k\}$ then imply that $\hat{a}_\epsilon$ and $\hat{b}_\epsilon^2$ converge weakly in L^2.

We turn now to the analysis of the maximization problem (41). To impose the integral constraint $\int \psi = \pi x$, we introduce a Lagrange multiplier,

$$\tilde{Q} \ = \ \frac{2}{\pi}(t\lambda - \theta(\lambda)\ ,\ \psi(\lambda)) + \frac{1}{2\pi^2}(L\psi(\lambda)\ ,\ \psi(\lambda)) \ + \ l\left[(\psi(\lambda)\ ,\ 1) - \pi x\right].$$

The associated Euler - Lagrange equations are

$$\frac{2}{\pi}(t\lambda - \theta(\lambda)) \ + \ \frac{1}{\pi^2}L\psi(\lambda) \ + \ l \ = \ 0 \ , \tag{45}$$

$$(\psi(\lambda)\ ,\ 1) \ = \ \pi x \ .$$

Now as we are solving a constrained maximization problem, the function ψ in general solves the equation (45) on a subset of the spectral domain, which we call I_1, and the complement of this set is split further into subsets on which ψ attains one or the other of the constraints $\psi = 0$ or $\psi = \phi$.

Now we use the elementary fact (see, for example, [**13**], [**22**]) that the operator L is the anti-derivative of the Hilbert transform operator,

$$L\psi(\lambda) \ = \ L\psi(\alpha) \ + \ 2\pi \int_\alpha^\lambda H\psi(\mu)\, d\mu \ ,$$

to differentiate the Euler - Lagrange equation (45) on I_1, and obtain a Riemann - Hilbert problem:

$$\begin{aligned} H\psi &= \theta'(\lambda) - t \quad , \quad \lambda \in I_1 \quad , \\ \psi &= 0 \quad , \quad \lambda \in I_2 \quad , \\ \psi &= \phi(\lambda) \quad , \quad \lambda \in I_3 \quad . \end{aligned}$$

To describe the continuum limit of the Toda lattice, we construct solutions to this Riemann - Hilbert problem, parameterized by x and t, and show that the function produced, $\psi(\lambda\ ;\ x, t)$, solves the maximization problem (41).

CHAPTER 3

An Example

For the purpose of illustration, this section is devoted to the analysis of the continuum limit in the particular case

$$a^0(x) = 0 \ ,$$
$$b^0(x) = 1/2 \ ,$$

for which we are able to compute the limit explicitly.

Remark: The function $b^0(x)$ does not satisfy the natural boundary condition at $x = 0$ and $x = 1$. In this connection, however, see the remark following (21).

For these simple functions, the spectral data is explicit and one finds

$$\lambda_j = \cos j\pi\epsilon \ ,$$
$$f_j(1,0) = \sin j\pi\epsilon \ ,$$

or, equivalently, the spectral function is

$$\Phi(\lambda) = \cos^{-1}(\lambda)$$

and hence

$$\phi(\lambda) = \frac{1}{\sqrt{1-\lambda^2}} \ ,$$

and also the associated norming constant is

$$\theta(\lambda) \ = \ \epsilon\frac{1}{2}\log(1-\lambda^2) \ .$$

The functional Q_s has the form

$$Q_s = \frac{2t}{\pi}\Big(\lambda \ , \ \psi_s\Big) + \frac{1}{2\pi^2}\Big(L\psi_s \ , \ \psi_s\Big) + \frac{2\epsilon}{\pi}\Big(\log{(1-\lambda^2)}^{1/2} \ , \ \psi_s\Big)$$

where ψ_s and L are as described in previous sections. From this we obtain as before (see (40)) the maximization problem which we will study in depth:

$$Q(x,t) \ = \ \max_{\psi\in A} \ \left[\frac{2}{\pi}(t\lambda \ , \psi) \ + \ \frac{1}{2\pi^2}(L\psi \ , \psi)\right] \ . \tag{46}$$

Remark: Observe that there is no contribution from the asymptotic norming constant $\theta(\lambda)$. This reflects the special nature of the above initial data (in particular, $b^0(0) \neq 0$, $b^0(1) \neq 0$), which leads to norming constants which are *not* exponentially small.

Following the previous section, we obtain from (46) the Euler - Lagrange equations for ψ:

$$\frac{1}{\pi^2} L\psi(\lambda) + \frac{2}{\pi} t\lambda + l = 0 \ , \qquad (\psi \, , \, 1) = \pi x \ , \tag{47}$$

and differentiating with respect to λ yields the Riemann - Hilbert problem

$$H\psi = -t \ , \tag{48}$$

where we have yet to specify the subset I_1 of the spectral domain, on which this is to be satisfied, as well as the subsets I_2 and I_3 on which the constraints $\psi = 0$ and $\psi = \phi$, respectively, are to be achieved. We remark that in [**13**] the original Riemann - Hilbert problem is differentiated with respect to the independent parameter x to obtain the simple Riemann - Hilbert problem

$$H\psi_x = 1 \ ,$$

with right hand side independent of λ. For the continuum limit of the Toda lattice with the above initial data, we see that a similar simple situation arises in which the right hand side is independent of λ, even without differentiating with respect to x or t.

In the succeeding chapters we will present a systematic approach to the solution of the Riemann - Hilbert problems that arise from the maximization problem for Q. However, for the particular simple problem at hand, we will proceed, in this chapter, in an ad hoc fashion.

The maximizer for $t = 0$

We begin with the problem for $t = 0$, and make the ansatz that the set I_1 on which ψ is to satisfy (48) is the entire spectral domain:

$$H\psi = 0 \ , \quad \lambda \in I_1 \ , \qquad I_1 = [-1, 1] \ . \tag{49}$$

As we will now show, this problem has a simple solution which satisfies the constraints

$$0 \le \psi \le \phi \ , \qquad \int \psi = \pi x \ ,$$

and also solves the maximization problem (46). The solution is given by

$$\psi^* = Re \, \frac{ix}{(\lambda^2 - 1)^{1/2}} \ , \tag{50}$$

where the branch of the root is chosen such that $(\lambda^2 - 1)^{1/2} > 0$ for $\lambda > 0$. Indeed, ψ^* as defined in (50) can be extended to an $L^p(-\infty, \infty)$ function, as $Re\frac{ix}{(\lambda^2-1)^{1/2}} = 0$ for $|\lambda| > 1$. General theory [12] then implies that

$$\begin{aligned} H\psi^* &= Im\,\frac{ix}{(\lambda^2-1)^2} \\ &= \; = 0 \quad for\ \lambda \in (-1,1)\ , \end{aligned} \tag{51}$$

i.e. that ψ^* solves (49). Observe that at $x = 1$ ψ^* achieves the constraint $\psi^* = \phi$, while at $x = 0$ it achieves the other constraint $\psi^* = 0$.

LEMMA 3.1. *The function ψ^* solves the maximization problem (46).*

Proof. Let χ be any function such that $\psi^* + \chi \in A$, where A is (as before) the set of all Lebesgue measurable functions on $[-1,1]$ which satisfy $0 \leq \psi \leq \phi$. Thus

$$0 \leq \psi^* + \chi \leq \phi\ ,$$
$$\int \chi d\lambda = 0\ .$$

We will define $Q(\psi)$ as follows,

$$Q(\psi) = \frac{2}{\pi}(t\lambda\ ,\ \psi) + \frac{1}{2\pi^2}(L\psi\ ,\ \psi)\ . \tag{52}$$

We evaluate

$$Q(\psi^* + \chi) = Q(\psi^*) + \frac{1}{2\pi^2}\Big(L\chi\ ,\ \chi\Big) + \frac{1}{\pi^2}\Big(L\psi^*\ ,\ \chi\Big)\ .$$

Now we use the representation for $L\psi^*$,

$$\begin{aligned} L\psi^* &= L\psi^*(0) + 2\pi\int_0^\lambda H\psi^*\, d\mu \\ &= L\psi^*(0)\quad ,\quad for\ \lambda \in [-1,1]\ , \end{aligned}$$

where the last equality follows from the fact that $H\psi^* = 0$ for $\lambda \in [-1,1]$, to obtain

$$Q(\psi^* + \chi) = Q(\psi^*) + \frac{1}{2\pi^2}\Big(L\chi\ ,\ \chi\Big) + \frac{1}{\pi^2}\Big(L\psi^*(0),\ \chi\Big)\ .$$

Since χ is mean zero, we have

$$\Big(L\psi^*(0),\ \chi\Big) = 0\ ,$$

and since the operator L is negative definite on mean zero functions [13],[22], we also have

$$\Big(L\chi\ ,\ \chi\Big) < 0\ .$$

Thus we conclude that

$$Q(\psi^* + \chi) < Q(\psi^*)$$

for all χ (not identically zero) such that $\psi^* + \chi \in A$. □

A simple calculation shows that $Q(\psi^*)$ has the following explicit form:

$$Q(\psi^*) = x^2 \log(1/2)\ .$$

From this it is clear that we have reconstructed the initial data: by (42),

$$\log \hat{b}_\epsilon^2 \;\to\; Q_{xx} \;=\; 2\log\frac{1}{2}\;,$$

which is consistent with $\log b^0 \;=\; \log\frac{1}{2}$.

The region $0 \le t \le \frac{1}{2}$, $x \in\; (t, 1-t)$

The maximization problem (46) for t positive is solved by the same general representation, but on a shrinking spatial interval. In fact, if we suppose that for $t > 0$ the set I_1 is still the full spectral interval $[-1,1]$, the Riemann - Hilbert problem (48) has a solution given by

$$\psi^* = Re\;\; i\left[\frac{t\lambda + x}{(\lambda^2-1)^{1/2}} - t\right] \tag{53}$$

$$where \quad (\lambda^2-1)^{1/2} \;>\; 0 \;\; for \;\; |\lambda| > 1\;.$$

Indeed, ψ^* can once again be extended to an $L^p(-\infty,\infty)$ function, as

$$Re\;\; i\left[\frac{t\lambda + x}{(\lambda^2-1)^{1/2}} - t\right] \;=\; 0 \quad for\; |\lambda| > 1\;.$$

As above, we have

$$\begin{aligned} H\psi^* \quad &= \quad Im\; i\left[\frac{t\lambda + x}{(\lambda^2-1)^{1/2}} - t\right] \\ &= \quad -t \quad , \quad for\; \lambda \in (-1,1)\;. \end{aligned}$$

However, ψ^* no longer belongs to the set A for all values of x. Indeed, the conditions $0 \le \psi^* \le \phi$ imply

$$t \le x \le 1-t\;,$$

i.e. that (53) belongs to the admissible set A on a shrinking spatial domain, which collapses at $t = 1/2$.

LEMMA 3.2. *The function* $\psi^*(\lambda)$ *defined in (53) solves the maximization problem (46), provided* $0 \le t \le \frac{1}{2}$, *and* $x \in [t, 1-t]$.

Proof.

As in lemma (3.1), let χ be any function such that $\psi^* + \chi \;\in A$. We evaluate

$$Q(\psi^*+\chi) = Q(\psi^*) + \frac{1}{2\pi^2}\Big(L\chi\;,\;\chi\Big) + \frac{2}{\pi}(t\lambda\;,\;\chi) + \quad + \;\frac{1}{\pi^2}\Big(L\psi^*\;,\;\chi\Big)\;.$$

Now we use the representation for $L\psi^*$,

$$\begin{aligned} L\psi^* \quad &= \quad L\psi^*(0) + 2\pi\int_0^\lambda H\psi^*\,d\mu \\ &= \quad L\psi^*(0)\; - 2\pi t\lambda\;,\quad for\; \lambda \in [-1,1]\;, \end{aligned}$$

to obtain

$$Q(\psi^*+\chi) = Q(\psi^*) + \frac{1}{2\pi^2}\Big(L\chi\;,\;\chi\Big) \;+\; \frac{1}{\pi^2}\Big(L\psi^*(0),\;\chi\Big)\;. \tag{54}$$

The fact that χ is mean zero shows that the second term in (54) is strictly negative, that the third vanishes, and hence that

$$Q(\psi^* + \chi) < Q(\psi^*) .$$

□

Now using the definition of the solution $Q(x,t)$ of the maximization problem (46), and also the fact that $Q(x,t) = Q(\psi^*)$, we have

$$\frac{\partial}{\partial t} Q(x,t) = \frac{\partial}{\partial t}\left[\frac{2}{\pi}(t\lambda , \psi^*) + \frac{1}{2\pi^2}(L\psi^* , \psi^*)\right] \tag{55}$$

$$= \frac{2}{\pi}(\lambda , \psi^*) + \frac{2}{\pi}(t\lambda , \psi_t^*) + \frac{1}{\pi^2}(L\psi^* , \psi_t^*) , \tag{56}$$

The Euler - Lagrange equation satisfied by ψ^* (47) implies

$$Q_t = \frac{2}{\pi}(\lambda , \psi^*) + l(1 , \psi_t^*) \tag{57}$$

$$= \frac{2}{\pi}(\lambda , \psi^*) , \tag{58}$$

where the last equality is because

$$(1 , \psi^*) = \pi x \Rightarrow (1 , \psi_t^*) = 0 .$$

Thus we see that the weak limit of $\hat{b}_\epsilon(x,t)$ is

$$w - \lim_{\epsilon\to 0} \hat{b}_\epsilon^2(x,t) = \frac{1}{4}Q_{tt} = \frac{1}{\pi}(\lambda , \psi_t^*) ,$$

and evaluating the right hand side we obtain

$$w - \lim_{\epsilon\to 0} \hat{b}_\epsilon^2(x,t) = \frac{1}{4} ,$$

and similarly we see that the weak limit of $\hat{a}_\epsilon(x,t)$ is

$$w - \lim_{\epsilon\to 0} \hat{a}_\epsilon(x,t) = \frac{1}{2}Q_{xt} = \frac{1}{\pi}(\lambda , \psi^*{x})$$
$$= 0 .$$

Thus on the shrinking spatial domain $[t, 1-t]$ the continuum limit is at rest.

We note that at the boundaries of the domain, the function ψ^* attains a constraint. That is, we see that $\psi^*(\lambda ; x,t)$ vanishes for $\lambda = -1$, $x = t$, and also that

$$\lim_{\lambda\to 1}[\psi^*(\lambda ; 1-t, t) - \phi(\lambda)] = 0 .$$

In fact, this indicates how to solve the maximization problem on the exterior of this shrinking set. We will consider values of x in the region $[0,t]$ first.

The maximizer for $0 < t \le \frac{1}{2}$, $x \in [0,t]$.

Fix $0 < t \le \frac{1}{2}$. For values of x in the subinterval $[0,t]$, we make the ansatz that the solution to the maximization problem solves the Riemann -

Order ID: 113-4338963-3298650

Thank you for buying from hendersonbooks98225 on Amazon Marketplace.

Shipping Address:	Order Date:	Thu, Oct 22, 2026
AMZ APTP ORDER	Shipping Service:	Standard
800 AVONDALE AVE	Buyer Name:	Lucy
COLUMBUS, OH 43212-3473	Seller Name:	hendersonbooks98225

Quantity	Product Details
1	**A Continuum Limit of the Toda Lattice (Memoirs of the American Mathematical Society) [Paperback] [1998] Deift, Percy; McLaughlin, Kenneth T-R** **SKU:** 73-7WJK-KEDQ **ASIN:** 0821806912 **Condition:** Used - Very Good **Order Item ID:** 173103521300521 **Condition note:** Clean--tight binding--text Clean--L

Returning your item:

Go to "Your Account" on Amazon.com, click "Your Orders" and then click the "seller profile" link for this order to get information about the return and refund policies that apply.

Visit https://www.amazon.com/returns to print a return shipping label. Please have your order ID ready.

Thanks for buying on Amazon Marketplace. To provide feedback for the seller please visit www.amazon.com/feedback. To contact the seller, go to Your Orders in Your Account. Click the seller's name under the appropriate product. Then, in the

Hilbert problem

$$(59)\qquad \begin{aligned} H\psi &= -t\ , \quad \lambda \in (\alpha, 1)\ , \\ \psi &= 0\ , \quad \lambda \in (-1, \alpha)\ , \end{aligned}$$

for some $\alpha = \alpha(x,t) \in (-1,1)$. Note that at $x = t$ we must have $\alpha = -1$.

As we now show, equation (59) is solved by $\psi^* = Ref$, where

$$f = -it\left[1 - \frac{\lambda + a}{[(\lambda-\alpha)(\lambda-1)]^{1/2}}\right],$$

$$\textit{with} \qquad [(\lambda-\alpha)(\lambda-1)]^{1/2} > 0 \ \ \textit{for}\ \lambda > 1\ ,$$

and we pick the (real) parameters a, α to satisfy the constraints $0 \le \psi^* \le \phi$ and $\int \psi^* = \pi x$. Observe that $Ref(\lambda) = 0$ for $|\lambda| > 1$, and that we can extend ψ^* to an $L^p(-\infty,\infty)$ function for any $1 < p < 2$ by setting $\psi^*(\lambda) = Ref(\lambda)$ for all λ. General theory [**12**] then implies that

$$(60)\ H\psi^* = Imf = \left\{ \begin{array}{ll} -t\ , & \lambda \in (\alpha, 1) \\ -t\left[1 - \frac{\lambda+a}{[(\lambda-\alpha)(\lambda-1)]^{1/2}}\right]\ , & \lambda \in \mathbf{R}\backslash(\alpha,1) \end{array} \right\} .$$

Also, $\psi^* = Ref = 0$ for $\lambda \in (-1, \alpha)$. Thus ψ^* solves (59).

For $\lambda \in (\alpha, 1)$, $\left[(\lambda-\alpha)(\lambda-1)\right]^{1/2} = i\left|(\lambda-\alpha)(\lambda-1)\right|^{1/2}$, and hence

$$\psi^* = t\frac{\lambda + a}{\left|(\lambda-\alpha)(\lambda-1)\right|^{1/2}} .$$

In order that $\psi^* \le \phi$ for $\lambda \in (\alpha, 1)$, we must have that $a = -\alpha$, hence

$$(61)\qquad \psi^* = t\left[\frac{\lambda-\alpha}{1-\lambda}\right]^{1/2}\ , \quad \lambda \in (\alpha, 1)\ ,$$

which is less than $\phi(\lambda) = \frac{1}{(\lambda^2-1)^{1/2}}$, for $t < \frac{1}{2}$. Clearly $\psi^* \ge 0$ for $\lambda \in (\alpha, 1)$.

Finally the constraint $\int \psi^* = \pi x$ implies that

$$\frac{\pi t}{2}(1-\alpha) = \pi x\ ,$$

and hence

$$(62)\qquad \alpha(x,t) = 1 - \frac{2x}{t}\ .$$

From (62) it is clear that $\alpha \in [-1, 1]$, as $0 \le x \le t$.

LEMMA 3.3. *The candidate (61) with α defined in (62) solves the maximization problem (46), for $0 < t \le \frac{1}{2}$ and $x \in [0, t]$.*

Proof. We take χ such that $\psi^* + \chi \in A$, and obtain

$$(63)\ Q(\psi^* + \chi) = Q(\psi^*) + \frac{2}{\pi}(t\lambda\ , \chi) + \frac{1}{2\pi^2}\Big(L\chi\ , \chi\Big) + \frac{1}{\pi^2}\Big(L\psi^*, \chi\Big)\ .$$

We use a slightly different representation for $L\psi^*$,

$$L\psi^* = L\psi^*(\alpha) + 2\pi \int_\alpha^\lambda H\psi^* \, d\mu \tag{64}$$
$$= L\psi^*(\alpha) \; - 2\pi t(\lambda - \alpha) \;, \quad for \; \lambda \in [\alpha, 1] \;,$$

to obtain, from (63),

$$Q(\psi^* + \chi) = Q(\psi^*) + \frac{1}{2\pi^2}\Big(L\chi \,, \, \chi\Big) + \Big(\frac{1}{\pi^2}L\psi^* + \frac{2}{\pi}t\lambda \,, \, \chi\Big)_{(-1,\alpha)}$$
$$+ \Big(\frac{1}{\pi^2}L\psi^*(\alpha) + \frac{2}{\pi}t\alpha\Big)\Big(1 \,, \, \chi\Big)_{(\alpha,1)}$$
$$= Q(\psi^*) + \frac{1}{2\pi^2}\Big(L\chi \,, \, \chi\Big)$$
$$+ \Big(\frac{1}{\pi^2}(L\psi^*(\lambda) - L\psi^*(\alpha)) + \frac{2}{\pi}t(\lambda - \alpha) \;, \, \chi\Big)_{(-1,\alpha)} \;, \tag{65}$$

where the notation $\Big(,\Big)_{(a,b)}$ is used to signify that the integral is over the subinterval (a, b), and the last equality is a consequence of the fact that χ has mean zero, i.e. that

$$\Big(1 \,, \, \chi\Big)_{(-1,\alpha)} + \Big(1 \,, \, \chi\Big)_{(\alpha,1)} = 0 \;.$$

We again use (64) to re - write the expression (65) for $Q(\psi^* + \chi)$,

$$Q(\psi^* + \chi) = Q(\psi^*) + \frac{1}{2\pi^2}\Big(L\chi \,,\chi\Big) \; + \; \frac{2}{\pi}\Big(\int_\alpha^\lambda H\psi^*(\mu) + t d\mu, \chi\Big)_{(-1,\alpha)} \;.$$

As we must have $\chi \geq 0$ on $(-1, \alpha)$ to enforce $0 \; \leq \; \psi^* + \chi$, the proof of the lemma will be complete upon establishing

$$\int_\alpha^\lambda H\psi^*(\mu) \; + \; t d\mu \; \leq \; 0 \quad for \;\; \lambda \in (-1, \alpha) \;. \tag{66}$$

But this follows from the fact (60) that

$$H\psi^*(\mu) + t \;=\; t\left[\frac{\mu - \alpha}{\mu - 1}\right]^{1/2} \; > \; 0 \;\; for \; \mu \in (-1, \alpha) \;,$$

along with the observation that the integration in (66) is in the negative direction.

□

The derivation of

$$Q_t \;=\; \frac{2}{\pi}(\lambda \,, \, \psi^*)$$

presented in equations (55 - 58) above is general and applies to the case of the maximizer (61) with α defined in (62), and so we see using formula (43) that the weak limit of $\hat{b}_\epsilon^2(x, t)$ is

$$w - \lim_{\epsilon \to 0} \hat{b}_\epsilon^2(x, t) \;=\; \frac{1}{4}Q_{tt} \;=\; \frac{1}{\pi}(\lambda \,, \, \psi_t^*) \;,$$

and using (61) to evaluate the right hand side, we obtain

$$w - \lim_{\epsilon \to 0} \hat{b}_\epsilon^2(x,t) = \left(\frac{x}{2t}\right)^2 ,$$

and similarly we see from formula (44) that the weak limit of $\hat{a}_\epsilon(x,t)$ is

$$\begin{aligned} w - \lim_{\epsilon \to 0} \hat{a}_\epsilon(x,t) &= \frac{1}{2} Q_{xt} = \frac{1}{\pi} (\lambda , \psi_x^*) \\ &= 1 - \frac{x}{t} . \end{aligned}$$

The maximizer for $0 \le t \le \frac{1}{2}$, $x \in [1-t,1]$.

For the other subinterval, $[1-t,1]$, the ansatz we make for the Riemann - Hilbert problem is

$$H\psi = -t \quad \lambda \in I_1 = (-1,\beta) ,$$
$$\psi = \phi , \ \lambda \in I_3 = (\beta, 1) .$$

We are led to this because the maximizer valid in the interval $[t , 1-t]$ satisfies the constraint

$$\lim_{\lambda \to 1} (\psi^*(\lambda;\, x,t) - \phi) = 0 ,$$

for $x = 1-t$.

By making the substitution $\psi = \phi - p$, $Q(\psi)$ becomes

$$\begin{aligned} Q(\psi) &= Q(\phi) - \frac{1}{\pi^2}(L\phi , p) - \frac{2}{\pi}(t\lambda , p) + \frac{1}{2\pi^2}\Big(Lp , p\Big) \\ &= Q(\phi) - \frac{1}{\pi} L\phi(0)(1-x) - \frac{2}{\pi}(t\lambda , p) + \frac{1}{2\pi^2}(Lp , p) , \end{aligned}$$

where we have used that

$$\lambda \in (-1,1) \Rightarrow L\phi(\lambda) = L\phi(0) .$$

We will solve the original maximization problem (46) by constructing a solution $p^*(\lambda)$ to the new maximization problem,

$$\tilde{Q} = \max_{p \in \tilde{A}} -\frac{2}{\pi}(t\lambda , p) + \frac{1}{2\pi^2}(Lp , p) ,$$

where $\tilde{A}$ is the set of all Lebesgue measurable functions satisfying

$$0 \le p \le \phi ,$$
$$\int p d\lambda = \pi(1-x) .$$

The Riemann - Hilbert problem arising out of the Euler - Lagrange equations associated to this maximization problem is

$$Hp = t \quad for \ \lambda \in I_1 , \tag{67}$$
$$p = \phi \quad for \ \lambda \in I_2 , \tag{68}$$
$$p = 0 \quad for \ \lambda \in I_3 . \tag{69}$$

For $0 < t \leq \frac{1}{2}$, we have

$$\begin{aligned} I_1 &= (-1,\beta) \ , \\ I_3 &= (\beta,1) \ , \end{aligned} \tag{70}$$

and I_2 is empty. The solution to the Riemann - Hilbert problem (67) with I_j specified in (70) and satisfying the constraints $0 \leq p \leq \phi$ is

$$\begin{aligned} p^* &= Re\ it\left[1 - \left(\frac{\lambda-\beta}{\lambda+1}\right)^{1/2}\right] \\ &= t\left(\frac{\beta-\lambda}{\lambda+1}\right)^{1/2} \ . \end{aligned} \tag{71}$$

The fact that p^* solves (67) follows once again from extending p^* to an $L^p(-\infty,\infty)$ function, and invoking the general theory of the Hilbert transform operator.

Enforcing $\int p^* = \pi(1-x)$ implies that

$$\beta(x,t) = -1 + \frac{2}{t}(1-x) \ . \tag{72}$$

From (72) it is clear that $\beta(x,t) \in [-1,1]$, as $x \in (1-t,1)$. We have thus established that $p^* \in \tilde{A}$, and hence that $\psi^* = \phi - p^*$ is a valid candidate for the original maximization problem (46). The proof of the following lemma follows from arguments very similar to those presented to prove lemmas (3.1 - 3.3), and we omit it.

LEMMA 3.4. *For $0 < t \leq \frac{1}{2}$, $x \in [1-t,1]$, the function $\psi^* = \phi - p^*$, where p^* is defined in (71) and β is defined in (72), solves the maximization problem (46). Moreover,*

$$\begin{aligned} w - \lim_{\epsilon\to 0} \hat{b}_\epsilon^2(x,t) &= \left(\frac{1-x}{2t}\right) \ , \\ w - \lim_{\epsilon\to 0} \hat{a}_\epsilon(x,t) &= -1 + \frac{1-x}{t} \ . \end{aligned}$$

Thus we see that the Riemann invariant quantities $\alpha(x,t)$ and $\beta(x,t)$, and the continuum limit of $\hat{a}_\epsilon(x,t)$ and $\hat{b}_\epsilon^2(x,t)$ are as pictured in figure (1).

The Continuation Past $t = \frac{1}{2}$

At time $t = 1/2$ the middle spatial interval $[t, 1-t]$ collapses, but the functions ψ^* and p^* defined above, which solve the maximization problems for the exterior regions $[0,t]$ and $[1-t,1]$ respectively, continue to solve the maximization problem on spatial regions which begin to shrink at $t = 1/2$. This can be seen by taking the function ψ^* which solves the problem on $[0,t]$ (for $t < \frac{1}{2}$), and enforcing, for $t \geq 1/2$, that

$$\psi^* \leq \phi = [(1-\lambda)(1+\lambda)]^{-1/2} \ . \tag{73}$$

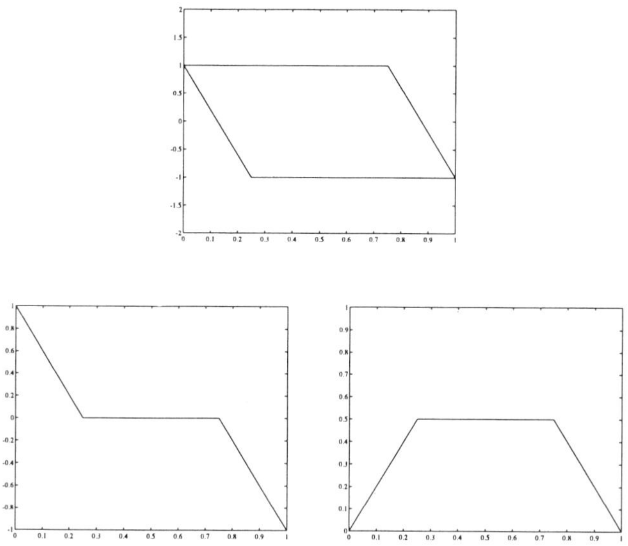

FIGURE 1. The Riemann invariants $\beta(x,t) \geq \alpha(x,t)$ (above), and the continuum limit quantities $\hat{a}(x,t)$ (left), and $\hat{b}(x,t)$ (right), $t < 1/2$. The corners occur at $x = t$ and $x = 1-t$.

From (61) we have for $\alpha < \lambda < 1$,

$$t^2\left(\frac{\lambda-\alpha}{1-\lambda}\right) \leq \frac{1}{(1-\lambda)(1+\lambda)}$$

$$\Longleftrightarrow (\lambda-\alpha)(\lambda+1) \leq t^{-2} \ . \tag{74}$$

Now the left hand side is largest when $\lambda = 1$, and so the condition (73) becomes

$$2(1-\alpha) \leq t^{-2} \ ,$$

and we use the definition (62) to establish

$$x \leq \frac{1}{4t} \ .$$

Similar calculations apply to the continuation of the interval $[1-t, 1]$, and show from (71), (72), that

$$p^* \leq \phi \ \Rightarrow x \geq 1 - \frac{1}{4t} \ .$$

Thus we see that we have solved the maximization problem on the *unshaded* region of the $x-t$ plane shown in Figure (2) below.

The maximizer for the region $t > \frac{1}{2}$, $x \in (\frac{1}{4t}, 1-\frac{1}{4t})$.

We must understand the behavior of the maximizer in the shaded region of Figure (2). This region emerges from the point $x = t = \frac{1}{2}$. At this point we have 2 candidates for the maximizer, $\psi^*(\lambda\ ;\ x,t)$, determined by allowing

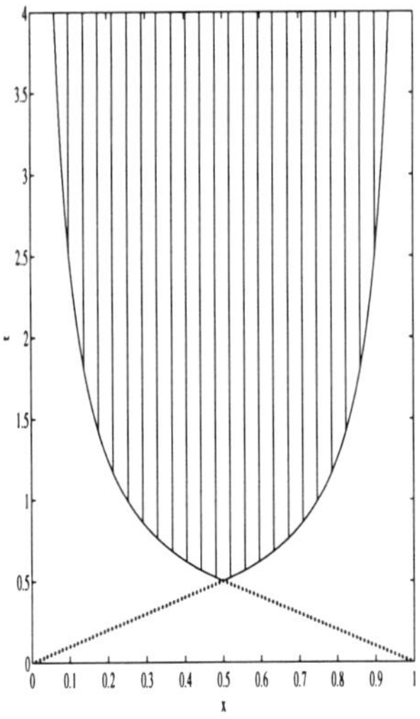

FIGURE 2. x-t plot of regions (unshaded) where maximizer is given by (1.11) and (1.22). The solid boundary curve is given by $|x - \frac{1}{2}| = \frac{1}{2} - \frac{1}{4t}$, $t \geq \frac{1}{2}$.

the lower endpoint of the band I_1 to evolve, and $\phi - p^*(\lambda\ ;\ x, t)$, determined by allowing the upper endpoint to evolve (cf. top diagram in Figure (1)). In fact the two candidates are identical at $x = t = \frac{1}{2}$, and so intuition tells us that we should seek a solution in this new interior spatial region for which *both* endpoints of the band are evolving.

We consider the Riemann - Hilbert problem with ψ already extended by zero to be an $L^p(-\infty, \infty)$ function,

$$\begin{aligned} H\psi &= -t \ , \ \lambda \in I_1 = (\alpha(x,t), \beta(x,t)) \\ \psi &= 0 \ , \ \lambda \in I_2 = (-\infty, \alpha(x,t)) \ and \ (1, \infty) \\ \psi &= \phi(\lambda) \ , \ \lambda \in I_3 = (\beta(x,t), 1), \end{aligned}$$

which has the explicit solution $\psi^*(\lambda) = Re\ f(\lambda)$, $H\psi^*(\lambda) = Im\ f(\lambda)$, $\lambda \in \mathbf{R}$, where $f(z)$ is analytic in $\mathbf{C}^+$,

$$f(z) = \left[\frac{z-\alpha}{z-\beta}\right]^{1/2} \frac{-t}{\pi} \left(\int_\alpha^\beta \left[\frac{\lambda-\beta}{\lambda-\alpha}\right]^{1/2} \frac{d\lambda}{\lambda - z} + \int_\beta^1 \frac{\left[\frac{\lambda-\beta}{\lambda-\alpha}\right]^{1/2}}{(1-\lambda^2)^{1/2}} \frac{d\lambda}{\lambda - z} \right),$$

for $z \in \mathbf{C}^+$, and

$$f(\lambda) = \lim_{\epsilon \downarrow 0} f(\lambda + i\epsilon) \ , \ \lambda \in \mathbf{R}.$$

Carrying out the first integral explicitly, we derive that ψ^* has the following form:

$$(75)\ \psi^* = \left\{ \begin{array}{ll} \frac{1}{(1-z^2)^{1/2}} , & z \in (\beta, 1) \\ t\left[\frac{z-\alpha}{\beta-z}\right]^{1/2} + \left[\frac{z-\alpha}{z-\beta}\right]^{1/2} \frac{1}{\pi i}\int_\beta^1 \frac{\left[\frac{\lambda-\beta}{\lambda-\alpha}\right]^{1/2}}{(1-\lambda^2)^{1/2}} \frac{d\lambda}{\lambda - z} , & z \in (\alpha, \beta) \\ 0 , & z \in (-1, \alpha) \end{array} \right\} .$$

In order that $\psi^* \leq \phi$, we must have

$$(76) \qquad T(\alpha,\beta) \equiv \int_\beta^1 \frac{1}{[(\lambda-\beta)(\lambda-\alpha)(1-\lambda^2)]^{1/2}} d\lambda = \pi t \ .$$

On the other hand, as we now show, enforcing the second condition $\int \psi^* = \pi x$, yields

$$(77) \qquad X(\alpha,\beta) \equiv \int_\beta^1 \frac{\lambda - m}{[(\lambda-\beta)(\lambda-\alpha)(1-\lambda^2)]^{1/2}} d\lambda = \pi x \ ,$$

where $m = \frac{\beta+\alpha}{2}$. We note for future reference that it is easy to show that $T(\alpha,\beta), X(\alpha,\beta)$ are smooth functions of α, β, for $\alpha < \beta$.

Consider

$$(78) \int_\alpha^\beta \psi^* =$$

$$\int_\alpha^\beta t\left(\frac{z-\alpha}{\beta-z}\right)^{1/2} dz + \int_\alpha^\beta dz \left(\frac{z-\alpha}{z-\beta}\right)^{1/2} \frac{1}{\pi i}\int_\beta^1 \frac{\left(\frac{\lambda-\beta}{\lambda-\alpha}\right)^{1/2}}{(1-\lambda^2)^{1/2}} \frac{d\lambda}{\lambda - z} \ .$$

The first integral on the right hand side of (78) can be calculated explicitly:

$$(79) \qquad \int_\alpha^\beta t\left(\frac{z-\alpha}{\beta-z}\right)^{1/2} dz = \pi t \frac{\beta-\alpha}{2} \ .$$

In the case of the second integral in (78), we have, interchanging orders of integration,

$$(80) \qquad \int_\alpha^\beta dz \left(\frac{z-\alpha}{z-\beta}\right)^{1/2} \frac{1}{\pi i}\int_\beta^1 \frac{\left(\frac{\lambda-\beta}{\lambda-\alpha}\right)^{1/2}}{(1-\lambda^2)^{1/2}} \frac{d\lambda}{\lambda - z} =$$

$$-\int_\beta^1 d\lambda \frac{\left(\frac{\lambda-\beta}{\lambda-\alpha}\right)^{1/2}}{(1-\lambda^2)^{1/2}} \frac{1}{\pi i}\int_\alpha^\beta \left(\frac{z-\alpha}{z-\beta}\right)^{1/2} \frac{dz}{z-\lambda} \ .$$

Now the inner integral on the right hand side of (80) is given by $H\zeta(\lambda)$, where

$$\zeta(\lambda) = Re\ i \left[\left(\frac{\lambda-\alpha}{\lambda-\beta}\right)^{1/2} - 1 \right] \ .$$

Therefore, for $\lambda > \beta$, we have

$$\int_\alpha^\beta dz \left(\frac{z-\alpha}{z-\beta}\right)^{1/2} \frac{1}{\pi i}\int_\beta^1 \frac{\left(\frac{\lambda-\beta}{\lambda-\alpha}\right)^{1/2}}{(1-\lambda^2)^{1/2}} \frac{d\lambda}{\lambda - z}$$

$$= -\int_\beta^1 d\lambda \frac{\left(\frac{\lambda-\beta}{\lambda-\alpha}\right)^{1/2}}{(1-\lambda^2)^{1/2}} \left[\left(\frac{\lambda-\alpha}{\lambda-\beta}\right)^{1/2} - 1\right]$$

(81) $$= \int_\beta^1 \frac{1}{(1-\lambda^2)^{1/2}} d\lambda + \int_\beta^1 \frac{\left(\frac{\lambda-\beta}{\lambda-\alpha}\right)^{1/2}}{(1-\lambda^2)^{1/2}} d\lambda \ ,$$

and so, using (76) to express the right hand side of (79) as

$$\frac{\beta-\alpha}{2}\pi t = \frac{\beta-\alpha}{2}\int_\beta^1 \frac{1}{[(\lambda-\alpha)(\lambda-\beta)(1-\lambda^2)]^{1/2}} d\lambda \ ,$$

and using (81) and (75), we see

(82) $$\int_\alpha^1 \psi^* = \int_\beta^1 \frac{\lambda - m}{[(\lambda-\alpha)(\lambda-\beta)(1-\lambda^2)]^{1/2}} d\lambda \ ,$$

where $m = \frac{\beta+\alpha}{2}$.

We will now show that the two conditions (76) and (77) express a complete linearization of the flow, by showing that these conditions determine $\beta(x,t) \geq \alpha(x,t)$ uniquely. Establishing that ψ^* as defined in (75) belongs to the set A is a straightforward calculation, which we omit.

In what follows we will take $t > 1/2$ fixed, and we consider values of x in the interval $[\frac{1}{4t}, 1 - \frac{1}{4t}]$. At the left endpoint of this interval the value of α and β are known from (62),

$$\beta(\frac{1}{4t}, t) = 1 \ ,$$

$$\alpha(\frac{1}{4t}, t) = 1 - \frac{1}{2t^2} \ .$$

Note that $\beta(x,t)$ is fixed, as $I_1 = (\alpha(x,t), 1)$. At the right endpoint, we have from (72)

$$\beta(1 - \frac{1}{4t}, t) = \frac{1}{2t^2} - 1 \ ,$$

$$\alpha(1 - \frac{1}{4t}, t) = -1 \ ,$$

where again we note that $\alpha(x,t)$ is determined by $I_1 = (-1, \beta(x,t))$. Thus we would like to show that there is a unique solution to the system (76 - 77) for $x \in (\frac{1}{4t}, 1 - \frac{1}{4t})$, with $\alpha \in (-1, 1 - \frac{1}{2t^2})$ and $\beta \in (\frac{1}{2t^2} - 1, 1)$ and $\beta \geq \alpha$. We will show this by first establishing that for each β in the interval $(\frac{1}{2t^2} - 1, 1)$, we can find a unique $\alpha(\beta, t)$ such that (76) is satisfied. Then we will show that $X(\alpha(\beta,t),\beta)$ is a decreasing function of β, and so by varying β we will be able to solve (77).

LEMMA 3.5. *Suppose that* $\beta \in (\frac{1}{2t^2}-1,1)$, *and that* $t > 1/2$ *is fixed. Then there is a unique* $\alpha(\beta,t)$ *such that*

$$T(\alpha(\beta,t),\beta) = \pi t \ .$$

Proof.
T is clearly increasing in α,

$$T_\alpha = \int_\beta^1 \frac{1}{[(\lambda-\beta)(\lambda-\alpha)(1-\lambda^2)]^{1/2}} \frac{\frac{1}{2}}{\lambda-\alpha} d\lambda \quad > \quad 0 \ . \tag{83}$$

It is also clear that as α approaches β, T tends to infinity. At $\alpha = -1$, T attains its minimum value,

$$T(-1,\beta) = \frac{\pi}{[2(\beta+1)]^{1/2}} \ .$$

Now for β in the interval $(\frac{1}{2t^2}-1,1)$, we have

$$\min_{-1\le\alpha\le\beta} T(\alpha,\beta) \ = \ \frac{\pi}{[2(\beta+1)]^{1/2}} < \pi t \tag{84}$$

as

$$\beta > \frac{1}{2t^2} - 1 \Rightarrow \ \frac{1}{2(\beta+1)} < t^2 \ . \tag{85}$$

Thus, for $\beta \in (\frac{1}{2t^2}-1\ ,\ 1)$ we see by varying α that

$$\pi t \ \in \ (\min_\alpha T(\alpha,\beta)\ ,\ \max_\alpha T(\alpha,\beta)) \ .$$

By continuity we can find a (unique) $\alpha(\beta,t) < \beta$ that solves $T(\alpha,\beta) = \pi t$.
□

We now take this solution $\alpha(\beta,t)$ and substitute it into the function $X(\alpha,\beta)$, and show that the function obtained is then monotone decreasing in β.

LEMMA 3.6. *The function defined by*

$$X(\beta,t) \equiv X(\alpha(\beta,t),\beta)$$

is decreasing in β *for* $\beta \in (\frac{1}{2t^2}-1,1)$.

Proof.
We first show that the following formulae are true:

$$\frac{\partial}{\partial\beta} X(\alpha,\beta) = \frac{(\beta-\alpha)}{2} T_\beta \ , \tag{86}$$

$$\frac{\partial}{\partial\alpha} X(\alpha,\beta) = -\frac{(\beta-\alpha)}{2} T_\alpha \ . \tag{87}$$

To establish the first, we write

$$X = \int_\beta^1 \frac{\lambda - \beta}{R^{1/2}} d\lambda + (\beta - m) T(\alpha, \beta) \ ,$$

where

$$R(\lambda) = (\lambda - \alpha)(\lambda - \beta)(1 - \lambda^2) \ ,$$

which yields

$$\frac{\partial}{\partial \beta} X(\alpha, \beta) = \frac{-1}{2} T(\alpha, \beta) + \frac{1}{2} T(\alpha, \beta) + \frac{\beta - \alpha}{2} T_\beta,$$

which establishes (86). The second formula, (87), follows from similar calculations. The signature table for

$$R(\lambda)^{1/2} = \lim_{\epsilon \downarrow 0} R(\lambda + i\epsilon)^{1/2}$$

is given by

$$\begin{aligned} R(\lambda)^{1/2} &\in \mathbf{R}_+ \ , \ \lambda > 1 \ , \\ R(\lambda)^{1/2} &\in i\mathbf{R}_+ \ , \ \lambda \in (\beta, 1) \ , \\ R(\lambda)^{1/2} &\in \mathbf{R}_- \ , \ \lambda \in (\alpha, \beta) \ , \\ R(\lambda)^{1/2} &\in i\mathbf{R}_- \ , \ \lambda \in (-1, \alpha) \ , \\ R(\lambda)^{1/2} &\in \mathbf{R}_+ \ , \ \lambda < -1 \ . \end{aligned}$$

Using formulae (86 - 87), and the fact that

(88) $$\frac{\partial}{\partial \beta} \alpha(\beta, t) = -\frac{T_\beta}{T_\alpha} \ ,$$

we obtain

(89) $$\begin{aligned} \frac{\partial}{\partial \beta} X(\beta, t) &= \frac{\partial}{\partial \alpha} X \cdot \frac{\partial \alpha(\beta, t)}{\partial \beta} + \frac{\partial}{\partial \beta} X \\ &= -\frac{\beta - \alpha}{2} T_\alpha \left(-\frac{T_\beta}{T_\alpha} \right) + \frac{\beta - \alpha}{2} T_\beta \\ &= (\beta - \alpha) T_\beta \ , \end{aligned}$$

and so we see that to prove the lemma, we must show that $T_\beta(\alpha, \beta)$ is negative. This is done by contour integration:

(90) $$\int_{-1}^\alpha \frac{d\lambda}{R(\lambda)^{1/2}} + \int_\beta^1 \frac{d\lambda}{R(\lambda)^{1/2}} = \frac{1}{2} \int_c \frac{d\lambda}{R(\lambda)^{1/2}} \ ,$$

where c is a closed, clockwise contour encircling $(-1,1)$. Letting the diameter of c go to infinity, we see that the integral is zero. Hence

$$\begin{aligned} T &= \int_\beta^1 \frac{i}{[(\lambda-\alpha)(\lambda-\beta)(\lambda^2-1)]^{1/2}}\, d\lambda \\ &= -\int_{-1}^{\alpha} \frac{i}{[(\lambda-\alpha)(\lambda-\beta)(\lambda^2-1)]^{1/2}} d\lambda \ , \end{aligned}$$

from which we see explicitly that

(91) $$T_\beta \ = \ -\frac{1}{2}\int_{-1}^{\alpha} \frac{i}{R(\lambda)^{1/2}} \frac{d\lambda}{\lambda-\beta} \ < \ 0 \ ,$$

as $R(\lambda)^{1/2} \in i\mathbf{R}_-$ for $-1 < \lambda < \alpha$ □

Summarizing, we have shown that there is a unique $\alpha(\beta, t) < \beta$ solving the first integral condition (76), and that the function $X(\beta, t)$ defined above is decreasing in β. In particular, letting $\beta \uparrow 1$ in (76) we obtain

$$\frac{\pi}{\sqrt{2(1-\alpha)}} \ = \ \pi t \ ,$$

and hence

(92) $$\alpha \ = \ 1 - \frac{1}{2t^2} \ .$$

On the other hand, from (84) and (85), as $\beta \to \frac{1}{2t^2} - 1$,

(93) $$\alpha \to -1 \ .$$

Simple calculations now show from (77) that

$$\begin{aligned} X(\beta = 1, t) \ &= \ \frac{1}{4t} \ , \\ X(\beta = \frac{1}{2t^2} - 1, t) \ &= \ 1 - \frac{1}{4t} \ , \end{aligned}$$

which shows that we can solve (uniquely) the pair of equations for all $x \in [\frac{1}{4t}, 1 - \frac{1}{4t}]$, for any $t > 1/2$.

We have proved

THEOREM 3.1. *The conditions*

$$\begin{aligned} T(\alpha,\beta) &= \pi t \ , \\ X(\alpha,\beta) &= \pi x \end{aligned}$$

determine two continuously differentiable functions of x and t, $\alpha(x,t), \beta(x,t)$ for $t > \frac{1}{2}$, and $x \in [\frac{1}{4t}, 1 - \frac{1}{4t}]$.

We note here that the two conditions (76) and (77) imply that $\alpha(x,t)$ and $\beta(x,t)$ solve the Riemann invariant form of the continuous Toda equations,

(94) $$\frac{\partial}{\partial t}\beta \ = \ \frac{\beta-\alpha}{2}\frac{\partial}{\partial x}\beta$$

(95) $$\frac{\partial}{\partial t}\alpha \ = \ -\frac{\beta-\alpha}{2}\frac{\partial}{\partial x}\alpha \ .$$

Indeed, differentiating the two conditions with respect to x and t, we obtain a matrix equation for α_t, α_x and β_t, β_x:

$$\begin{pmatrix} T_\alpha & T_\beta \\ X_\alpha & X_\beta \end{pmatrix} \begin{pmatrix} \alpha_t & \alpha_x \\ \beta_t & \beta_x \end{pmatrix} = \begin{pmatrix} \pi & 0 \\ 0 & \pi \end{pmatrix} ,$$

and using (86 - 87),

$$\begin{pmatrix} \alpha_t & \alpha_x \\ \beta_t & \beta_x \end{pmatrix} = \frac{1}{(\beta-\alpha)T_\alpha T_\beta} \begin{pmatrix} \frac{\beta-\alpha}{2}T_\beta & -T_\beta \\ \frac{\beta-\alpha}{2}T_\alpha & T_\alpha \end{pmatrix} \begin{pmatrix} \pi & 0 \\ 0 & \pi \end{pmatrix} .$$

Because the derivatives of T with respect to α and β are nonzero (see (91), (83), we see that

$$\beta_t = \frac{\pi}{2T_\beta} = \frac{\beta-\alpha}{2}\beta_x ,$$

and

$$\alpha_t = \frac{\pi}{2T_\alpha} = -\frac{\beta-\alpha}{2}\alpha_x$$

which proves (94) and (95).

Remark: Note that the above calculations show that if α, β solve $T = \pi t$, $X = \pi x$, then necessarily α and β solve (94) and (95). This fact is true in great generality, as we will see in the succeeding chapters.

Finally we note that the conditions arising out of the Riemann - Hilbert problem allow us to establish rigorous long-time asymptotics for the continuous Toda equations, and hence for the continuum limit of the Toda lattice.

THEOREM 3.2.

$$\begin{aligned} \lim_{t\to\infty}\lim_{\epsilon\to 0} \hat{b}_\epsilon(x,t) &= \lim_{t\to\infty} \frac{\beta(x,t)-\alpha(x,t)}{4} &= 0 \\ \lim_{t\to\infty}\lim_{\epsilon\to 0} \hat{a}_\epsilon(x,t) &= \lim_{t\to\infty} \frac{\beta(x,t)+\alpha(x,t)}{2} &= \Phi^{-1}(\pi x) = \cos(\pi x) . \end{aligned}$$

Proof.

From the equation

$$T(\alpha,\beta) \equiv \int_\beta^1 \frac{1}{[(\lambda-\beta)(\lambda-\alpha)(1-\lambda^2)]^{1/2}} d\lambda = \pi t ,$$

we see that as $t \to \infty$, we must have $\beta - \alpha \to 0$. This proves the first claim.

As α approaches β, the function $X(\beta,t)$ tends, by dominated convergence, to

$$\int_\beta^1 \frac{d\lambda}{[1-\lambda^2]^{1/2}} = \cos^{-1}\beta .$$

From this we see that $X(\beta,t) = \pi x$ implies, in the long time limit, that

$$\lim_{\epsilon\to 0} \hat{a}_\epsilon(x,t) = \frac{\beta(x,t)+\alpha(x,t)}{2} \to \cos \pi x .$$

□

This is the continuum analog of the fact that the finite dimensional Toda lattice with zero boundary conditions converges to the eigenvalues of the Toda matrix $L(t=0)$ [**16**]. Here we see that the continuum limit of the Toda lattice converges to the asymptotic spectral measure as t tends to infinity.

This simple example sheds some light on the work to come. We have shown that the continuum limit of the Toda lattice, with this initial data, converges to the solution of the continuous Toda equations. Further, we have seen that the continuous Toda equations are equivalent to a Riemann - Hilbert problem in complex function theory, and that analysis of this problem allows us to prove a variety of facts about the underlying PDE system, and the continuum limit of the lattice.

We have shown that for this data the continuous Toda equations possess a solution for all $t > 0$, that this solution has two corners whose evolution is explicit, and we have established rigorous long-time asymptotics for the solution.

We shall see that these results remain true for a larger class of data, and that our methods enable us to derive explicit information concerning shock formation, and the evolution of the continuum limit of the Toda lattice beyond shock time.

Remark: We see explicitly, for example from (61), that the maximizer ψ^* lies in $L^p(\mathbf{R})$ for $1 \leq p < 2$, but not in $L^2(\mathbf{R})$. For this reason, in succeeding chapters, we seek the maximzer ψ^* in $L^p(\mathbf{R})$, with $p < 2$.

CHAPTER 4

Monotone Initial Data

In this chapter we consider the general class of data, for which the Riemann invariants of the inital data, $\alpha^0(x)$ and $\beta^0(x)$, defined by

$$\begin{aligned} \alpha^0(x) &= a^0(x) - 2b^0(x) \ , \\ \beta^0(x) &= a^0(x) + 2b^0(x) \end{aligned}$$

are arbitrary smooth monotone **decreasing** functions, with $\alpha^0(x) < \beta^0(x)$ for $x \in (0,1)$, $\alpha^0(0) = \beta^0(0)$, $\alpha^0(1) = \beta^0(1)$. For this data, our WKB analysis indicates that the eigenvalues $\{\lambda_j\}$ of the matrix $L(t=0)$ satisfy the equation

$$\Phi(\lambda_j) \equiv \int_{x_-(\lambda_j)}^{x_+(\lambda_j)} \cos^{-1}(\frac{\lambda_j - a}{2b})\ dx \quad = \quad (j\pi + \pi/2)\epsilon \tag{96}$$

asymptotically, and the associated norming constant is given (asymptotically) by

$$f_{\lambda_j}(1) \approx \exp\left[-\frac{\theta(\lambda_j)}{\epsilon}\right]$$

where θ has the form

$$\theta(\lambda) = \int_0^{x_-(\lambda)} \cosh^{-1}(\frac{\lambda - a}{2b})\ dx \ , \tag{97}$$

and hence

$$\theta'(\lambda) \ = \ \int_0^{\lambda} \frac{1}{[(\lambda - \alpha^0(y))(\lambda - \beta^0(y))]^{1/2}}\ dy \ . \tag{98}$$

As described in Chapter 2, we consider the maximization problem

$$Q(x,t) \ = \ \max_{\psi \in A} \frac{2}{\pi}(t\lambda - \theta \ , \ \psi) \ + \ \frac{1}{2\pi^2}(L\psi \ , \ \psi) \ , \tag{99}$$

where A is the set of all Lebesgue measurable functions ψ satisfying

$$\begin{aligned} 0 \le \psi \le \phi \ , \\ \int \psi \ = \ \pi x \ , \end{aligned}$$

with $\phi(\lambda)$ given by

$$\phi(\lambda) \ = \ -\frac{d}{d\lambda}\Phi(\lambda) \ = \ \int_{x_-(\lambda)}^{x_+(\lambda)} \frac{i}{[(\lambda - \alpha^0(y))(\lambda - \beta^0(y))]^{1/2}} dy \tag{100}$$

The Euler - Lagrange equations arising from the maximization problem (99) become

$$
\begin{aligned}
(101) \qquad \frac{2}{\pi}(t\lambda - \theta(\lambda)) + \frac{1}{\pi^2}L\psi(\lambda) + l &= 0 \ , \quad \lambda \in I_1 \ , \\
\psi(\lambda) &= 0 \ , \ \lambda \in I_2 \ , \\
\psi(\lambda) &= \phi(\lambda) \ , \quad , \lambda \in I_3 \ .
\end{aligned}
$$

Differentiating, we obtain the Riemann - Hilbert problem as outlined in Chapter 2,

$$
\begin{aligned}
(102) \qquad H\psi &= \theta'(\lambda) - t \ , \quad \lambda \in I_1 \ , \\
(103) \qquad \psi(\lambda) &= 0 \ , \quad \lambda \in I_2 \ , \\
(104) \qquad \psi(\lambda) &= \phi(\lambda) \ , \quad \lambda \in I_3 \ .
\end{aligned}
$$

The results of Chapter 3 lead us to the following ansatz for the Riemann - Hilbert problem satisfied by the maximizer $\psi = \psi^*$, again extended to an $L^p(-\infty, \infty)$ function by setting $\psi^*(\lambda) = 0$ for $|\lambda| > 1$. We seek a function $\psi^* \in L^p(\mathbf{R})$, satisfying the properties (102 - 104), where the I_j's are chosen as follows,

$$
\begin{aligned}
(105) \qquad I_1 &= (\alpha(x,t), \beta(x,t)) \\
(106) \qquad I_2 &= (-\infty, \alpha(x,t)) \cup (1, \infty) \\
(107) \qquad I_3 &= (\beta(x,t), 1) \ .
\end{aligned}
$$

The conditions (102 - 104), together with (105 - 107) comprise the so - called "zero gap" ansatz, and the solution is obtained as follows.

Using the general theory of the Hilbert transform operator, the Riemann - Hilbert problem can be re - written in terms of the function

$$
f(\lambda) = \psi^*(\lambda) + iH\psi^*(\lambda) \in L^p(\mathbf{R}) \ , 1 \le p < 2 \ ,
$$

which is the boundary value of a function (which we will also call $f(\lambda)$) analytic in the upper half plane, and which satisfies

$$
\begin{aligned}
Im\ f(\lambda) &= \theta'(\lambda) - t \ , \ \lambda \in I_1 \ , \\
Re\ f(\lambda) &= 0 \ , \ \lambda \in I_2 \ , \\
Re\ f(\lambda) &= \phi(\lambda) \ , \ \lambda \in I_3 \ .
\end{aligned}
$$

We consider the induced problem for the function

$$
g(\lambda) = \frac{f(\lambda)}{R(\lambda)^{1/2}} \ ,
$$

with $R(\lambda)$ defined by

$$
R(\lambda) = (\lambda - \alpha)(\lambda - \beta) \ ,
$$

which is given by

$$
\begin{aligned}
(108)\qquad Re\ g(\lambda) &= i\frac{\theta' - t}{R(\lambda)^{1/2}} \quad , \lambda \in I_1 \ , \\
Re\ g(\lambda) &= 0 \quad , \lambda \in I_2 \ , \\
Re\ g(\lambda) &= -\frac{\phi(\lambda)}{R(\lambda)^{1/2}} \quad , \lambda \in I_3 \ ,
\end{aligned}
$$

as we take $R(\lambda)^{1/2} > 0$ for $\lambda > \beta$. The solution of the problem (108) is explicit,

$$
g(z) = \frac{1}{\pi i}\left(\int_\alpha^\beta \frac{i(\theta'(\lambda) - t)}{R(\lambda)^{1/2}}\frac{d\lambda}{\lambda - z} + \int_\beta^1 \frac{\phi(\lambda)}{R(\lambda)^{1/2}}\frac{d\lambda}{\lambda - z}\right),
$$

and hence we have constructed the function $f(z)$, and $\psi^* = Re\ f(z)$,

$$
(109)\quad f(z) = R(z)^{1/2}\frac{1}{\pi i}\left(\int_\alpha^\beta \frac{i(\theta'(\lambda) - t)}{R(\lambda)^{1/2}}\frac{d\lambda}{\lambda - z} + \int_\beta^1 \frac{\phi(\lambda)}{R(\lambda)^{1/2}}\frac{d\lambda}{\lambda - z}\right) .
$$

By expanding the representation (109) about $z = \infty$, we see that the condition that $f \in L^p(\mathbf{R})$ becomes the integral condition

$$
(110)\qquad T(\alpha,\beta) \equiv \int_\alpha^\beta \frac{1}{R(\lambda)^{1/2}}\, i\theta' d\lambda \ + \ \int_\beta^1 \frac{1}{R(\lambda)^{1/2}}\phi d\lambda = \pi t \ .
$$

The condition $f \in L^p(\mathbf{R})$ arises because the formula (109) does not necessarily tend to zero as $\lambda \to \infty$. However, the integral condition (110) can also be interpreted in terms of the constraints

$$
0 \leq \psi^* \leq \phi \ .
$$

Indeed, if we consider the induced Riemann - Hilbert problem for the function $\tilde{g}$ defined by

$$
\tilde{g} = \left(\frac{\lambda - \beta}{\lambda - \alpha}\right)^{1/2} f(\lambda) \ ,
$$

we find that the solution is

$$
\begin{aligned}
(111)\qquad \tilde{g}(z) &= \frac{1}{\pi i}\int_\alpha^\beta i(\theta'(\lambda) - t)\left(\frac{\lambda - \beta}{\lambda - \alpha}\right)^{1/2}\frac{d\lambda}{\lambda - z} \\
&\quad + \int_\beta^1 \phi(\lambda)\left(\frac{\lambda - \beta}{\lambda - \alpha}\right)^{1/2}\frac{d\lambda}{\lambda - z} \ ,
\end{aligned}
$$

from which we obtain another representation for $f(z)$ which is clearly in $L^p(\mathbf{R})$ (for $\lambda \to \infty$):

$$
f(z) = \left(\frac{\lambda - \alpha}{\lambda - \beta}\right)^{1/2} \tilde{g}(z) \ ,
$$

but which may blow up at $z = \beta$. This would violate the constraint $Re\ f = \psi^* \leq \phi$, and so we must impose the constraint

$$\lim_{z \to \beta} \tilde{g}(z) = 0 \ ,$$

which becomes, from (111),

$$\int_{\alpha}^{\beta} \frac{1}{R(\lambda)^{1/2}} i\theta' d\lambda + \int_{\beta}^{1} \frac{1}{R(\lambda)^{1/2}} \phi d\lambda = \pi t \ .$$

As in Chapter 3, the condition $\int \psi^* = \pi x$ leads to

$$X(\alpha, \beta) \equiv \int_{\alpha}^{\beta} i \frac{\lambda - m}{R(\lambda)^{1/2}} \theta' d\lambda + \int_{\beta}^{1} \frac{\lambda - m}{R(\lambda)^{1/2}} \phi d\lambda = \pi x \ , \tag{112}$$

where $m = \frac{\beta + \alpha}{2}$.

Again we have obtained two condions, and expressed a complete linearization of the flow. The next step is to show that these two conditions uniquely determine the endpoints of the band α and β, at $t = 0$ first, and then for $t > 0$.

LEMMA 4.1. *Suppose $t = 0$. Then conditions (110) and (112) uniquely determine two continuously differentiable functions $\alpha(x)$, $\beta(x)$, which are precisely the Riemann invariants of the initial data,*

$$\alpha(x) = \alpha^0(x) \ ,$$
$$\beta(x) = \beta^0(x) \ .$$

Proof.

We first consider the condition for T. Using the definition (98) of $\theta'(\lambda)$, and (100) of $\phi(\lambda)$, we have

$$T(\alpha, \beta) = \int_{\alpha}^{\beta} \int_{0}^{x_-(\lambda)} \frac{i}{[R(\lambda) R^0(\lambda, y)]^{1/2}} dy d\lambda + \int_{\beta}^{1} \int_{x_-(\lambda)}^{x_+(\lambda)} \frac{i}{[R(\lambda) R^0(\lambda, y)]^{1/2}} dy d\lambda \ ,$$

where

$$R^0(\lambda, y) = (\lambda - \alpha^0(y))(\lambda - \beta^0(y)) \ .$$

Aided by figure (1), we interchange orders of integration,

$$(113)\quad T(\alpha,\beta) = \int_0^{x_-(\beta)}\int_\alpha^\beta \frac{i}{[R(\lambda)R^0(\lambda,y)]^{1/2}}\,d\lambda dy + \int_{x_-(\beta)}^{x_-(\alpha)}\int_\alpha^{\alpha_0(y)} \frac{i}{[R(\lambda)R^0(\lambda,y)]^{1/2}}\,d\lambda dy + \int_0^{x_-(\beta)}\int_{\alpha_0(y)}^{\beta_0(y)} \frac{i}{[R(\lambda)R^0(\lambda,y)]^{1/2}}\,d\lambda dy + \int_{x_-(\beta)}^{x_+(\beta)}\int_\beta^{\beta_0(y)} \frac{i}{[R(\lambda)R^0(\lambda,y)]^{1/2}}\,d\lambda dy\ ,$$

Now we can write (113) as

$$(114)\quad T(\alpha,\beta) = \int_0^{x_-(\beta)}\left\{\int_\alpha^\beta + \int_{\alpha_0(y)}^{\beta_0(y)}\right\} \frac{i}{[R(\lambda)R^0(\lambda,y)]^{1/2}}\,d\lambda dy + \int_{x_-(\beta)}^{x_-(\alpha)}\left\{\int_\alpha^{\alpha_0(y)} + \int_\beta^{\beta_0(y)}\right\} \frac{i}{[R(\lambda)R^0(\lambda,y)]^{1/2}}\,d\lambda dy + \int_{x_-(\alpha)}^{x_+(\beta)}\int_\beta^{\beta_0(y)} \frac{i}{[R(\lambda)R^0(\lambda,y)]^{1/2}}\,d\lambda dy\ ,$$

and we see that the first two integrals vanish by contour deformation (cf. (90)). Thus the condition for T is expressed as

$$(115)\quad T(\alpha,\beta) = \int_{x_-(\alpha)}^{x_+(\beta)}\int_\beta^{\beta_0(y)} \frac{i}{[R(\lambda)R^0(\lambda,y)]^{1/2}}\,d\lambda dy = \pi t,$$

where the integration is over the shaded region shown in Figure (1). Noting that the integrand is of one sign over the region of integration, we see that the only way to satisfy the condition (110) at $t = 0$ is if the region itself vanishes. This implies that

$$(116)\qquad x_-(\alpha) = x_+(\beta)\ .$$

The steps used to obtain the representation (114) for the function T can also be used for the function X, and so we can write

$$(117)\ X(\alpha,\beta) = \int_0^{x_-(\beta)}\left\{\int_\alpha^\beta + \int_{\alpha_0(y)}^{\beta_0(y)}\right\} \frac{i(\lambda-m)}{[R(\lambda)R^0(\lambda,y)]^{1/2}}\,d\lambda dy + \int_{x_-(\beta)}^{x_-(\alpha)}\left\{\int_\alpha^{\alpha_0(y)} + \int_\beta^{\beta_0(y)}\right\} \frac{i(\lambda-m)}{[R(\lambda)R^0(\lambda,y)]^{1/2}}\,d\lambda dy + \int_{x_-(\alpha)}^{x_+(\beta)}\int_\beta^{\beta_0(y)} \frac{i(\lambda-m)}{[R(\lambda)R^0(\lambda,y)]^{1/2}}\,d\lambda dy\ .$$

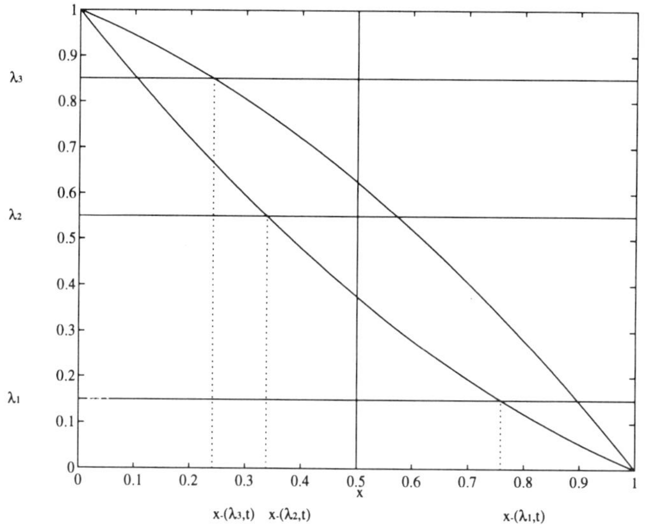

FIGURE 1. The Riemann Invariants $\alpha^0(x)$ and $\beta^0(x)$ (note $\alpha^0(x) \le \beta^0(x)$), and a particular choice of α and β, along with locations of $x_-(\alpha)$, and $x_+(\beta)$, and the region of integration for $T(\alpha,\beta)$ in (3.20).

By contour integration, we have the following identities

$$\left\{\int_\alpha^\beta + \int_{\alpha_0(y)}^{\beta_0(y)}\right\} \frac{i(\lambda - m)d\lambda}{[R(\lambda)R^0(\lambda,y)]^{1/2}} = \pi \text{ , } for\ y \in (0, x_-(\beta)) \text{ ,}$$

$$\left\{\int_\alpha^{\alpha_0(y)} + \int_\beta^{\beta_0(y)}\right\} \frac{i(\lambda - m)d\lambda}{[R(\lambda)R^0(\lambda,y)]^{1/2}} = \pi, \ for\ y \in (x_-(\beta), x_-(\alpha)) \text{ ,}$$

and hence the condition (112) becomes

$$(118) \quad X(\alpha,\beta) = \pi x_-(\alpha) + \int_{x_-(\alpha)}^{x_+(\beta)} \int_\beta^{\beta_0(y)} \frac{i\,(\lambda - m)\,d\lambda dy}{[R(\lambda)R^0(\lambda,y)]^{1/2}} = \pi x \text{ ,}$$

but then by (116), we obtain

$$\pi x_-(\alpha) = \pi x \text{ ,}$$

and so we have reconstructed the initial data, $x_-(\alpha) = x$, $x_+(\beta) = x$, i.e. $\alpha = \alpha^0(x)$, $\beta^0(x)$. □

We are now ready to show that we can solve this system for all $t > 0$. This is accomplished, as in Chapter 3, by showing that the function $T(\alpha,\beta)$ is, for fixed β, increasing in α for all $\alpha < \beta$. This will establish the existence of a unique $\alpha(\beta,t)$ such that condition (110) is satisfied. Substituting this solution into the function $X(\alpha,\beta)$, we obtain a function of β alone, which we call $X(\beta, t)$. By showing that $X(\beta, t)$ is decreasing in β, we will, as in Chapter 3, establish the result.

LEMMA 4.2. *The function $T(\alpha, \beta)$ defined in (110) satisfies*

$$T_\alpha > 0 \ .$$

Proof.

Using the formula (115), and by studying the region of integration (see figure (1)), one sees that if we increase α, the region of integration increases, and over the original region, the integrand has increased. Thus $T(\alpha, \beta)$ is increasing in α. We can, of course differentiate the formula (115), and doing so we obtain

$$\begin{aligned} T_\alpha &= \frac{\pi \partial_\alpha x_-(\alpha)}{[(\alpha-\beta)(\alpha-\beta_0(x_-(\alpha)))]^{1/2}} + \\ &+ \int_{x_-(\alpha)}^{x_+(\beta)} \int_\beta^{\beta_0(y)} \frac{i}{[R(\lambda)R^0(\lambda, y)]^{1/2}} \frac{\frac{1}{2}}{\lambda - \alpha} \, d\lambda dy \ , \end{aligned}$$

where the first term arises from evaluating the inner integral in (115) at $y = x_-(\alpha)$, as follows

$$\begin{aligned} &\int_\beta^{\beta^0(x_-(\alpha))} \frac{i}{[R(\lambda)R^0(\lambda, x_-(\alpha))]^{1/2}} \, d\lambda \\ &= \int_\beta^{\beta^0(x_-(\alpha))} \frac{i}{[(\lambda-\beta)(\lambda-\beta^0(x_-(\alpha)))]^{1/2}} \frac{d\lambda}{\lambda-\alpha} \quad \left(as \ \alpha^0(x_-(\alpha)) = \alpha\right) \\ &= -\frac{\pi}{[(\alpha-\beta)(\alpha-\beta^0(x_-(\alpha)))]^{1/2}} \ , \end{aligned}$$

by contour integration. But both $[(\alpha-\beta)(\alpha-\beta_0(x_-(\alpha)))]^{1/2}$ and $\partial_\alpha x_-(\alpha)$ are negative, and hence $T_\alpha > 0$. □

Now from this we easily see that we have the existence of a solution to the equation $T(\alpha, \beta) = \pi t$, for all t. Indeed, we have already seen that we can solve this equation for $t = 0$, and it is easy to see from (110) that, for fixed β, if we let α approach β, then $T(\alpha, \beta) \to \infty$. Thus we have a solution, which is unique, by the above lemma.

LEMMA 4.3. *Set*

$$X(\beta, t) = X(\alpha(\beta, t), \beta) \ . \tag{119}$$

Then

$$\frac{\partial}{\partial \beta} X(\alpha(\beta, t), \beta) < 0 \ .$$

Proof.

We first establish the following relations (cf. (86,87)),

$$(120)\qquad \frac{\partial}{\partial\alpha}X(\alpha,\beta) = -\frac{\beta-\alpha}{2}\,\frac{\partial}{\partial\alpha}T(\alpha,\beta)\ ,$$

$$(121)\qquad \frac{\partial}{\partial\beta}X(\alpha,\beta) = \frac{\beta-\alpha}{2}\,\frac{\partial}{\partial\beta}T(\alpha,\beta)\ .$$

From formulae (110 - 112), we have

$$X = \int_\alpha^\beta \frac{i\theta'}{R^{1/2}}(\lambda-\beta)d\lambda + \int_\beta^1 \frac{\phi}{R^{1/2}}(\lambda-\beta)d\lambda + \frac{\beta-\alpha}{2}T\ ,$$

and thus

$$X_\beta = -\frac{1}{2}T + \frac{1}{2}T + \frac{\beta-\alpha}{2}T_\beta\ ,$$

which establishes (121). Relationship (120) follows similarly.

As in (88, 89) of Chapter 3, these formulae imply

$$\frac{\partial}{\partial\beta}X(\beta,t) = (\beta-\alpha)\ \left[\frac{\partial}{\partial\beta}T(\alpha,\beta)\right]_{\alpha=\alpha(\beta,t)}\ .$$

So to prove the lemma it is sufficient to verify that for all $\alpha<\beta$,

$$\frac{\partial}{\partial\beta}T(\alpha,\beta) < 0\ .$$

To show this, we use another version of the formula (115),

$$(122)\qquad T(\alpha,\beta) = -\int_{x_-(\alpha)}^{x_+(\beta)}\int_{\alpha_0(y)}^{\alpha}\frac{i}{[R(\lambda)R^0(\lambda,y)]^{1/2}}\,d\lambda dy\ .$$

which follows from the contour integral

$$\left(\int_{\alpha^0(y)}^{\alpha} + \int_\beta^{\beta^0(y)}\right)\frac{i}{[R(\lambda)R^0(\lambda,y)]^{1/2}} = \frac{1}{2}\int_\gamma \frac{d\lambda}{R(\lambda)^{1/2}}\ ,$$

where γ is a closed, clockwise contour encircling $(-1,1)$, by letting the diameter of γ go to infinity.

Once again, it is easy to see that if we decrease β, then we are gaining a region of integration, and over the old region, the integrand is increasing. So T is decreasing in β. Once again, similar calculations to those used to differentiate T with respect to α show that

$$\begin{aligned} T_\beta &= \frac{\pi\partial_\beta x_+(\beta)}{[(\beta-\alpha)(\beta-\alpha_0(x_+(\beta)))]^{1/2}} + \\ &\quad - \int_{x_-(\alpha)}^{x_+(\beta)}\int_{\alpha_0(y)}^{\alpha}\frac{i}{[R(\lambda)R^0(\lambda,y)]^{1/2}}\,\frac{\frac{1}{2}}{\lambda-\beta}\,d\lambda dy\ , \end{aligned}$$

which is clearly negative.

□

Remark: as in Chapter 3, we have

$$\beta_x = \frac{\pi}{(\beta-\alpha)T_\beta} < 0 ,$$

and

$$\alpha_x = -\frac{\pi}{(\beta-\alpha)T_\alpha} < 0 .$$

Thus we have established the following theorem.

THEOREM 4.1. *For all $t \geq 0$ the system (110), (112) uniquely determines the endpoints of the band I_1, $\alpha(x,t)$ and $\beta(x,t)$ as continuously differentiable functions of x, t. Moreover $\alpha(x,t)$ and $\beta(x,t)$ are monotone decreasing functions of x.*

Again, just as for the simple data studied in the previous section, these endpoints evolve so as to solve the Riemann invariant form of the continuous Toda equations,

$$\begin{aligned} \frac{\partial}{\partial t}\beta &= \frac{\beta-\alpha}{2}\frac{\partial}{\partial x}\beta \\ \frac{\partial}{\partial t}\alpha &= -\frac{\beta-\alpha}{2}\frac{\partial}{\partial x}\alpha . \end{aligned}$$

Specifically, this shows that the continuous Toda equations with initial data from the class of data being considered, are equivalent to a Riemann - Hilbert problem. In future chapters we will show this to be true not only for the continuous Toda equations, but also for the Whitham modulational equations arising from Toda, as well as those for KdV.

THEOREM 4.2. *The function*

$$\psi^*(z) = Re\ f(z) = Re \lim_{\epsilon\downarrow 0} f(z+i\epsilon) ,$$

where $f(z)$ is defined in (109), solves the maximization problem (99).

Proof. We must show that $\psi^* \in A$, and that ψ^* maximizes $Q(\psi)$. We suppose for the moment that $\psi^* \in A$, and consider an arbitrary candidate for the maximizer,

$$\psi^* + \chi ,$$

where $\psi^* + \chi \in A$. This imposes the constraints on the function χ,

$$\begin{aligned} 0 < \psi^* &+ \chi < \phi , \\ \int \chi d\lambda &= 0. \end{aligned}$$

We have

$$
\begin{aligned}
Q(\psi^* + \chi) &= Q(\psi^*) + \frac{1}{2\pi^2}\Big(L\chi, \chi\Big) + \frac{2}{\pi}\Big(t\lambda - \theta\ ,\ \chi\Big)\ + \\
&\quad \frac{1}{\pi^2}\Big(L\psi^*\ ,\ \chi\Big) \\
&= Q(\psi^*) + \frac{1}{2\pi^2}\Big(L\chi, \chi\Big) + \frac{2}{\pi}\Big(t\lambda - \theta\ ,\ \chi\Big)_{(-1,\alpha)\cup(\beta,1)}\ + \\
&\quad -l\Big(1\ ,\ \chi\Big)_{(\alpha,\beta)} + \frac{1}{\pi^2}\Big(L\psi^*\ ,\ \chi\Big)_{(-1,\alpha)\cup(\beta,1)} \\
(123) &= Q(\psi^*) + \frac{1}{2\pi^2}\Big(L\chi, \chi\Big) + \Big(\frac{1}{\pi^2}L\psi^* + \frac{2}{\pi}(t\lambda - \theta) + l\ ,\ \chi\Big)_{(-1,\alpha)\cup(\beta,1)}\ ,
\end{aligned}
$$

where we have used that ψ^* solves the Euler - Lagrange equations (101), and that χ has mean zero.

Now using the representations for $L\psi^*(\lambda)$,

$$
L\psi^*(\lambda) = L\psi^*(\beta) + 2\pi \int_\beta^\lambda H\psi^*(\mu)\, d\mu\ , \tag{124}
$$

$$
L\psi^*(\lambda) = L\psi^*(\alpha) + 2\pi \int_\alpha^\lambda H\psi^*(\mu)\, d\mu\ , \tag{125}
$$

the right hand side of (123) becomes

$$
\begin{aligned}
&Q(\psi^*) + \frac{1}{2\pi^2}\Big(L\chi, \chi\Big) + \\
&+ \Big(\frac{1}{\pi^2}L\psi^*(\alpha) + l - (\theta(\alpha) - t\alpha) + \frac{2}{\pi}\int_\alpha^\lambda H\psi^* - (\theta'(\mu) - t)d\mu, \chi\Big)_{(-1,\alpha)} \\
&+ \Big(\frac{1}{\pi^2}L\psi^*(\beta) + l - (\theta(\beta) - t\beta) + \frac{2}{\pi}\int_\beta^\lambda H\psi^* - (\theta'(\mu) - t)d\mu, \chi\Big)_{(\beta,1)}\ .
\end{aligned}
$$

From the variational equation (101), we see that

$$
\begin{aligned}
\frac{1}{\pi^2}L\psi^*(\alpha) + l - (\theta(\alpha) - t\alpha) &= 0\ , \\
\frac{1}{\pi^2}L\psi^*(\beta) + l - (\theta(\beta) - t\beta) &= 0\ .
\end{aligned}
$$

and we obtain, finally,

$$
\begin{aligned}
Q(\psi^* + \chi) &= Q(\psi^*) + \frac{1}{2\pi^2}(L\chi, \chi) + \\
&\quad \frac{2}{\pi}\Big(\int_\alpha^\lambda H\psi^*(\mu) - \Big(\theta'(\mu) - t\Big)\ d\mu\ ,\ \chi\Big)_{(-1,\alpha)}\ + \\
&\quad \frac{2}{\pi}\Big(\int_\beta^\lambda H\psi^*(\mu) - \Big(\theta'(\mu) - t\Big)\ d\mu\ ,\ \chi\Big)_{(\beta,1)}
\end{aligned}
$$

Now to see that we are solving the maximization problem, we must show that the last terms of this formula are negative. The first is an inner product over the region $[-1, \alpha]$, where the function χ must be positive, as $\psi^* = 0$. The second term is an inner product over the region $[\beta, 1]$, where the function χ must be negative, as $\psi^* = \phi$. Thus we would like to show that

$$\int_{\alpha}^{\lambda} H\psi^*(\mu) - \Big(\theta'(\mu) - t\Big) \; d\mu \quad < \quad 0 \;,\quad \lambda \in (-1, \alpha)\;, \tag{126}$$

$$\int_{\beta}^{\lambda} H\psi^*(\mu) - \Big(\theta'(\mu) - t\Big) \; d\mu \quad > \quad 0 \;,\quad \lambda \in (\beta, 1)\;. \tag{127}$$

This is true because the integrand has the correct sign over the intervals in question, as we now show.

LEMMA 4.4.

$$H\psi^*(\mu) - \Big(\theta'(\mu) - t\Big) \quad > \; 0\;,\quad \lambda \in (-1, \alpha) \cup (\beta, 1)\;.$$

Proof.

To prove this claim, we will use the derivatives of the functions ψ^* and $H\psi^*$ with respect to x. One can show that $f(\lambda)$, defined in (109), is differentiable in α and β, for $\lambda \neq \alpha$, $\lambda \neq \beta$, and hence so is $\psi^*(\lambda)$. Since $\alpha(x,t)$ and $\beta(x,t)$ are continuously differentiable functions of x, we can compute the Riemann - Hilbert problem satisfied by the differentiated quantity $\psi_x^*(\lambda; x, t)$,

$$H\psi_x^* \;=\; 0 \;,\quad \lambda \in \Big(\alpha(x,t), \beta(x,t)\Big) \tag{128}$$

$$\psi_x^* \;=\; 0 \;,\quad \lambda \in \Big(-\infty, \alpha(x,t)\Big) \cup \Big(\beta(x,t), \infty\Big)\;, \tag{129}$$

$$H\psi_x^* \quad \to 0 \;\; as \;\lambda \to \infty\;, \tag{130}$$

$$\int \psi_x^* \;=\; \pi\;. \tag{131}$$

One can also obtain from (109) that $f_x(\lambda)$ has square root singularities at $\lambda = \alpha$ and $\lambda = \beta$. The function ψ_x^* defined by

$$\psi_x^*(\lambda; x, t) = Re \left[\frac{i}{[(\lambda - \alpha(x,t))(\lambda - \beta(x,t))]^{1/2}}\right]\;, \tag{132}$$

$$H\psi_x^*(\lambda; x, t) = Im \left[\frac{i}{[(\lambda - \alpha(x,t))(\lambda - \beta(x,t))]^{1/2}}\right]\;, \tag{133}$$

is the unique solution to (128 - 131) with square root singularites at $\lambda = \alpha$ and $\lambda = \beta$.

Since we have, for any fixed $t \geq 0$, smooth functions $\alpha(x,t), \beta(x,t)$, we may integrate formulae (132) and (133) to obtain a representation for the function ψ^*, as follows.

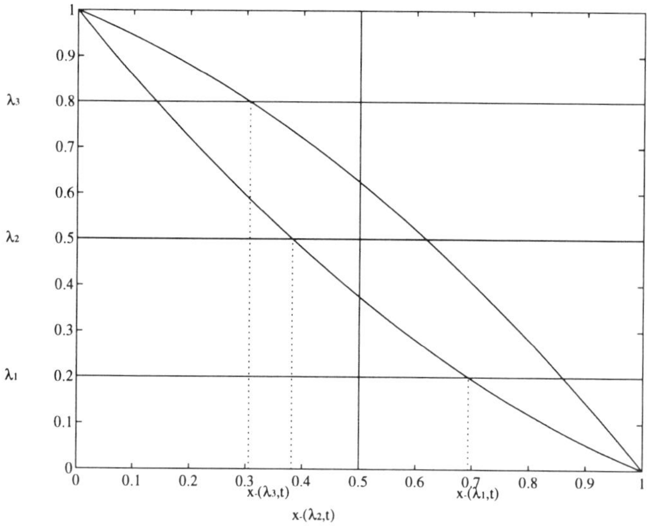

FIGURE 2. Riemann invariants $\alpha(x,t) \leq \beta(x,t)$, and a value of x, along with three choices for λ, λ_1, λ_2, and λ_3, and the corresponding values $x_-(\lambda_2)$, $x_-(\lambda_2)$, and $x_+(\lambda_3)$

We refer to Figure (2), and consider $\psi^*(\lambda;\ x,t)$, for $\lambda = \lambda_2$, and x as pictured. Clearly $\lambda_2 \in \Big(\alpha(x,t), \beta(x,t)\Big)$, and we write

$$\psi^*(\lambda_2;\ x,t) = \psi^*(\lambda_2;\ x_2,t) \ + \int_{x_2}^{x} \psi_y^*(\lambda_2;\ ,y,t)dy \ ,$$

where we have chosen x_2 as pictured, such that $\lambda_2 = \alpha(x_2,t)$. But by (101), $\psi^*(\lambda_2;\ x_2,t) = 0$, and hence

$$\psi^*(\lambda_2;\ x,t) = \int_{x_2}^{x} \psi_y^*(\lambda_2; y,t)dy \ .$$

We can, of course, replace λ_2 by any λ, provided we replace x_2 by the value of $x = x_-(\lambda,t)$ corresponding to λ,

$$\alpha(x_-(\lambda,t),t) \ = \ \lambda \ .$$

Thus for any $0 < \lambda < 1$, we have

$$\begin{aligned} \psi^*(\lambda;\ x,t) \ &= \ \int_{x_-(\lambda,t)}^{x} \psi_y^*(\lambda; y,t)dy \\ &= \ \int_{x_-(\lambda,t)}^{x} \frac{i}{R^t(\lambda,y)^{1/2}} dy \ > \ 0, \end{aligned} \tag{134}$$

where we have used (132), and we define

$$R^t(\lambda,y) = \Big(\lambda-\alpha(y,t)\Big)\Big(\lambda-\beta(y,t)\Big)\,, \qquad (135)$$
$$where\;\; R^t(\lambda,y)^{1/2}>0 \;\; for\;\; \lambda>\beta(y,t)\ .$$

We define the associated quantity $x_+(\lambda,t)$ by

$$\beta(x_+(\lambda,t),t) = \lambda\ .$$

Again by (101),

$$\psi^*(\lambda;x_+(\lambda,t),t) = \phi(\lambda)\,,$$

which implies

$$\int_{x_-(\lambda,t)}^{x_+(\lambda,t)} \psi_y^*(\lambda;y,t)dy = \phi(\lambda)\ . \qquad (136)$$

Thus, for $\lambda\in\Big(\alpha(x,t),\beta(x,t)\Big)$, we have

$$\psi^*(\lambda;\ x,t) = \phi(\lambda) - \int_x^{x_+(\lambda,t)} \frac{i}{R^t(\lambda,y)^{1/2}}\,dy < \phi(\lambda)\ . \qquad (137)$$

Similarly, using (133) (see, for illustration, Figure (2), taking $\lambda=\lambda_3$) and (102), we have the following representation, for $\lambda>\beta(x,t)$, for the Hilbert transform of ψ^*,

$$H\psi^*(\lambda;\ x,t) = \int_{x_+(\lambda,t)}^{x} \frac{1}{R^t(\lambda,y)^{1/2}}\,dy + \theta'(\lambda)-t\ .$$

From this last formula, we see that $\theta'-t-H\psi^*<0$, for $\lambda>\beta(x,t)$.

If we take $\lambda<\alpha(x,t)$, we obtain (see Figure (2), with $\lambda=\lambda_1$),

$$\Big(\theta'-t\Big) - H\psi^*(\lambda;\ x,t) = \int_x^{x_-(\lambda,t)} \frac{1}{R^t(\lambda,y)^{1/2}}\,dy < 0\ ,$$

as $x<x_-(\lambda,t)$, and so we have proven the lemma. □

We note that a consequence of this lemma (see (134), (137)), is that $0\le\psi^*\le\phi$, and hence $\psi^*\in A$. This completes the proof of the theorem. □

We have established that the solution to the maximization problem is obtained from a "single band guess". First, we showed that the Riemann - Hilbert problem has a solution provided the two conditions (110) and (112) on the functions $\alpha(x,t),\beta(x,t)$ are satisfied. Then we established that these two conditions determine the functions uniquely, and, as a consequence, that the functions solve the continuous Toda equations (in Riemann invariant form), with the required initial data. Finally we established that the solution to the Hilbert problem so constructed solves the maximization problem.

We calculate the derivative of $Q(x,t) = Q(\psi^*)$ with respect to t,

$$\begin{aligned}\frac{\partial}{\partial t}Q(x,t) &= \frac{\partial}{\partial t}\left[\frac{2}{\pi}\Big(t\lambda - \theta(\lambda)\ ,\ \psi^*\Big) + \frac{1}{2\pi^2}\Big(L\psi^*\ ,\ \psi^*\Big)\right] \\ &= \frac{2}{\pi}\Big(\lambda\ ,\ \psi^*\Big) + \frac{2}{\pi}\Big(t\lambda - \theta(\lambda)\ ,\ \psi_t^*\Big) + \frac{1}{\pi^2}\Big(L\psi^*\ ,\ \psi_t^*\Big)\ .\end{aligned}$$

As in the case of ψ_x^*, the quantity ψ_t^* solves an analogous Riemann - Hilbert problem,

$$\begin{aligned} H\psi_t^* &= -1\ ,\ \lambda \in \Big(\alpha(x,t), \beta(x,t)\Big)\ , \\ \psi_t^* &= 0\quad ,\ \lambda \notin \Big(\alpha(x,t), \beta(x,t)\Big)\ , \\ \int \psi_t^* d\lambda &= 0\ . \end{aligned}$$

We find

$$\begin{aligned} \psi_t^* &= Re\ i\left[\frac{\lambda - m}{R(\lambda)^{1/2}} - 1\right] \\ &= i\frac{\lambda - m}{R(\lambda)^{1/2}}\ ,\ \lambda \in \Big(\alpha(x,t), \beta(x,t)\Big)\ . \end{aligned} \tag{138}$$

Using (138) Euler - Lagrange equations (101) satisfied by ψ^* , we have, by a computation similar to (55 - 58),

$$Q_t = \frac{2}{\pi}\Big(\lambda\ ,\ \psi^*\Big)\ .$$

Thus we have, from (43-44),

$$\begin{aligned} w - \lim_{\epsilon\to 0} \hat{b}_\epsilon^2(x,t) &= \frac{1}{4}Q_{tt} = \frac{2}{\pi}\Big(\lambda\ ,\ \psi_t^*\Big)\ , \\ w - \lim_{\epsilon\to 0} \hat{a}_\epsilon(x,t) &= \frac{1}{2}Q_{xt} = \frac{2}{\pi}\Big(\lambda\ ,\ \psi_x^*\Big)\ . \end{aligned}$$

At this point we note, in analogy with the zero dispersion limit of the KdV equation considered in [**13**] for example, that prior to shock formation, (and hence, for the initial data considered in this chapter, for all $t > 0$), the weak limits above can be replaced by strong L^2 limits.

Using (138) and (132), we obtain

$$\begin{aligned} \lim_{\epsilon\to 0} \hat{b}_\epsilon^2(x,t) &= b(x,t)^2\ , \\ \lim_{\epsilon\to 0} \hat{a}_\epsilon(x,t) &= a(x,t)\ , \end{aligned}$$

where we define

$$\begin{aligned} a(x,t) &= \frac{\beta(x,t) + \alpha(x,t)}{2}\ , \\ b(x,t) &= \frac{\beta(x,t) - \alpha(x,t)}{4}\ . \end{aligned} \tag{139}$$

Remark: The identity

$$\int_{x_-(\lambda,t)}^{x_+(\lambda,t)} \frac{i}{R^t(\lambda,y)^{1/2}}\,dy \;=\; \phi(\lambda) \tag{140}$$

(see 136) illustrates the **isospectral** nature of the continuous Toda equations. More precisely, define

$$\Phi(\lambda,t) \;=\; \int_{x_-(\lambda,t)}^{x_+(\lambda,t)} \cos^{-1}\left(\frac{\lambda - a(x,t)}{2b(x,t)}\right)\,dy\ .$$

This function is clearly the analog at $t > 0$ of the spectral measure $\Phi(\lambda)$ defined in (96) at $t = 0$. Using the continuous Toda equations

$$\begin{aligned} a_t &= 2(b^2)_x\ , \\ b_t &= b a_x\ , \end{aligned}$$

which are a consequence of $\alpha_t = -\frac{\beta-\alpha}{2}\alpha_x$, $\beta_t = \frac{\beta-\alpha}{2}\beta_x$, and (139), note that

$$\begin{aligned} \Phi_t &= \int_{x_-(\lambda,t)}^{x_+(\lambda,t)} \frac{-2ba_t - 2(\lambda-a)b_t}{2b\sqrt{(2b)^2-(\lambda-a)^2}}dy \\ &= -\frac{1}{2}\int_{x_-(\lambda,t)}^{x_+(\lambda,t)} \frac{8bb_x + 2(\lambda-a)a_x}{\sqrt{(2b)^2-(\lambda-a)^2}}\,dy \\ &= 0\ . \end{aligned}$$

Hence

$$\phi(\lambda,t) \;=\; -\frac{\partial}{\partial\lambda}\Phi(\lambda,t)$$

is independant of t. This is the content of (140).

For future reference, we note that a similar calculation shows

$$\theta(\lambda) - t\lambda = \int_0^{x_-(\lambda,t)} \cosh^{-1}\left(\frac{\lambda - a(y,t)}{2b(y,t)}\right)\,dy\ . \tag{141}$$

Comparing with (97), we see then that

$$\theta(\lambda,t) \equiv \theta(\lambda) - t\lambda$$

has the meaning of an asymptotic norming constant at time t.

Finally we compute the long - time asymptotics of the continuum limit of the Toda lattice.

THEOREM 4.3. *Let the initial data for the Toda lattice (6 - 7) be given by monotone decreasing functions* $\alpha^0(x,t) \le \beta^0(x,t)$, *as above.*

Then

$$\begin{aligned} \lim_{t\to\infty}\lim_{\epsilon\to 0} \hat{b}_\epsilon^2(x,t) &= \left(\frac{\beta(x,t)-\alpha(x,t)}{4}\right)^2 &= 0\ , \\ \lim_{t\to\infty}\lim_{\epsilon\to 0} \hat{a}_\epsilon(x,t) &= \frac{\beta(x,t)+\alpha(x,t)}{2} &= \Phi^{-1}(\pi x)\ . \end{aligned}$$

Proof.

It is evident from the condition for T that as $t \to \infty$, we must have that $\beta - \alpha(\beta, t) \to \infty$. Analyzing at the function $X(\alpha(\beta, t), \beta)$, we see by dominated convergence that

$$X(\alpha(\beta,t),\beta) \to \Phi(\beta) ,$$

and so the condition $X(\alpha, \beta) = \pi x$ implies that

$$\frac{\beta(x,t) + \alpha(x,t)}{2} \longrightarrow \Phi^{-1}(\pi x) .$$

□

One of the goals of this thesis is to understand the evolution of the continuum limit of the Toda lattice beyond the time of shock formation for the underlying PDE system, the continuous Toda equations. We note that in the work of Lax and Levermore, the underlying PDE system is the inviscid Burger's equation $u_t - 6uu_x = 0$, which is explicitly solvable by the method of characteristics. The explicit solution afforded by this method enables one to understand the formation of a shock in this equation completely. The work in this section plays the role of the method of characteristics for the case of the continuum limit of the Toda lattice. As we will show, the pair of equations (110), (112) presented above enable us to obtain detailed information about shock formation for the continuous Toda PDE system. In the next chapter we will use these equations to produce initial data which forms a shock in finite time. We will begin these considerations by presenting data for which a shock forms, but for which the continuum limit of the Toda lattice continues to be governed by the continuous Toda equations after shock formation.

CHAPTER 5

Shock 1

Here we consider a further generalization of the class of data (see Figure (1)), for which the upper Riemann invariant $\beta^0(x)$ possesses one nondegenerate critical point, a maximum. For definiteness we assume that $\alpha^0(0) = \beta^0(0) = 0$, $\alpha^0(1) = \beta^0(1) = -1$, , and that $\beta^0(x_c) = \max \beta^0(x) = 1$. It will be clear from the analysis that follows, that the qualitative behavior of the continuum limit does not depend on these specific choices, which are made only for convenience. The lower Riemann invariant $\alpha^0(x)$ however, is still assumed to be strictly monotone decreasing, i.e. $\alpha^0_x < 0$ for all $x \in [0,1]$. Note that as before we have

$$\alpha^0(0) = \beta^0(0)\ ,$$
$$\alpha^0(1) = \beta^0(1)$$

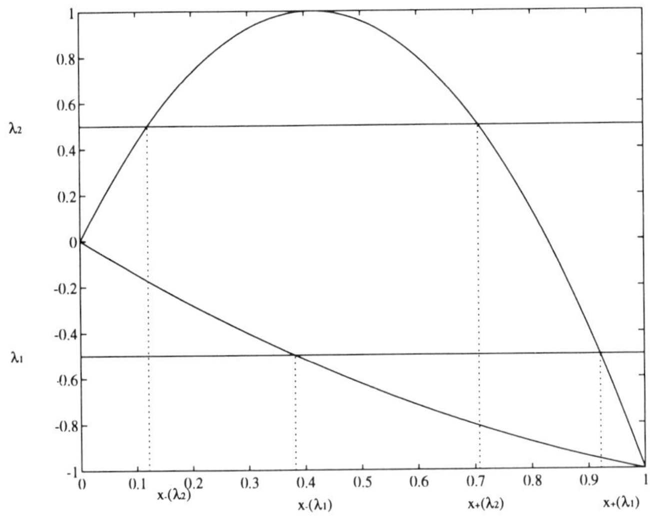

FIGURE 1. Riemann invariants $\alpha^0(x) \le \beta^0(x)$, and representative choices of λ, along with corresponding values of $x_-(\lambda)$ and $x_+(\lambda)$.

For this data, the norming constant $\theta(\lambda)$ has two separate definitions (see Appendix I), one for $\lambda < 0$, and another for $\lambda > 0$:

$$\theta(\lambda) \equiv \int_0^{x_-(\lambda)} \cosh^{-1}(\frac{\lambda - a}{2b})\, dy \ , \quad for\ \lambda > 0 \ , \tag{142}$$

$$\theta(\lambda) \equiv \int_0^{x_-(\lambda)} \cosh^{-1}(\frac{a - \lambda}{2b})\, dy \ , \quad for\ \lambda < 0 \ ,$$

but the spectral function $\Phi(\lambda)$ is defined, as before, by

$$\Phi(\lambda) \equiv \int_{x_-(\lambda)}^{x_+(\lambda)} \cos^{-1}(\frac{\lambda - a}{2b})\, dy \tag{143}$$

for all $-1 < \lambda < 1$. The derivatives of these functions are, for all $-1 < \lambda < 1$, $\lambda \neq 0$,

$$\theta'(\lambda) \ = \ \int_0^{x_-(\lambda)} \frac{1}{R^0(\lambda, y)^{1/2}} dy \ , \tag{144}$$

$$\phi(\lambda) \ = \ -\frac{d}{d\lambda}\Phi(\lambda) = \int_{x_-(\lambda)}^{x_+(\lambda)} \frac{i}{R^0(\lambda, y)^{1/2}} dy \ , \tag{145}$$

where we define $R^0(\lambda, y)$ in the usual way, $R^0(\lambda, y) = (\lambda - \alpha^0(y))(\lambda - \beta^0(y))$, and we take

$$R^0(\lambda, y)^{1/2} > 0 \quad for \lambda > \beta^0(y) \ .$$

The maximization problem is, as before,

$$Q(x,t) \ = \ \max_{\psi \in A} \ \frac{2}{\pi}(t\lambda - \theta \ , \ \psi) \ + \ \frac{1}{2\pi^2}(L\psi \ , \ \psi) \ , \tag{146}$$

where A is the set of all Lebesgue measurable functions ψ satisfying

$$0 \leq \psi \leq \phi \ ,$$

$$\int \psi \ = \ \pi x \ .$$

We re-write the Euler - Lagrange equations,

$$\begin{aligned} \frac{2}{\pi}(t\lambda - \theta(\lambda)) \ + \ \frac{1}{\pi^2}L\psi(\lambda) + l \ &= \ 0 \ , \quad \lambda \in I_1 \ , \\ \psi(\lambda) \ &= \ 0 \quad , \lambda \in I_2 \ , \\ \psi(\lambda) \ &= \ \phi(\lambda) \ , \quad , \lambda \in I_3 \ , \end{aligned} \tag{147}$$

and the Riemann - Hilbert problem,

$$H\psi \ = \ \theta'(\lambda) - t \ , \ \ \lambda \in I_1 \ , \tag{148}$$

$$\psi(\lambda) \ = \ 0 \ , \ \ \lambda \in I_2 \ , \tag{149}$$

$$\psi(\lambda) \ = \ \phi(\lambda) \ , \ \ \lambda \in I_3 \ . \tag{150}$$

On the new region $[0, x_c]$ on which the upper Riemann invariant is *increasing*, we can no longer solve the maximization problem with the ansatz used for the analysis of monotone decreasing initial data in Chapter 4. To see why, we first solve the maximization problem on the complementary region $(x_c, 1)$, with the ansatz used in previous sections.

The maximizer for $t = 0$, $x \in (x_c, 1)$.

We consider the Riemann - Hilbert problem for a function $\psi(\lambda) \in L^p(\mathbf{R})$, $1 \le p < 2$,

$$H\psi = \theta'(\lambda) \quad on \quad I_1 = (\alpha(x,t), \beta(x,t)) \tag{151}$$

$$\psi = 0 \quad on \quad I_2 = (-\infty, \alpha(x,t)) \cup (1, \infty) \tag{152}$$

$$\psi = \phi(\lambda) \quad on \quad I_3 = (\beta(x,t), 1), \tag{153}$$

(note $t = 0$ in (151)).

The solution to this problem is explicit (cf. (109), from Chapter 4),

$$\psi_R^*(z) = Re \lim_{\epsilon \downarrow 0} f_R(z + i\epsilon) \ , \tag{154}$$

where $f_R(z)$ is given by,

$$f_R(z) = R(z)^{1/2} \frac{1}{\pi i} \left(\int_\alpha^\beta \frac{i(\theta'(\lambda))}{R(\lambda)^{1/2}} \frac{d\lambda}{\lambda - z} + \int_\beta^1 \frac{\phi(\lambda)}{R(\lambda)^{1/2}} \frac{d\lambda}{\lambda - z} \right) \ , \tag{155}$$

where once again we define $R(\lambda) = (\lambda - \alpha)(\lambda - \beta)$, and take $R(\lambda)^{1/2} > 0$ for $\lambda > \beta$. Again enforcing that $\psi_R^* + iH\psi_R^* \in H^p$, and that $\int \psi_R^* = \pi x$, we obtain the two conditions

$$T_R(\alpha, \beta) \equiv \int_\alpha^\beta \frac{i}{[(\lambda - \alpha)(\lambda - \beta)]^{1/2}} \theta' d\lambda + \\ + \int_\beta^1 \frac{1}{[(\lambda - \alpha)(\lambda - \beta)]^{1/2}} \phi d\lambda = 0 \ , \tag{156}$$

$$X_R(\alpha, \beta) \equiv \int_\alpha^\beta i \frac{\lambda - m}{[(\lambda - \alpha)(\lambda - \beta)]^{1/2}} \theta' d\lambda + \\ + \int_\beta^1 \frac{\lambda - m}{[(\lambda - \alpha)(\lambda - \beta)]^{1/2}} \phi d\lambda = \pi x \ . \tag{157}$$

As in the previous section, the first task is to reconstruct the initial data from these conditions. We use the definitions (145) and (144) of ϕ and θ', along with Figure (2) to interchange orders of integration, and write (cf. (113) - (115) of Chapter 4)

$$T_R(\alpha, \beta) = \int_0^{x_-(\beta)} \left\{ \int_\alpha^{\alpha^0(y)} + \int_{\beta^0(y)}^\beta \right\} \frac{i}{[R(\lambda) R_0(\lambda, y)]^{1/2}} d\lambda dy \quad + \\ + \int_{x_-(\beta)}^{x_-(\alpha)} \left\{ \int_\alpha^{\alpha^0(y)} + \int_\beta^{\beta^0(y)} \right\} \frac{i}{[R(\lambda) R_0(\lambda, y)]^{1/2}} d\lambda dy \\ + \int_{x_-(\alpha)}^{x_+(\beta)} \int_\beta^{\beta^0(y)} \frac{i}{[R(\lambda) R_0(\lambda, y)]^{1/2}} d\lambda dy \qquad ,$$

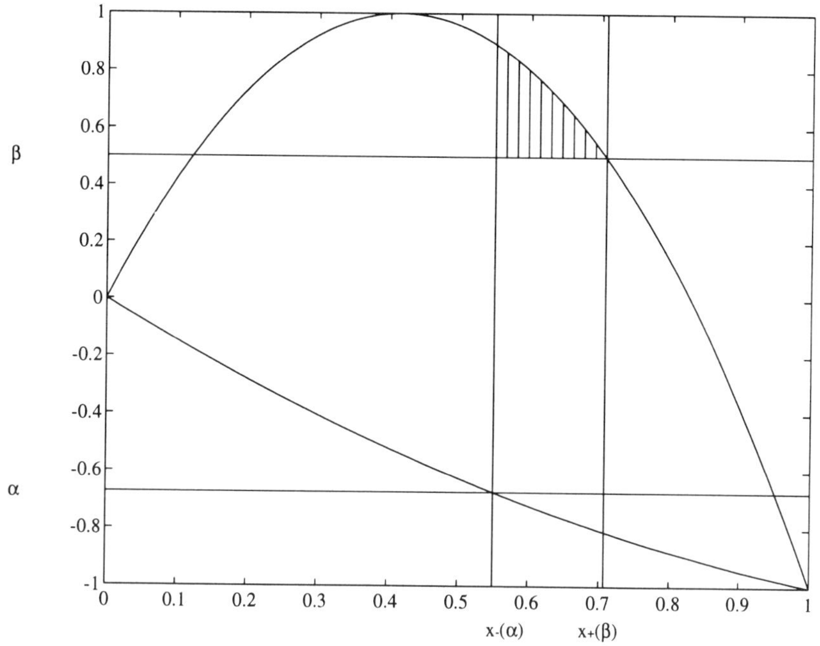

FIGURE 2. The Riemann Invariants $\alpha^0(x)$ and $\beta^0(x)$ (note $\alpha^0(x) \leq \beta^0(x)$), and a particular choice of α and β, along with locations of $x_-(\alpha)$, and $x_+(\beta)$, and the region of integration for $T_R(\alpha,\beta)$ in (5.17).

The first two terms are identically zero, by contour deformation, and so we have the representation for T_R,

$$(158)\qquad T_R(\alpha,\beta) \;=\; \int_{x_-(\alpha)}^{x_+(\beta)} \int_{\beta}^{\beta^0(y)} \frac{i}{[R(\lambda)R_0(\lambda,y)]^{1/2}}\, d\lambda dy \;.$$

Also, similar computations for (158) lead to

$$(159)\quad X_R(\alpha,\beta) = \int_{x_-(\alpha)}^{x_+(\beta)} \int_{\beta}^{\beta^0(y)} \frac{i(\lambda-m)}{[R(\lambda)R_0(\lambda,y)]^{1/2}}\, d\lambda dy \;+\pi x_-(\alpha) \;=\; \pi x \;.$$

Clearly then, to enforce condition (156), we must pick α and β such that the region of integration vanishes,

$$x_-(\alpha) = x_+(\beta) \;.$$

but then, by (159),

$$x_-(\alpha) = x_+(\beta) = x \;,$$

and so we have reconstructed the initial data, for $x_c < x$.

The proof that the function ψ_R^* indeed solves the maximization problem (146) is given in Theorem (5.1) below.

Observe that $x_+(\beta) \geq x_c$, and hence the condition $x_-(\alpha) = x_+(\beta) = x$ can only be solved for $x \geq x_c$. Hence the ansatz (151), (152) at $t = 0$ breaks down for $x < x_c$. Note also that at $x = x_c$, the solution of $x_-(\alpha) = x_+(\beta) = x_c$ is given by $\alpha = \alpha^0(x_c)$, $\beta = 1$ and hence the set I_3 collapses to the empty set. This leads us to the following ansatz.

The maximizer for $t = 0$, $x \in (0, x_c)$.

Consider the Riemann - Hilbert problem ,

$$\begin{aligned} H\psi^* &= \theta'(\lambda) \;\; on \;\; I_1 = (\alpha(x,t), \beta(x,t)) \\ \psi^* &= 0 \;\; on \;\; I_2 = \mathbf{R} \backslash [\alpha(x,t), \beta(x,t)] \; . \end{aligned} \tag{160}$$

The solution to this problem can again be obtained by formula,

$$\psi_L^*(\lambda) = Re \lim_{\epsilon \downarrow 0} f_L(\lambda + i\epsilon) \;\; , \tag{161}$$

where

$$f_L(z) = R(z)^{1/2} \frac{1}{\pi i} \int_\alpha^\beta \frac{i(\theta'(\lambda))}{R(\lambda)^{1/2}} \frac{d\lambda}{\lambda - z} \;\; , \tag{162}$$

and by enforcing the condition $\psi_L^* + iH\psi_L^* \in H^p$, along with $\int \psi_L^* = \pi x$, we again obtain two conditions,

$$T_L(\alpha, \beta) \equiv \int_\alpha^\beta \frac{i}{[(\lambda - \alpha)(\lambda - \beta)]^{1/2}} \theta' d\lambda = 0 \;\; , \tag{163}$$

$$X_L(\alpha, \beta) \equiv \int_\alpha^\beta i \frac{\lambda - m}{[(\lambda - \alpha)(\lambda - \beta)]^{1/2}} \theta' d\lambda = \pi x \;\; . \tag{164}$$

Carrying out a similar analysis to that used to derive formulae (158) and (159), we obtain the representations:

$$T_L(\alpha, \beta) = \int_{x_-(\alpha)}^{x_-(\beta)} \int_{\beta^0(y)}^\beta \frac{i}{[R(\lambda) R_0(\lambda, y)]^{1/2}} d\lambda dy \tag{165}$$

$$X_L(\alpha, \beta) = \int_{x_-(\alpha)}^{x_-(\beta)} \int_{\beta^0(y)}^\beta i \frac{\lambda - m}{[R(\lambda) R_0(\lambda, y)]^{1/2}} d\lambda dy \; + \pi x_-(\alpha) \;\; , \tag{166}$$

where the region of integration is as shaded in Figure (3). By contour integration, (165) can be written

$$T_L(\alpha, \beta) = - \int_{x_-(\alpha)}^{x_-(\beta)} \int_{\alpha^0(y)}^\alpha \frac{i}{[R(\lambda) R_0(\lambda, y)]^{1/2}} d\lambda dy \;\; , \tag{167}$$

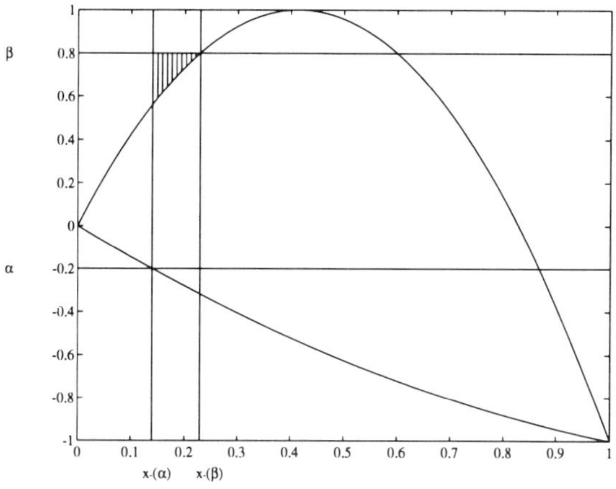

FIGURE 3. The Riemann Invariants $\alpha^0(x)$ and $\beta^0(x)$ (note $\alpha^0(x) \leq \beta^0(x)$), and a particular choice of α and β, along with locations of $x_-(\alpha)$, and $x_-(\beta)$, and the region of integration for $T_L(\alpha, \beta)$ in (1.24).

So, for this ansatz to satisfy the conditions (163), (164) at $t = 0$, we must have

$$x_-(\alpha) = x_-(\beta) = x ,$$

which is solvable if (and only if) $x < x_c$. Thus we see that our two ansatzes combine to reproduce the initial data, $\alpha = \alpha^0(x)$, $\beta = \beta^0(x)$, for all $x \in [0,1]$.

THEOREM 5.1. *The functions $\psi_R^*(\lambda;\ x)$ (for $0 < x < x_c$) and $\psi_L^*(\lambda;\ x)$ (for $x_c < x < 1$) defined by (154), and (161), respectively, solve the maximization problem for all $x \in [0,1]$.*

Proof. We must show that ψ_L^* and ψ_R^* belong to A, and that they maximizes $Q(\psi)$ on their spatial regions of definition.

Evaluating $Q(\psi^* + \chi)$ (for either ψ_L^* or ψ_R^*), we obtain the now familiar formula (note $t = 0$),

$$(168) \qquad Q(\psi^* + \chi) = Q(\psi^*) + \frac{1}{2\pi^2}(L\chi, \chi) + \\ + \frac{2}{\pi}\left(\int_\alpha^\cdot H\psi^*(\mu) - \theta'(\mu)\ d\mu\ ,\ \chi(\cdot)\right)_{(-1,\alpha)} + \\ + \frac{2}{\pi}\left(\int_\beta^\cdot H\psi^*(\mu) - \theta'(\mu)\ d\mu\ ,\ \chi(\cdot)\right)_{(\beta,1)} .$$

Here we recall that

$$\int \chi = 0 \ ,$$

$$\lambda \notin (\alpha, \beta) \Rightarrow \chi(\lambda) \geq 0 \quad , \quad \mathit{for}\ x \in (0, x_c) \ ,$$

and, for $x \in (x_c, 1)$, we have

$$\lambda \in (-1, \alpha) \Rightarrow \chi(\lambda) \geq 0 \quad ,$$
$$\lambda \in (\beta, 1) \Rightarrow \chi(\lambda) \leq 0 \quad ,$$

Notation: If a statement is true for ψ_L^* **and** for ψ_R^*, we simply use the symbol ψ^* (see equations (169) and (172) directly below).

Thus to show that we have constructed the maximizer it is enough to verify that

$$\int_{\alpha^0(x)}^{\lambda} H\psi^*(\mu;\ x) - \theta'(\mu)\ \ d\mu < 0 \ , \ \mathit{for}\ x \in (0,1) \ , \tag{169}$$

$$\int_{\beta^0(x)}^{\lambda} H\psi_L^*(\mu;\ x) - \theta'(\mu)\ \ d\mu < 0 \ , \ \mathit{for}\ x \in (0,x_c) \ , \tag{170}$$

$$\int_{\beta^0(x)}^{\lambda} H\psi_R^*(\mu;\ x) - \theta'(\mu)\ \ d\mu > 0 \ , \ \mathit{for}\ x \in (x_c,1) \ , \tag{171}$$

$$0 \leq \psi^*(\lambda;\ x) \leq \phi \ . \tag{172}$$

As in Chapter 4, the verification of these inequalities uses the representation of the functions ψ_L^* and ψ_R^* obtained by integrating the simple expressions for the quantities ψ_x^*, $H\psi_x^*$,

$$\psi_x^* \quad = \quad Re\ \frac{i}{R^0(\lambda, x)^{1/2}} \ , \tag{173}$$

$$H\psi_x^* \quad = \quad Im\ \frac{i}{R^0(\lambda, x)^{1/2}} \ , \tag{174}$$

which may be obtained, as in Chapter 4, from the Riemann - Hilbert problem,

$$\begin{aligned} H\psi_x^* &= 0 \ , \ \ \lambda \in (\alpha^0(x), \beta^0(x)) \ , \\ \psi_x^* &= 0 \ , \ \ \lambda \notin (\alpha^0(x), \beta^0(x)) \ , \\ \psi_x^* &\in L^p(\mathbf{R}) \ , \ \ \mathit{for\ some}\ \ 1 \leq p < 2 \ , \\ \int \psi_x^* dx &= \pi \ , \end{aligned}$$

the unique solution of which is (173 - 174). Note that these formulae are true for *all* $x \in (0,1)$, and hence are true without subscripts L or R.

Using Figure (4), with $\lambda = \lambda_2$, we integrate the expression (173) to obtain, for $\lambda \in \left(\alpha^0(x), \beta^0(x)\right)$, and any $x \in (0,1)$

$$\psi^*(\lambda; x) \quad = \quad \int_{x_-(\lambda)}^{x} \frac{i}{R^0(\lambda, y)^{1/2}}\, dy \ , \tag{175}$$

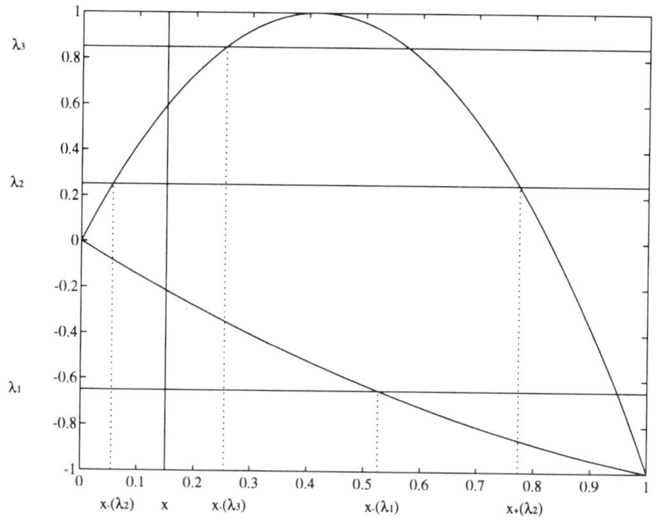

FIGURE 4. The Riemann invariants $\alpha^0(x)$ and $\beta^0(x)$, along with three choices of λ, and the associated values $x_-(\lambda_j)$, and $x_+(\lambda_2)$, used in the proof of Theorem 16.

as both $\psi_L^*(\lambda; x)$ and $\psi_R^*(\lambda; x)$ vanish for $x = x_-(\lambda)$, by (160) or (152). We note that (175) has no subscript, as it is true for both ψ_L^* and ψ_R^*. Thus we see that

- $\psi^*(\lambda;\ x) \geq 0\ ,\ for\ \lambda \in (\alpha^0(x)\ ,\ \beta^0(x))\ for\ all\ x \in (0,1)\ .$

From (153) we see

$$\psi_R^*(\lambda; x_+(\lambda), t)\ =\ \phi(\lambda)\ , \tag{176}$$

(177)

and hence, integrating (173) from $x_-(\lambda)$ to $x_+(\lambda)$, we obtain

$$\int_{x_-(\lambda)}^{x_+(\lambda)} \frac{i}{R^0(\lambda, x)^{1/2}} dy\ =\ \phi(\lambda)\ , \tag{178}$$

and we conclude, from (175) and (178),

- $0 \leq\ \psi^*(\lambda;\ x)\ \leq \phi(\lambda)\ ,\ \ for\ \lambda \in (\alpha^0(x)\ ,\ \beta^0(x))\ for\ all\ x \in (0,1)\ .$

Now observe that from (151-153) and (160), we have

- $\psi_L^*(\lambda;\ x) = 0\ ,\ for\ \lambda \in \mathbf{R}\backslash(\alpha^0(x), \beta^0(x))\ ,\ and\ x \in (0, x_c)\ \ ,$
- $\psi_R^*(\lambda;\ x) = \phi(\lambda)\ ,\ for\ \lambda \in (\beta^0(x), 1)\ and\ x \in (x_c, 1)\ ,$
- $\psi_R^*(\lambda;\ x) = 0\ ,\ for\ \lambda \in (-\infty, \alpha^0(x)) \cup (1, \infty)\ and\ x \in (x_c, 1)\ \ .$

Together, the inequalities preceded by a "•" above verify (172).

Similarly, we obtain, for $\lambda > \beta^0(x)$, and $x < x_c$ (again, see Figure (4), with $\lambda = \lambda_3$),

$$H\psi_L^*(\lambda; x) - \theta'(\lambda) = -\int_x^{x_-(\lambda)} \frac{1}{R^0(\lambda, y)^{1/2}}\, dy \ ,$$

which is clearly negative, and hence (170) follows.

For the region $x > x_c$, and for $\lambda > \beta^0(x)$, we use the representation

$$H\psi_R^*(\lambda; x) - \theta'(\lambda) = \int_{x_+(\lambda)}^{x} \frac{1}{R_0(\lambda, y)^{1/2}}\, dy \ > \ 0 \ ,$$

from which (171) follows.

Finally, for any $x \in (0,1)$, and for $\lambda < \alpha^0(x)$, we have

$$H\psi^*(\lambda;\ x) - \theta'(\lambda) = -\int_x^{x_-(\lambda)} \frac{1}{R_0(\lambda, y)^{1/2}}\, dy \ > \ 0 \ ,$$

which establishes (169).

This completes the proof of the Theorem. □

This Theorem justifies the replacement of the initial data by the approximate WKB data as described in Chapter 2 (see the discussion prior to equation (36)).

REMARK 5.1.

Observe that the proof that $\psi^*(\lambda;\ x)$ solves the maximization problem relies **only** on our ability to integrate the differentiated quantities ψ_x^* and $H\psi_x^*$, and on the general shape of the curves $\alpha^0(x)$ and $\beta^0(x)$. More precisely, a careful reading of the proof of Theorem 5.1 shows that **if** ψ^* solves the Riemann - Hilbert problem, and **if** the curves have the same general shape, then ψ^* **automatically** satisfies the side constraints, and hence solves the maximization problem.

We have established that at $t = 0$, the maximization problem is solved by a combination of two ansatzes,

$$\begin{aligned} \psi_L^*(\lambda) &= Re \lim_{\epsilon \downarrow 0} f_L(\lambda + i\epsilon) \ , \ for\ x \in (0, x_c) \\ \psi_R^*(\lambda) &= Re \lim_{\epsilon \downarrow 0} f_R(\lambda + i\epsilon) \ , \ for\ x \in (x_c, 1) \ , \end{aligned}$$

with f_L defined in (155), and f_R defined in (162). As we will now show, the analog of the functions ψ_L^* and ψ_R^*, for (small) $t > 0$, again combine to solve the maximization problem for all $x \in [0, 1]$.

The Riemann - Hilbert problem, for $t > 0$, (t small).

The Riemann - Hilbert problems we consider are,

for $x \in [0, x_L(t)]$,

$$H\psi_L^* = \theta'(\lambda) - t \ \ on\ \ I_1 = (\alpha(x,t), \beta(x,t)) \tag{179}$$

$$\psi_L^* = 0 \ \ on\ \ I_2 = \mathbf{R} \backslash [\alpha(x,t), \beta(x,t)] \ . \tag{180}$$

and for $x \in [x_R(t), 1]$,

$$H\psi_R^* = \theta'(\lambda) \quad on \quad I_1 = (\alpha(x,t), \beta(x,t)) \tag{181}$$

$$\psi_R^* = 0 \quad on \quad I_2 = (-\infty, \alpha(x,t)) \cup (1, \infty) \tag{182}$$

$$\psi_R^* = \phi(\lambda) \quad on \quad I_3 = (\beta(x,t), 1), \tag{183}$$

Here $x_L(t)$ and $x_R(t)$ are the upper and lower endpoints of the regions where we can solve the Riemann - Hilbert problems, and at this point they are unknown. The analysis below will show that $x_L(t) = x_R(t) \equiv x_c(t)$. The quantity $x_c(t)$ is a critical point for $\beta(x,t)$,

$$\beta(x_c(t), t) = \max_{0 \le x \le 1} \beta(x,t) .$$

The solution to the Riemann - Hilbert problems (179 - 180) and (181 - 183) are

$$\begin{aligned} \psi_L^*(\lambda; x, t) &= Re \lim_{\epsilon \downarrow 0} f_L(\lambda + i\epsilon) \ , \ for \ x \in (0, x_L(t)) \\ \psi_R^*(\lambda; x, t) &= Re \lim_{\epsilon \downarrow 0} f_R(\lambda + i\epsilon) \ , \ for \ x \in (x_R(t), 1) \ , \end{aligned} \tag{184}$$

where

$$f_L(z) = R(z)^{1/2} \frac{1}{\pi i} \int_\alpha^\beta \frac{i(\theta'(\lambda) - t)}{R(\lambda)^{1/2}} \frac{d\lambda}{\lambda - z} \ , \tag{185}$$

and

$$f_R(z) = R(z)^{1/2} \frac{1}{\pi i} \left(\int_\alpha^\beta \frac{i(\theta'(\lambda) - t)}{R(\lambda)^{1/2}} \frac{d\lambda}{\lambda - z} + \int_\beta^1 \frac{\phi(\lambda)}{R(\lambda)^{1/2}} \frac{d\lambda}{\lambda - z} \right) . \tag{186}$$

The conditions on $\psi_L^*(\lambda)$ and $\psi_R^*(\lambda)$, that both satisfy $\psi_L^* + iH\psi_L^* \in H^p$, $1 \le p < 2$, and $\int \psi_L^* = \pi x$, become

$$T_L(\alpha, \beta) \equiv \int_\alpha^\beta \frac{i}{[(\lambda - \alpha)(\lambda - \beta)]^{1/2}} \theta' d\lambda = \pi t \ , \tag{187}$$

$$X_L(\alpha, \beta) \equiv \int_\alpha^\beta i \frac{\lambda - m}{[(\lambda - \alpha)(\lambda - \beta)]^{1/2}} \theta' d\lambda = \pi x \ , \tag{188}$$

and

$$\begin{aligned} T_R(\alpha, \beta) \equiv & \int_\alpha^\beta \frac{i}{[(\lambda - \alpha)(\lambda - \beta)]^{1/2}} \theta' d\lambda + \\ & + \int_\beta^1 \frac{1}{[(\lambda - \alpha)(\lambda - \beta)]^{1/2}} \phi d\lambda = \pi t \ , \end{aligned} \tag{189}$$

$$\begin{aligned} X_R(\alpha, \beta) \equiv & \int_\alpha^\beta i \frac{\lambda - m}{[(\lambda - \alpha)(\lambda - \beta)]^{1/2}} \theta' d\lambda + \\ & \int_\beta^1 \frac{\lambda - m}{[(\lambda - \alpha)(\lambda - \beta)]^{1/2}} \phi d\lambda = \pi x \ . \end{aligned} \tag{190}$$

Remark on notation: Properly we should denote the endpoints α, β that appear in (179 - 180), and also in (187 - 188) by α_L, β_L, and similarly $\alpha = \alpha_R$, $\beta = \beta_R$ for the case (181 - 183) and (189 - 191). However, it will always be clear from the outset which pair (α, β) is under discussion, and we have chosen not to overburden the text with subscripts.

Remark: Observe that at $t = 0$, the systems of equations (187 - 188) and (189 - 191) reduce to the problems described above for the reconstruction of the initial data, and combine to solve the maximization problem on the full spatial interval $(0, 1)$.

To show that this remains true for (small) $t > 0$, we must first show that these systems can be evolved forward in time, and determine $\alpha(x,t)$ and $\beta(x,t)$ uniquely for all $x \in [0,1]$. This we will do via some lemmas.

LEMMA 5.1. *For $\alpha < 0$, $\alpha < \beta$, we have*

$$T_{L\alpha} > 0 \quad , \quad for\ \alpha^0(x_-(\beta)) < \alpha \ ,$$
$$T_{R\alpha} > 0 \quad , \quad for\ \alpha^0(x_+(\beta)) < \alpha \ .$$

Proof. The derivations of equations (158) and (167) presented above show, since the functions $T(\alpha,\beta), X(\alpha,\beta)$ (with either subscript) are independent of t and x, that we have

$$(191) \qquad T_L(\alpha,\beta) = \int_{x_-(\alpha)}^{x_-(\beta)} \int_{\beta^0(y)}^{\beta} \frac{i}{[R(\lambda)R_0(\lambda,y)]^{1/2}} \, d\lambda dy \ ,$$

$$(192) \qquad T_R(\alpha,\beta) = \int_{x_-(\alpha)}^{x_+(\beta)} \int_{\beta}^{\beta^0(y)} \frac{i}{[R(\lambda)R_0(\lambda,y)]^{1/2}} \, d\lambda dy \ .$$

Differentiating, we obtain

$$(193) T_{L\alpha} = \frac{\pi \partial_\alpha x_-(\alpha)}{[(\alpha - \beta)(\alpha - \beta^0(x_-(\alpha)))]^{1/2}} + \\ + \int_{x_-(\alpha)}^{x_-(\beta)} \int_{\beta^0(y)}^{\beta} \frac{i}{[R(\lambda)R_0(\lambda,y)]^{1/2}} \frac{\frac{1}{2}}{\lambda - \alpha} \, d\lambda dy > 0 \ ,$$

$$(194) T_{R\alpha} = \frac{\pi \partial_\alpha x_-(\alpha)}{[(\alpha - \beta)(\alpha - \beta^0(x_-(\alpha)))]^{1/2}} + \\ + \int_{x_-(\alpha)}^{x_+(\beta)} \int_{\beta^0(y)}^{\beta} \frac{i}{[R(\lambda)R_0(\lambda,y)]^{1/2}} \frac{\frac{1}{2}}{\lambda - \alpha} \, d\lambda dy > 0 \ ,$$

where the first terms appearing in each of (193) and (194) arise from the evaluation of the inner integrals in (191) and (192), respectively, using the identity

$$\int_{\beta^0(x_\alpha)}^{\beta} \frac{i}{[R(\lambda)R_0(\lambda,x_\alpha)]^{1/2}} \frac{1}{\lambda - \alpha} \, d\lambda = \pi \frac{1}{[(\alpha - \beta^0(x_\alpha))(\alpha - \beta)]^{1/2}} \ .$$

As all terms in (193) and (194) above are positive, the lemma is proved.
□

LEMMA 5.2. *Set* $t_m = \min_{\beta\in(0,1)} \frac{1}{\pi} T_L(0,\beta)$, *and define* $\alpha_L^0(\beta)$ *and* $\alpha_R^0(\beta)$ *to be the solution, for* β *fixed, to the equations*

$$\begin{aligned} T_L(\alpha,\beta) &= 0 \ , \\ T_R(\alpha,\beta) &= 0 \ , \end{aligned}$$

respectively. Then, for all $0 \le t < t_m$, *we have*

1. For each $\beta \in (0,1)$, *there exists a unique solution* $\alpha_L(\beta,t) \in (\alpha_L^0(\beta),0)$ *to the equation*

$$T_L(\alpha,\beta) = \pi t \ .$$

2. For each $\beta \in (-1,1)$ *there exists a unique solution* $\alpha_R(\beta,t) \in (\alpha_R^0(\beta),0)$ *to*

$$T_R(\alpha,\beta) = \pi t \ .$$

Proof. Note first that the functions $\alpha_L^0(\beta)$, $\alpha_R^0(\beta)$ indeed exist, by the arguments preceding Theorem 5.1, and hence

$$\begin{aligned} T_L(\alpha_L^0(\beta),\beta) &= 0 \ , \quad 0 \le \beta \le 1 \ , \\ T_R(\alpha_R^0(\beta),\beta) &= 0 \ , \quad -1 \le \beta \le 1 \ . \end{aligned} \tag{195}$$

The arguments also show that $\alpha_L^0(\beta)$, $\alpha_R^0(\beta) \le 0$. But for $0 < \beta < 1$, we have

$$T_L(0,\beta) \ \ge \ \pi t_m \ ,$$

$$T_R(0,\beta) = T_L(0,\beta) + \int_\beta^1 \frac{\phi}{R^{1/2}} d\lambda \tag{196}$$

$$> T_L(0,\beta) \ge \pi t_m \ , \tag{197}$$

and for $-1 < \beta < 0$ we have

$$\lim_{\alpha\to\beta} T_R(\alpha,\beta) = \infty \ ,$$

and we see that we have solutions $\alpha_L(\beta,t)$ and $\alpha_R(\beta,t)$ to $T_L(\alpha,\beta) = \pi t$ and $T_R(\alpha,\beta) = \pi t$, respectively, and since (from Lemma 5.1) $T_{R\alpha}$ and $T_{L\alpha}$ are positive, these solutions are unique. □

LEMMA 5.3. *For all* $\alpha < \beta$, *we have*

$$T_{R\beta} < 0 \ .$$

Proof. From the representation (192), we obtain, by contour integration

$$T_R(\alpha,\beta) = -\int_{x_-(\alpha)}^{x_+(\beta)} \int_{\alpha^0(y)}^{\alpha} \frac{i}{[R(\lambda)R_0(\lambda,y)]^{1/2}} d\lambda dy \ , \tag{198}$$

and hence

$$\begin{aligned} T_{R\beta} &= \frac{\pi \partial_\beta x_+(\beta)}{[(\beta - \alpha^0(x_+(\beta)))(\beta-\alpha)]^{1/2}} + \\ &- \int_{x_-(\alpha)}^{x_+(\beta)} \int_{\alpha^0(y)}^{\alpha} \frac{i}{[R(\lambda)R_0(\lambda,y)]^{1/2}} \frac{1/2}{\lambda-\beta} d\lambda dy \ , \end{aligned}$$

where as in (193) and (194), the first term arises from evaluating the inner integral in (198) at $y = x_+(\beta)$. Clearly both terms are negative, and the lemma is proved.

□

LEMMA 5.4. *For $t = 0$, and hence for small $t > 0$,*

$$T_{L\beta}(\alpha_L(\beta,t),\beta) \geq c > 0 \ , \ for\ 0 \leq \beta \leq 1 \ .$$

Proof.
From (191), we have, by contour integration

$$(199) \qquad T_L(\alpha,\beta) = -\int_{x_-(\alpha)}^{x_-(\beta)} \int_{\alpha^0(y)}^{\alpha} \frac{i}{[R(\lambda)R_0(\lambda,y)]^{1/2}}\, d\lambda dy \ ,$$

and hence,

$$\begin{aligned} T_{L\beta} &= \frac{\pi \partial_\beta x_-(\beta)}{[(\beta - \alpha^0(x_+(\beta)))(\beta-\alpha)]^{1/2}} + \\ &\quad - \int_{x_-(\alpha)}^{x_-(\beta)} \int_{\alpha^0(y)}^{\alpha} \frac{i}{[R(\lambda)R_0(\lambda,y)]^{1/2}} \frac{1/2}{\lambda-\beta} d\lambda dy \ , \end{aligned}$$

where, as in Lemma (5.3) above, the first term arises from evaluating the inner integral in (199) at $y = x_-(\beta)$ (see also (193), (194)).

Recall that, at $t = 0$, to enforce $T = 0$ we must have $x_-(\alpha) = x_-(\beta)$, and hence the second term in (199) vanishes, and so we have (at $t = 0$)

$$T_{L\beta}(\alpha(\beta,0),\beta) = \frac{\pi \partial_\beta x_-(\beta)}{[(\beta - \alpha^0(x_+(\beta)))(\beta-\alpha)]^{1/2}} \ ,$$

which is **strictly** positive for $\beta \in (0,1)$, by assumption. □

Note in particular that as $\beta \uparrow 1$, $\partial_\beta x_-(\beta) \to \infty$, and hence $T_{L\beta}(\alpha(\beta,t),\beta) \to \infty$ as well.

THEOREM 5.2. *Define*

$$\begin{aligned} X_L(\beta,t) &= X_L(\alpha_L(\beta,t),\beta) \ for\ \beta \in (0,1) \ , \\ X_R(\beta,t) &= X_R(\alpha_R(\beta,t),\beta) \ for\ \beta \in (-1,1) \ , \\ x_L(t) &= \frac{1}{\pi} \lim_{\beta \uparrow 1} X_L(\beta,1) \ , \\ x_R(t) &= \frac{1}{\pi} \lim_{\beta \uparrow 1} X_R(\beta,1) \ , \end{aligned}$$

Then, for (small) $t > 0$,
$X_L(\cdot,t)$ maps the interval $(0,1)$ onto $(0,x_L(t))$ in a 1-1 fashion, and $X_R(\cdot,t)$ maps the interval $(-1,1)$ onto $(x_R(t),1)$ in a 1-1 fashion.

Proof. First, observe that for the left region,

$$\lim_{\beta \to 0} \alpha_L(\beta,t) = 0 \ ,$$

as we have $\alpha_L^0(\beta) \le \alpha_L(\beta,t) < 0$, and

$$\lim_{\beta\to 0} \alpha_L^0(\beta) \quad = \quad 0 \ .$$

Next observe that

$$\lim_{\beta\to 0} X_L(\beta,t) \quad = \quad 0 \ , \tag{200}$$

by dominated convergence. Thus $X_L(\beta,t)$ maps $(0,1)$ onto $(0,x_L(t))$.

Similarly, note that

$$\lim_{\beta\to -1} \alpha_R(\beta,t) \quad = \quad -1 \ ,$$

as $-1 \le \alpha_R(\beta,t) < \beta$. Thus, by dominated convergence,

$$\lim_{\beta\to 0} X_R(\beta,t) \quad = \quad 1 \ ,$$

and so $X_R(\beta,t)$ maps $(-1,1)$ onto $(x_R(t),1)$.

Now we calculate

$$\begin{aligned} \partial_\beta X_L(\beta,t) &= (X_{L\alpha})\,\partial_\beta \alpha_L(\beta,t) \ + \ X_{L\beta} \\ &= \left(-\frac{\beta-\alpha}{2}T_{L\alpha}\right)\left(-\frac{T_{L\alpha}}{T_{L\beta}}\right) \ + \ \frac{\beta-\alpha}{2}T_{L\beta} \\ &= (\beta-\alpha)T_{L\beta} \ . \end{aligned} \tag{201}$$

Similarly, we have

$$\partial_\beta X_L(\beta,t) \quad = \quad (\beta-\alpha)T_{R\beta} \ .$$

Hence, from Lemmas 5.4 and 5.3,

$$\begin{aligned} \partial_\beta X_L(\beta,t) &> 0 \ , \ \mathit{for\ all}\ \beta \in (0,1) \ , \\ \partial_\beta X_R(\beta,t) &< 0 \ , \ \mathit{for\ all}\ \beta \in (-1,1) \ . \end{aligned}$$

this proves the theorem. □

We have constructed four functions $\alpha_L(x,t)$, $\beta_L(x,t), \alpha_R(x,t)$, $\beta_R(x,t)$ which solve the systems of equations

$$\begin{aligned} T_L(\alpha,\beta) &= \pi t \quad , \quad X_L(\alpha,\beta) = \pi x \quad \mathit{for}\ x \in (0,x_L(t)) \ , \\ T_R(\alpha,\beta) &= \pi t \quad , \quad X_R(\alpha,\beta) = \pi x \quad \mathit{for}\ x \in (x_R(t),1) \ . \end{aligned}$$

Once again, we observe the general relationship (independent of the subscript)

$$\begin{pmatrix} T_\alpha & T_\beta \\ X_\alpha & X_\beta \end{pmatrix}\begin{pmatrix} \alpha_t & \alpha_x \\ \beta_t & \beta_x \end{pmatrix} \quad = \quad \begin{pmatrix} \pi & 0 \\ 0 & \pi \end{pmatrix} ,$$

which, using the other general relationships (again independent of subscript),

$$\begin{aligned} \frac{\partial}{\partial\beta}X(\alpha,\beta) = \frac{(\beta-\alpha)}{2}T_\beta \ , \\ \frac{\partial}{\partial\alpha}X(\alpha,\beta) = -\frac{(\beta-\alpha)}{2}T_\alpha \ , \end{aligned}$$

imply

$$\alpha_x = -\frac{\pi}{(\beta-\alpha)T_\alpha(\alpha(x,t),\beta(x,t))},$$
$$\beta_x = \frac{\pi}{(\beta-\alpha)T_\beta(\alpha(x,t),\beta(x,t))}.$$

Inserting the proper subscripts, and using the facts

$$T_{L\alpha} > 0,$$
$$T_{L\beta} > 0,$$
$$T_{R\alpha} > 0,$$
$$T_{R\beta} < 0,$$

we see

- $\alpha_L(x,t)$ is a monotone decreasing function of x
- $\beta_L(x,t)$ is a monotone increasing function of x
- $\alpha_R(x,t)$ is a monotone decreasing function of x
- $\beta_R(x,t)$ is a monotone decreasing function of x.

THEOREM 5.3.

$$x_L(t) = x_R(t).$$

Proof. We use the expressions (198) and (199) to write, for β near 1 and $\alpha_L = \alpha_L(\beta,t) < 0$, $\alpha_R = \alpha_R(\beta,t) < 0$,

$$T_L(\alpha_L,\beta) = -\int_{x_-(\alpha_L)}^{x_-(\beta)} \int_{\alpha^0(y)}^{\alpha_L} \frac{i}{[R_L(\lambda)R_0(\lambda,y)]^{1/2}}\, d\lambda dy,$$
$$T_R(\alpha_R,\beta) = -\int_{x_-(\alpha_R)}^{x_+(\beta)} \int_{\alpha^0(y)}^{\alpha_R} \frac{i}{[R_R(\lambda)R_0(\lambda,y)]^{1/2}}\, d\lambda dy.$$

Now letting $\beta \to 1$, we obtain

$$T_L(\alpha_L,1) = -\int_{x_-(\alpha_L)}^{x_c} \int_{\alpha^0(y)}^{\alpha_L} \frac{i}{[R_L(\lambda)R_0(\lambda,y)]^{1/2}}\, d\lambda dy,$$
$$T_R(\alpha_R,1) = -\int_{x_-(\alpha_R)}^{x_c} \int_{\alpha^0(y)}^{\alpha_R} \frac{i}{[R_R(\lambda)R_0(\lambda,y)]^{1/2}}\, d\lambda dy.$$

We see that $T_L(\alpha_L,1)$ and $T_R(\alpha_R,1)$ are just two evaluations of the same function at different values of α, and hence we can only solve

$$T_L(\alpha_L(1,t),1) = \pi t,$$
$$T_R(\alpha_R(1,t),1) = \pi t,$$

by taking $\alpha_L(1,t) = \alpha_R(1,t)$.

Now, the proof of (166) and (159) at $t=0$ also applies to $t>0$, and we obtain, after an additional contour integration

$$X_L(\alpha_L(\beta,t),\beta) = \\ -\int_{x_-(\alpha_L(\beta,t))}^{x_-(\beta)} \int_{\alpha^0(y)}^{\alpha_L(\beta,t)} \frac{i(\lambda - m_L(\beta,t))}{[R_L(\lambda)R_0(\lambda,y)]^{1/2}}\, d\lambda dy \ + \pi x_-(\beta)\ ,$$

$$X_R(\alpha_R(\beta,t),\beta) = \\ -\int_{x_-(\alpha_R(\beta,t))}^{x_+(\beta)} \int_{\alpha^0(y)}^{\alpha_R(\beta,t)} \frac{i(\lambda - m_R(\beta,t))}{[R_R(\lambda)R_0(\lambda,y)]^{1/2}}\, d\lambda dy \ + \pi x_+(\beta)\ ,$$

and letting $\beta \uparrow 1$, we see that

$$X_L \to \quad -\int_{x_-(\alpha_L(1,t))}^{x_c} \int_{\alpha^0(y)}^{\alpha_L(1,t)} \frac{i(\lambda - m_L(1,t))}{[R_L(\lambda)R_0(\lambda,y)]^{1/2}}\, d\lambda dy \ + \pi x_c\ ,$$

$$X_R \to \quad -\int_{x_-(\alpha_R(1,t))}^{x_c} \int_{\alpha^0(y)}^{\alpha_R(1,t)} \frac{i(\lambda - m_R(1,t))}{[R_R(\lambda)R_0(\lambda,y)]^{1/2}}\, d\lambda dy \ + \pi x_c\ ,$$

and hence

$$x_L(t) = \lim_{\beta\to 1} \frac{1}{\pi} X_L(\beta,t) \ = \ \lim_{\beta\to 1} \frac{1}{\pi} X_R(\beta,t) = x_R(t) \ \equiv \ x_c(t)\ ,$$

and we use the above to define $x_c(t)$. □

Thus we have established that the systems (187 - 188) and (189 - 191) uniquely determine $\alpha(x,t)$ and $\beta(x,t)$ for all $x \in (0,1)$, and small $t > 0$. Just as noted in the text prior to Theorem 5.3, $\alpha(x,t)$ is monotone decreasing on $(0,1)$, and $\beta(x,t)$ is monotone increasing on $(0, x_c(t))$, and monotone decreasing on $(x_c(t), 1)$. Now as remarked in 5.1, the proof of Theorem 5.1 applies so long as the curves $\alpha(x,t)$ and $\beta(x,t)$ have the same same shape as $\alpha(x,0) = \alpha^0(x)$, $\beta(x,0) = \beta^0(x)$, as described above.

Thus we have established the following result.

THEOREM 5.4. *For small $t > 0$, the continuum limit of the Toda Lattice with initial data as described at the beginning of this Chapter, is governed by two zero gap ansatzes, (179 - 180) and (181 - 183). The endpoints of the associated band I_1, $\alpha(x,t), \beta(x,t)$ evolve so as to solve the continuous Toda equations, in Riemann invariant form,*

$$\beta_t \ = \ \frac{\beta - \alpha}{2}\beta_x \ , \ \beta(x,0) = \beta^0(x) \tag{202}$$

$$\alpha_t \ = \ \frac{\beta - \alpha}{2}\alpha_x \ , \ \alpha(x,0) = \alpha^0(x) \ . \tag{203}$$

Moreover,

$$\lim_{\epsilon\to 0} \hat{b}^2_\epsilon(x,t) \ = \ b^2(x,t)\ ,$$

$$\lim_{\epsilon\to 0} \hat{a}_\epsilon(x,t) \ = \ a(x,t)\ ,$$

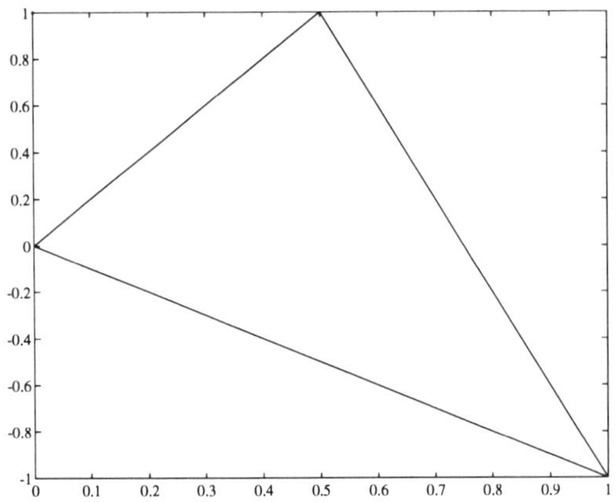

FIGURE 5. The Riemann Invariants $\alpha^0(x)$ and $\beta^0(x)$ (note $\alpha^0(x) \leq \beta^0(x)$), for the case of straight line initial data.Note that $x_c = 0.5$.

where we define

$$\begin{aligned} a(x,t) &= \frac{\beta(x,t)+\alpha(x,t)}{2} \quad , \\ b^2(x,t) &= \left(\frac{\beta(x,t)-\alpha(x,t)}{4}\right)^2 \quad . \end{aligned}$$

Observation: The above Theorem is proved only for small t and, in fact, cannot be true for all $t > 0$. Indeed, for fixed $\beta < 1$, one clearly cannot solve $T_L(\alpha, \beta) = \pi t$ for t sufficiently large, as

$$\begin{aligned} T &= \int_\alpha^\beta \frac{i\theta'}{R^{1/2}} d\lambda \\ &\leq \max_{\lambda \in (-1,\beta)} \theta'(\lambda)\, \pi \ . \end{aligned}$$

Furthermore, as we will see, $T_{L\beta}(\alpha_L(\beta,t),\beta) > 0$ only for t close to zero. Because of the explicit relationship between T_β and β_x,

$$\beta_x = \frac{\pi}{(\beta-\alpha)T_\beta}$$

we see that the ansatz breaks down because of the development of a shock ($\beta_x \to \infty \ \equiv \ T_\beta \to 0$) in the hyperbolic system (17 - 18).

As a model problem to understand the behavior of the continuum limit of the Toda lattice when shock formation is permitted by the initial data, we consider **straight line initial data** (see Figure (5)).

This data is a (nonsmooth) example of the general class of data described in the beginning of this chapter. Hence, the above Theorems and Lemmas apply (smoothness of the data at $x = x_c$ plays no role), and so we know that the maximization problem is solved, for small $t > 0$, by two zero gap ansatzes, one on $[0, x_c(t)]$, the other on $[x_c(t), 1]$. Also from the above, the endpoints $\alpha(x,t)$ and $\beta(x,t)$ of the spectral band I_1 associated to these problems define two functions of x, for $x \in [0,1]$, that are continuously differentiable on $[0, x_c(t),)$ and $(x_c(t), 1]$, and continuous across $x_c(t)$.

In our analysis of the continuum limit of the Toda lattice with this simple data, we will focus on the spatial region $[0, x_c(t))$, and show that as $t \to t_m$, where t_m is as defined in Lemma 5.2,

$$t_m = \min_{0\le\beta\le 1} T_L(0,\beta) ,$$

$\sup_{0\le x\le x_c(t)} \beta_x(x,t) \to \infty$. As will be seen, at shock time, $\alpha(x,t) < 0$ for all $x \in (0,1)$, and the curves $\alpha(x,t)$, $\beta(x,t)$ will be in the same general class as the initial data. Hence the continuum limit of the Toda lattice with this data is indeed described by the pair of zero gap ansatzes up to $t = t_m$, and therefore the continuum limit exhibits shock formation. For this simple data, it will be seen that the time of shock formation coincides with the vanishing of the "left" region, that is, that

$$x_c(t_m) = 0 .$$

The asymptotic norming constant for the simple data in Figure 5, is

$$\theta'(\lambda) = \left\{ \begin{array}{ll} +a, & \lambda > 0 \\ -a, & \lambda < 0 \end{array} \right\} , \tag{204}$$

where

$$a = \frac{1}{\sqrt{\beta^{0'}|\alpha^{0'}|}} \cos^{-1} \frac{\beta^{0'} - |\alpha^{0'}|}{\beta^{0'} + |\alpha^{0'}|} ,$$

where $\alpha^{0'}$ and $\beta^{0'}$ are the slopes of the lines on $[0, x_c)$.

We consider the Riemann - Hilbert problem

$$\begin{aligned} H\psi^* &= \theta'(\lambda) - t \;\; on \;\; I_1 = (\alpha(x,t), \beta(x,t)) , \\ \psi^* &= 0 \;\; on \;\; I_2 = \mathbf{R} \backslash (\alpha(x,t), \beta(x,t)) , \end{aligned} \tag{205}$$

with $\alpha(x,t) < 0 < \beta(x,t)$. We will focus only on the "left" ansatz, and so we drop the subscript "L", for ease of notation. As before,

$$\psi^*(\lambda) = Re \lim_{\epsilon\downarrow 0} f(\lambda + i\epsilon) , \tag{206}$$

where

$$f(z) = R(z)^{1/2} \frac{1}{\pi i} \int_\alpha^\beta \frac{i(\theta'(\lambda) - t)}{R(\lambda)^{1/2}} \frac{d\lambda}{\lambda - z} , \tag{207}$$

and the conditions $\psi^* + iH\psi^* \in H^p$, $\int \psi^* = \pi x$ become

$$(208) \qquad T(\alpha,\beta) \equiv \int_\alpha^\beta \frac{i}{[(\lambda-\alpha)(\lambda-\beta)]^{1/2}}\theta' d\lambda = \pi t \ ,$$

$$(209) \qquad X(\alpha,\beta) \equiv \int_\alpha^\beta i\frac{\lambda - m}{[(\lambda-\alpha)(\lambda-\beta)]^{1/2}}\theta' d\lambda = \pi x$$

(cf. (187 - 188)).

Inserting (204) into (208), we obtain,

$$(210) \qquad T = -2a\int_\alpha^0 \frac{i}{R(\lambda)^{1/2}}d\lambda + \pi a \ ,$$

or, by contour integration,

$$T = 2a\int_0^\beta \frac{i}{R(\lambda)^{1/2}}d\lambda - \pi a \ .$$

Thus we calculate , for this simple data,

$$(211) \qquad T_\beta = -2a\int_\alpha^0 \frac{i}{R(\lambda)^{1/2}}\frac{1/2}{\lambda-\beta}d\lambda > 0 \ ,$$

$$(212) \qquad T_\alpha = 2a\int_0^\beta \frac{i}{R(\lambda)^{1/2}}\frac{1/2}{\lambda-\alpha}d\lambda > 0 \ ,$$

provided $\alpha < 0 < \beta$.

As already noted, the general theory derived above applies, in particular, to the simple data in Figure 5. In what follows we will strengthen the results obtained from the genreal theory, in order to compute the shock time in this special case.

In fact, a careful reading of the forgoing proof shows that the assumption of smallness of t arises from the fact that we are not guaranteed that $T_{L\beta} > 0$ for all $\alpha < 0 < \beta$. Note, however, in (212) above, that for this simple data this inequality is in fact known. One can check then that the general theory applies for **all** $0 \le t \le t_m$.

Thus, Lemma (5.2) states, without modification, that for all $0 \le t \le t_m \equiv a$, there exists a solution $\alpha(\beta,t) < \beta$ to the equation

$$T(\alpha,\beta) = \pi t \ ,$$

and the following result strengthens Lemma 5.2, which is only true in general for small t.

LEMMA 5.5. *Fix $0 \le t \le t_m$, and take $\alpha(\beta,t)$ such that $T(\alpha,\beta) = \pi t$. Then there exists a unique β such that $X(\alpha(\beta,t),\beta) = \pi x$, for $x \in \left(0, \frac{1}{\pi}X(\alpha(\beta=1,t),1)\right)$.*

Proof.

First we observe that as in (200),

$$\lim_{\beta\downarrow 0} X(\alpha(\beta,t),\beta) = 0 \ .$$

Now,as we see that the range of $X(\alpha(\beta,t),\beta)$ for $\beta \in [0,1]$ clearly contains $\left(0, \frac{1}{\pi}X(\alpha(\beta=1,t),1)\right)$, we must only establish uniqueness of the solution β. But this follows from

$$\begin{aligned}
\partial_\beta X(\alpha(\beta,t),\beta) &= X_\alpha \alpha_\beta + X_\beta \\
&= \left(-\frac{\beta-\alpha}{2}T_\alpha\right)\left(\frac{-T_\beta}{T_\alpha}\right) + \frac{\beta-\alpha}{2}T_\beta \\
&= (\beta-\alpha)T_\beta \ , \qquad (213)
\end{aligned}$$

as (see (211)) $T_\beta > 0$. □

We have thus established, for fixed $t \in [0,t_m]$, a solution $\alpha(x,t)$, $\beta(x,t)$ to the pair of equations (208 - 209), for $x \in [0, x_L(t)]$, where

$$x_L(t) = \lim_{\beta \to 1} \frac{1}{\pi} X(\beta,t) \ .$$

We will not describe the region $(x_R(t),1)$ in such detail. However, we observe that for $0 \le t \le t_m$, the "Right" ansatz produces curves $\alpha_R(x,t), \beta_R(x,t)$ which are both monotone decreasing functions of x, for $x \in (x_R(t),1)$, and that $x_R(t) = x_L(t) \equiv x_c(t)$. The reason for this is that the assumption of smallness on t (used for the case of general initial data above, of which the data at hand is an example) comes solely from the "Left" region, to maintain $T_{L\beta} > 0$ (provided $\alpha_L(\beta,t)$, $\alpha_R(\beta,t) \le 0$). Here we see that $T_{L\beta} > 0$ for all $t < t_m$. Thus we conclude that for all $t < t_m$, the two ansatzes combine to produce $\alpha(x,t)$, $\beta(x,t)$ for $x \in (0,1)$, and they have the same general shape as the initial data described in the beginning of this chapter. Thus these ansatzes describe the continuum limit of the Toda lattice,

$$\begin{aligned}
\lim_{\epsilon\to 0} \hat{b}_\epsilon^2(x,t) &= \left(\frac{\beta(x,t)-\alpha(x,t)}{4}\right)^2 \ , \\
\lim_{\epsilon\to 0} \hat{a}_\epsilon(x,t) &= \frac{\beta(x,t)+\alpha(x,t)}{2} \ .
\end{aligned}$$

LEMMA 5.6. *For all $\beta \in (0,1]$,*

$$\lim_{t\to t_m} X(\alpha(\beta,t),\beta) = 0 \ .$$

Proof.

First observe from (204) and (208) that $T(0,\beta) = \pi a$, independent of β. Thus, as T is increasing in α,

$$T(\alpha,\beta) < T(0,\beta) = \pi t_m \ , \quad for\ \alpha < 0 \ .$$

But $T(\alpha(\beta,t),\beta) = \pi t$. Hence $\alpha(\beta,t) \to 0$ as $t \to t_m$. However, from (204) and (209),

$$X(\alpha=0,\beta) = a \int_0^\beta \frac{\lambda - m}{R(\lambda)^{1/2}} d\lambda \ ,$$

where $m = \frac{\alpha+\beta}{2} = \frac{\beta}{2}$, and hence

$$\lim_{t \to t_m} X(\alpha(\beta, t), \beta) = 0 \ .$$

□

Qualitatively, the "left" region (on which $\beta_x > 0$) is being pulled into the boundary, and hence, as $\beta(x,t)$ must cover the range of values $(0,1)$ as x varies over the region $(0, x_c(t))$, a shock must be forming. The time of shock formation is explicit,

$$t_m = a = \frac{1}{\sqrt{\beta^{0'}|\alpha^{0'}|}} \cos^{-1} \frac{\beta^{0'} - |\alpha^{0'}|}{\beta^{0'} + |\alpha^{0'}|} \ .$$

In addition, one sees that the "right" ansatz (151 - 153) solves the maximization problem for every $x \in (0,1]$, at $t = t_m$, and calculations of $T_{R\alpha}$ and $T_{R\beta}$ show that one can continue to solve *this* problem for t beyond shock time, and that the curves so defined remain monotone decreasing. Thus, we conclude that for the straight line initial data considered, the zero gap ansatz (151) solves the maximization problem (146) **for t beyond shock time**.

To see that this phenomenon is not restricted to straight line initial data, we will show that under less restrictive assumptions a shock will form at the boundary, and that the maximization problem (146) is solved, beyond shock time, by a zero gap ansatz. We consider general initial data as described above, such that $\alpha^0(x)$ is strictly monotone decreasing, and $\beta^0(x)$ has one (nondegenerate) critical point, a maximum, at x_c,

$$\begin{aligned} \alpha^0(0) &= \beta^0(0) = 0 \\ \alpha^0(1) &= \beta^0(1) = -1 \\ \beta^0(x_c) = 1 \ , & \end{aligned}$$

however, we will only consider the "left" Riemann - Hilbert problem (160), as we are interested in behavior at the boundary $x = 0$, and we will assume that on the "right" the maximization problem is solved by the appropriate zero gap ansatz. That is, that the continuation to $t > 0$ of the maximizer obtained from the ansatz (151) continues to solve the maximization problem on $(x_c(t), 1)$, where the location of x_c evolves in time, and that the connection at $x_c(t)$ to the ansatz (160) is made as in the case of the special data of Figure 5. Finally we assume that

$$\theta''(\lambda) > 0 \tag{214}$$

for all values of $\lambda \neq 0$ in the spectral domain $(-1,1)$. In Lemma 5.11 below, we will identify a class of initial data $\alpha^0(x), \beta^0(x)$ for which (214) is true.

We consider the Riemann - Hilbert problem (160),that is,

$$\begin{aligned} H\psi^* &= \theta'(\lambda) - t \ \ on \ \ I_1 = (\alpha(x,t), \beta(x,t)) \\ \psi^* &= 0 \ \ on \ \ I_2 = \mathbf{R} \backslash [\alpha(x,t), \beta(x,t)] \ . \end{aligned} \tag{215}$$

It's solution is (again, dropping the subscript L)

$$\psi^*(\lambda) = Re \lim_{\epsilon \downarrow 0} f(\lambda + i\epsilon) \ ,$$

where

$$f(z) \quad = \quad R(z)^{1/2}\frac{1}{\pi i}\int_\alpha^\beta \frac{i(\theta'(\lambda)-t)}{R(\lambda)^{1/2}}\frac{d\lambda}{\lambda - z} \ .$$

The conditions $\psi^* + iH\psi^* \in H^p$, $\int \psi^* = \pi x$ become

$$(216) \qquad T(\alpha,\beta) \equiv \int_\alpha^\beta \frac{i}{[(\lambda-\alpha)(\lambda-\beta)]^{1/2}}\theta' d\lambda = \pi t \ ,$$

$$(217) \qquad X(\alpha,\beta) \equiv \int_\alpha^\beta i\frac{\lambda - m}{[(\lambda-\alpha)(\lambda-\beta)]^{1/2}}\theta' d\lambda = \pi x \ .$$

We remark that these condition reconstruct the initial data at $t = 0$, as we have shown previously in this chapter, and that the function $\psi^*(\lambda; x, 0)$ solves the maximization problem (146) (of course, extending to $(x_c, 1)$ by the "right" ansatz). Further, as remarked above, the derivation of the formula

$$(218) \quad T_\alpha = \frac{\pi \partial_\alpha x_-(\alpha)}{[(\alpha-\beta)(\alpha - \beta^0(x_-(\alpha)))]^{1/2}} + \\ + \int_{x_-(\alpha)}^{x_-(\beta)} \int_{\beta^0(y)}^{\beta} \frac{i}{[R(\lambda)R_0(\lambda,y)]^{1/2}} \frac{\frac{1}{2}}{\lambda - \alpha} \, d\lambda dy \ > 0 \ , \quad \alpha < 0 \ ,$$

(c.f. proof of Lemma 5.1) is general and applies in this case as well.

LEMMA 5.7. *For $\beta > 0$, $\alpha < 0$, we have*

$$T_\beta > 0 \ .$$

Proof. We write

$$(219) \qquad T \ = \ \int_{-1}^{-\frac{\beta+\alpha}{\beta-\alpha}} \frac{1}{\sqrt{1-\mu^2}}\theta'\left(\frac{\beta-\alpha}{2}\mu + \frac{\beta+\alpha}{2}\right) d\mu \\ + \int_{-\frac{\beta+\alpha}{\beta-\alpha}}^{1} \frac{1}{\sqrt{1-\mu^2}}\theta'\left(\frac{\beta-\alpha}{2}\mu + \frac{\beta+\alpha}{2}\right) d\mu \ ,$$

and hence

$$(220) T_\beta = \sqrt{\frac{-4\alpha}{\beta}\frac{\theta'_+(0)}{\beta-\alpha}} + \frac{1}{2}\int_{-1}^{1} \frac{\mu+1}{\sqrt{1-\mu^2}}\theta''\left(\frac{\beta-\alpha}{2}\mu + m\right) d\mu \ ,$$

where $\theta'_+(0) = \lim_{\lambda \downarrow 0} \theta'(\lambda)$, and the first term above arises because of the discontinuity at $\lambda = 0$ in the function $\theta'(\lambda)$ (see (144)). As both terms in (220) are positive, the lemma is proved. □

Remark: Note that the Lemma is also true for $0 < \alpha < \beta$; inded (220) is still valid, and in fact simpler, as the first term is absent.

LEMMA 5.8. *Fix $\beta > 0$. Then, for all $0 \leq t \leq t_m$, where t_m is the shock time calculated for the case of the straight line data,*

$$t_m = \frac{1}{\pi} \lim_{\beta \downarrow 0} T(0,\beta) = \frac{1}{\sqrt{\beta^{0'}|\alpha^{0'}|}} \cos^{-1} \frac{\beta^{0'} - |\alpha^{0'}|}{\beta^{0'} + |\alpha^{0'}|} ,$$

$$\beta^{0'} = \beta^0_x(x=0) \quad \alpha^{0'} = -\alpha^0_x(x=0) ,$$

there exists a unique solution $\alpha(\beta, t)$ to the equation

$$T(\alpha, \beta) = \pi t . \tag{221}$$

Proof. We know that we can find $\alpha = \alpha^0(\beta)$ such that $T(\alpha, \beta) = 0$, as the ansatz reconstructs the initial data. As T is increasing in α (see (218)), we can solve (221) uniquely for all

$$t \leq \min_{0 \leq \beta \leq 1} T(\alpha = 0, \beta) .$$

However, T is increasing in β , and hence the minimum is achieved at $\beta = 0$. We have

$$\begin{aligned} \lim_{\beta \downarrow 0} T(0,\beta) &= \lim_{\beta \downarrow 0} \int_0^\beta \frac{i\theta'}{R^{1/2}} d\lambda = \pi \theta'(0) , \\ &= \pi t_m , \end{aligned}$$

which proves the lemma. □

LEMMA 5.9. *Fix $0 \leq t \leq t_m$, and take $\alpha(\beta, t)$ to solve $T(\alpha, \beta) = \pi t$. Then there exists a unique β such that $X(\alpha(\beta, t), \beta) = \pi x$ for each $x \in \left(0, \frac{1}{\pi} X(\alpha(\beta = 1, t), 1)\right)$.*

Proof.
As in (200),

$$\lim_{\beta \downarrow 0} X(\alpha(\beta, t), \beta) = 0 .$$

Thus the range of $X(\alpha(\beta, t), \beta)$ clearly contains $\left(0, \frac{1}{\pi} X(\alpha(\beta = 1, t), 1)\right)$. As above, we calculate

$$\begin{aligned} \partial_\beta X(\alpha(\beta, t), \beta) &= X_\alpha \alpha_\beta + X_\beta \qquad (222) \\ &= \left(-\frac{\beta - \alpha}{2} T_\alpha\right) \left(\frac{-T_\beta}{T_\alpha}\right) + \frac{\beta - \alpha}{2} T_\beta \\ &= (\beta - \alpha) T_\beta > 0 . \qquad (223) \end{aligned}$$

This proves the lemma. □

Thus we have shown that for $0 \leq t \leq t_m$, the conditions

$$\begin{aligned} T(\alpha, \beta) &= \pi t , \\ X(\alpha, \beta) &= \pi x , \end{aligned}$$

determine $\alpha(x,t)$, $\beta(x,t)$ as continuously differentiable functions of x,t. We can compute the equations for the quantities α_x, β_x, α_t, β_t, and we obtain

$$\begin{pmatrix} T_\alpha & T_\beta \\ X_\alpha & X_\beta \end{pmatrix} \begin{pmatrix} \alpha_t & \alpha_x \\ \beta_t & \beta_x \end{pmatrix} = \begin{pmatrix} \pi & 0 \\ 0 & \pi \end{pmatrix} . \tag{224}$$

Using the relations,

$$X_\alpha = -\frac{\beta-\alpha}{2} T_\alpha \ ,$$
$$X_\beta = \frac{\beta-\alpha}{2} T_\beta \ ,$$

(224) implies

$$\alpha_x = -\frac{\pi}{(\beta-\alpha)T_\alpha} < 0 \ ,$$
$$\beta_x = \frac{\pi}{(\beta-\alpha)T_\beta} > 0 \ ,$$

where we have used that $T_\alpha > 0$, and $T_\beta > 0$ (see (218) and Lemma 5.7). Thus we see that the curves $\alpha(x,t)$ and $\beta(x,t)$ are monotone functions.

As the initial data may be approximated near $x = 0$ by straight lines, we expect, from the previous calculations for the model problem with data in Figure 5, that some kind of shock should develop near $x = 0$, precisely at $t = t_m$. In fact, as we now show, at time $t = t_m$ a "boundary shock" develops at $x = 0$, i.e. $\beta_x(0,t) \to \infty$ as $t \uparrow t_m$, and $\lim_{x\downarrow 0} \beta_x(x,t) = \infty$ for $t > t_m$. However, $\beta(x,t_m)$ and $\alpha(x,t_m)$ remain continuous on $[0,1]$ for all $t \geq t_m$.

To prove this we show how to evolve the zero gap ansatz beyond $t = t_m$.

LEMMA 5.10. *For each fixed $\beta > 0$, there exists a unique $\alpha^0(\beta) < \alpha(\beta,t) < \beta$ such that $T(\alpha(\beta,t),\beta) = \pi t$ for all $0 \leq t \leq \theta'(\beta)$.*

Proof. For $\alpha < 0$, we have already established

$$T_\alpha > 0 \ .$$

We now show the same is true for all $0 \leq \alpha < \beta$. Indeed, in analogy to (219), we have for $\alpha \geq 0$,

$$T(\alpha,\beta) = \int_{-1}^{1} \frac{1}{\sqrt{1-\mu^2}} \theta'\left(\frac{\beta-\alpha}{2}\mu + \frac{\beta+\alpha}{2}\right) d\mu \ , \tag{225}$$

and hence

$$T_\alpha = \int_{-1}^{1} \frac{1}{\sqrt{1-\mu^2}} \theta''\left(\frac{\beta-\alpha}{2}\mu + \frac{\beta+\alpha}{2}\right) \frac{1-\mu}{2} d\mu > 0 \ .$$

Thus we see that T is increasing in α for all $\alpha < \beta$. Also, we observe, from (225), that

$$\lim_{\alpha\to\beta} T(\alpha,\beta) = \pi\theta'(\beta) \ ,$$

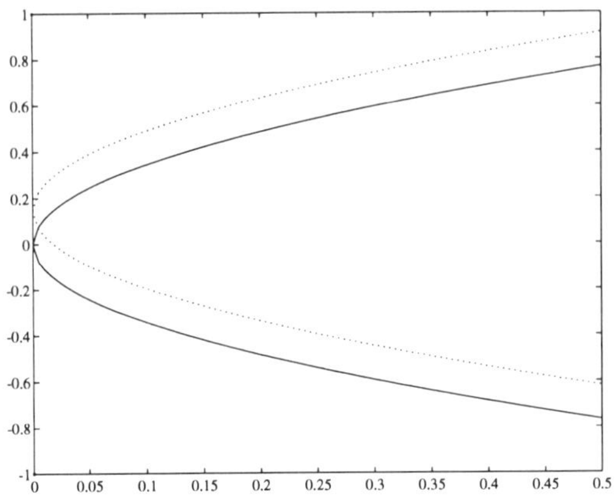

FIGURE 6. Evolution of the Riemann invariants $\alpha^0(x)$ and $\beta^0(x)$ past shock time. The solid line is $t = 0$, and the dotted line is $t > 0$. Note that $\alpha(0,t), \beta(0,t) \neq 0$.

and recall from Lemma 5.9 that $T(\alpha^0(\beta), \beta) = 0$. Thus, for each fixed $\beta > 0$, we can find a unique $\alpha(\beta, t)$ such that $T(\alpha, \beta) = \pi t$ for all $t \leq \theta'(\beta)$. □

Remark: As $\theta''(\lambda) > 0$, the above lemma implies that we can solve $T(\alpha, \beta) = \pi t$ for $\alpha = \alpha(\beta, t) \leq \beta$ provided $\beta > (\theta')^{-1}(t)$. Thus if $t > t_m$, then necessarily $\beta > (\theta')^{-1}(t_m) = 0$. This means that when we solve $X(\alpha(\beta,t), \beta)$ below, the curves $\alpha(\beta(\cdot,t)), \beta(\cdot,t)$ do **not** go to zero as $x \to 0$ (see Figure 6).

For a fixed $t > t_m$, we consider values of β that satisfy

$$\beta > (\theta')^{-1}(t) ,$$

and we note from (222), say, that $\partial_\beta X(\alpha(\beta,t), \beta) > 0$ as $T_\alpha > 0$ and $T_\beta > 0$

Further, as $\beta \downarrow (\theta')^{-1}(t)$, we see from

$$T(\alpha, (\theta')^{-1}(t)) < \lim_{\alpha \uparrow (\theta')^{-1}(t)} T(\alpha, (\theta')^{-1}(t)) = \pi t ,$$

that $\alpha(\beta, t)$ also converges to $(\theta')^{-1}(t)$, and hence from (217)

$$\lim_{\beta \downarrow (\theta')^{-1}(t)} X(\beta, t) = 0 . \tag{226}$$

Thus, for $t > t_m$, we still have a unique solution to the "zero gap" ansatz $(\alpha(x,t), \beta(x,t))$ on $\left[0, \frac{1}{\pi}X(\alpha(\beta = 1, t) , 1)\right]$; the solution in monotone in the sense that $\beta(x,t)$ is an increasing function of x, and $\alpha(x,t)$ is a decreasing function of x on the interval.

Note from (226) that

$$\beta(x = 0, t) = \alpha(x = 0, t) = (\theta')^{-1}(t) .$$

The functions $\alpha(\cdot,t),\beta(\cdot,t)$ are clearly continuous on $[0,1]$. However, as we now show for $t>t_m$, $\beta_x(x,t)$ blows up as $x\downarrow 0$. Indeed, for $x>0$, we have

$$\beta_x(x,t) = \frac{\pi}{(\beta-\alpha)T_\beta(\alpha(x,t),\beta(x,t)}\ ,$$

and

$$T_\beta = \frac{1}{2}\int_{-1}^{1}\frac{\mu+1}{\sqrt{1-\mu^2}}\theta''\left(\frac{\beta-\alpha}{2}\mu+m\right)d\mu\ .$$

Thus $|T_\beta|\le \pi\sup_{0\le\lambda\le\beta(x,t)}$. Thus, as $x\downarrow 0$, and hence α, $\beta\to(\theta')^{-1}(t)$, we see that $\beta_x(x,t)\to\infty$. A similar calculation shows that $\alpha_x(x,t)$ also blows up as $x\downarrow 0$. *This establishes that a shock develops at* $t=t_m$.

REMARK 5.2.

For $t<t_m$, $\alpha(0,t)=\beta(0,t)=0$ (see proof of Lemma 5.9). For $t>t_m$, we have that

$$\beta(x=0,t) = \alpha(x=0,t) = (\theta')^{-1}(t)\ ,$$

which is now time dependent. These facts are consistent with the dynamical equations

$$\alpha_t = -\frac{\beta-\alpha}{2}\alpha_x\ ,$$
$$\beta_t = \frac{\beta-\alpha}{2}\beta_x\ .$$

As long as α_x, β_x are bonded (as in the case $t<t_m$), we must have $\alpha_t(0,t)=\beta_t(0,t)=0$. But for $t>t_m$, the singularities of $\alpha_x(x,t)$ and $\beta_x(x,t)$ compensate the convergence $\alpha(x,t)-\beta(x,t)\to 0$, as $x\to 0$, and hence $\alpha_t(0,t),\beta_t(0,t)\neq 0$.

Finally we show that we can construct data for which $\theta''>0$.

LEMMA 5.11. *Suppose for some* $0<x^0<1$, *that* α^0 *and* β^0 *satisfy*

$$-\alpha^0(x)=\beta^0(x)\ ,$$
$$\beta''(x,0)<0\ , \tag{227}$$

for $0\le x\le x^0$, *and* $\beta^0(x^0)=1$. *Then* $\theta''>0$.

Proof. We have by simple changes of variables in (144),

$$\theta''(\lambda)=-\int_0^1\frac{\mu}{\sqrt{1-\mu^2}}\frac{\beta''\left(\beta^{-1}(\lambda\mu)\right)}{\left[\beta'\left(\beta^{-1}(\lambda\mu)\right)\right]^3}d\mu$$

from which the claim is clear. □

It is amusing to check the analysis of this chapter in the case where

$$\beta^0(x) = \sqrt{\frac{x}{x^0}}\ ,\quad x<x^0\ ,$$
$$\alpha^0(x) = -\sqrt{\frac{x}{x^0}}\ ,\quad x>x^0\ ,$$

for which we can explicitly solve the inversion problem.

Indeed, we find

$$
\begin{aligned}
\theta'(\lambda) &= \int_0^{x_-(\lambda)} \frac{1}{[(\lambda-\alpha^0(x))(\lambda-\beta^0(x))]^{1/2}}\,dx \\
&= \int_0^{x_-(\lambda)} \frac{1}{\left(\lambda^2-\frac{x}{x^0}\right)^{1/2}}dx \\
&= 2\lambda x^0 \ ,
\end{aligned}
$$

and, the expressions for T and X become

$$
\begin{aligned}
T &= \int_\alpha^\beta \frac{2\lambda x^0}{R(\lambda)^{1/2}}d\lambda = (\beta+\alpha)x^0 \ , \\
X &= \int_\alpha^\beta \frac{\lambda-m}{R(\lambda)^{1/2}}2\lambda x^0\, d\lambda = \pi\left(\frac{\beta-\alpha}{2}\right)^2 x^0 \ .
\end{aligned}
$$

which implies that

$$
\begin{aligned}
\beta(x,t) &= \sqrt{\frac{x}{x^0}} + \frac{t}{2x^0} \\
\alpha(x,t) &= -\sqrt{\frac{x}{x^0}} + \frac{t}{2x^0} \ ,
\end{aligned}
\tag{228}
$$

from which we see that for $t > 0$, we have a singularity at $x = 0$, with height given by $\frac{t}{2x^0}$.

From the analysis in the section, we see that the shock time is

$$t_m = \theta'(0) = 0$$

as it should be, and we see indeed that

$$\beta(0,t) = (\theta')^{-1}(t) = \frac{t}{2x^0} \ .$$

CHAPTER 6

Shock 2

So far we have solved the maximization problem by satisfying a one band ansatz, i.e. for $t \geq 0$ and for some $0 \leq x_c(t) \leq 1$ we solve for $\alpha(x,t), \beta(x,t)$ such that

$$T_L(\alpha,\beta) = \pi t \ , \tag{229}$$

$$X_L(\alpha,\beta) = \pi x \ , \tag{230}$$

for $0 \leq x \leq x_c(t)$, and

$$T_R(\alpha,\beta) = \pi t \ , \tag{231}$$

$$X_R(\alpha,\beta) = \pi x \ , \tag{232}$$

for $x_c(t) \leq x \leq 1$. If we can solve these equations for functions $\alpha(x,t)$ and $\beta(x,t)$ that are continuous and monotone (in the sense described in the Remark following Theorem 5.1), for all $0 \leq x \leq 1$, then ψ^* constructed from (206 - 207) solves the maximization problem (146), i.e. the side conditions (169 - 172) are **automatically** satisfied.

Thus the one band ansatz on the full interval $(0,1)$ can fail at a time t, **only** if the equations (229 - 232) fail to produce continuous and monotone functions $\alpha(x,t), \beta(x,t)$. Furthermore the method shows that we will be successful in constructing continuous and monotone functions $\alpha(x,t), \beta(x,t)$ once we know that $T_\alpha \neq 0$, $T_\beta \neq 0$. Thus the one - band ansatz on $(0,1)$ can break down only if $T_\alpha = 0$ or $T_\beta = 0$. Because of the explicit relationship between derivatives of the pair $(\alpha(x,t) \ , \ \beta(x,t))$ and T,

$$\begin{aligned} \alpha_x &= -\frac{\pi}{(\beta-\alpha)T_\alpha} \ , \\ \beta_x &= \frac{\pi}{(\beta-\alpha)T_\beta} \end{aligned}$$

we see that the vanishing of one of T_α or T_β is equivalent to the onset of a shock in the hyperbolic system in the case $\alpha < \beta$.

Remark: It is, of course, entirely possible to consider a solution $\alpha(x,t)$, $\beta(x,t)$ for x in some given subinterval $(x_1(t), x_2(t))$ of $(0,1)$. The function ψ^* can then be constructed for x in the interval, but the proof that ψ^* satisfies the side conditions is now no longer automatic, and must be checked by hand for each $x \in (x_1(t), x_2(t))$.

Our first goal in this section is to present a general class of data for which a shock occurs in the interior of the x-interval $(0,1)$, rather than at the boundary as in the previous chapter. Once we have established the formation of a shock, we will take data with a shock at $t = 0$, and show that we can indeed continue to solve $T_L(\alpha, \beta) = \pi t$, $X_L(\alpha, \beta) = \pi x$ for α, β beyond $t = 0$. Moreover, we will show that the solution is not unique, and that there is a region $(x_1(t), x_2(t)$ where the solution is three-sheeted, i.e. for $x \in (x_1(t), x_2(t))$ there are 3 pairs of solutions $(\alpha_j(x,t), \beta_j(x,t))$, $0 \le j \le 2$ of (229 - 230). In view of the previous Remark, it is not clear which ψ_j^*, if any, constructed from $(\alpha_j(x,t), \beta_j(x,t))$, $0 \le j \le 2$, solves the side conditions, and hence the maximization problem on $(x_1(t), x_2(t))$.

Remark: The existence of 3 pairs of solutions $(\alpha_j(x,t), \beta_j(x,t))$, $0 \le j \le 2$, is to be compared to the evolution by characteristics of the inviscid Burger's equation beyond shock time - the solution folds over and becomes three sheeted.

In succeeding chapters we will show that for small times $t > 0$ the one band ansatz continues to solve the maximization problem on a union of intervals $(0, x_L(t)) \cup (x_R(t), 1)$ in the following sense: the solutions $(\alpha(x,t), \beta(x,t))$ of (229 - 230) for $x \in (0, x_c(t))$, and of (231 - 232) for $x \in (x_c(t), 1)$, produce functions $\psi^*(\lambda;\ x, t)$ for $x \in (0, x_L(t)) \cup (x_R(t))$ which solve the side conditions and hence the maximization problem. The points $x_L(t)$, $x_R(t)$ interlace $x_1(t)$, $x_2(t)$, i.e. $x_1(t) < x_L(t) < x_2(t) < x_R(t)$. Moreover for any $x \in (x_L(t), x_R(t))$ the zero gap ansatz fails in the sense that the functions ψ^* constructed from any solution of (229 - 230) does not satisfy the side conditions. To compute the maximizer ψ^* for $x \in (x_L(t), x_R(t))$, we need a Riemann - Hilbert problem with gaps, and indeed for $x \in (x_L(t), x_R(t))$ and near $x_L(t)$ we will show that the maximization problem is solved by means of a one gap Riemann - Hilbert problem.

So our first task is to present data for which a shock forms in the interior of the interval $(0,1)$. Recall from Chapter 5 that if we consider data for which $\theta'' > 0$, then a shock forms at the boundary. We will take a class of data for which the opposite is true, i.e. $\theta'' < 0$. More precisely, we take initial data as in Chapter 5 for which the upper Riemann invariant $\beta^0(x)$ possesses one nondegenerate critical point, a maximum. Also, we assume that $\alpha^0(0) = \beta^0(0) = 0$, $\alpha^0(1) = \beta^0(1) = -1$, , and that $\beta^0(x_c) = \max_{x\in(0,1)} \beta^0(x) = 1$, for definiteness. Again as in Chapter 5, it will be clear from the analysis that follows, that the qualitative behavior of the continuum limit does not depend on these specific choices of the constants, which are made only for convenience. Furthermore, the lower Riemann invariant $\alpha^0(x)$ is still assumed to be negative and strictly monotone decreasing, i.e. $\alpha_x^0 < 0$ for all $x \in [0,1]$, but now we assume that for some $\bar{x} < x_c$,

$$-\alpha^0(x) = \beta_0(x) ,$$

and

$$(\beta^0)'' > 0 ,$$

for $0 \le x < \bar{x}$.

LEMMA 6.1. *For data as above, and for $\lambda \in (0, \beta^0(\bar{x})) \cup (\alpha^0(\bar{x}), 0)$, we have*

$$\theta''(\lambda) < 0 .$$

Proof. We take $\lambda > 0$. By making the change of variables $\lambda z = \beta^0(y)$ in (144), we can express (c.f. Lemma 5.11)

$$\theta'(\lambda) = \int_0^1 \frac{g'(\lambda z)}{\sqrt{1-z^2}} dz ,$$

where g is the inverse function of $\beta^0(x)$,

$$\beta^0(g(\mu)) = \mu .$$

Hence

$$\theta''(\lambda) = \int_0^1 \frac{g''(\lambda z)}{\sqrt{1-z^2}} dz < 0 ,$$

as $(\beta^0)'' > 0 \Rightarrow g'' < 0$. Note that $\theta'(\lambda)$ is odd about $\lambda = 0$, and differentiable away from $\lambda = 0$. Therefore $\theta''(\lambda)$ is even about $\lambda = 0$, and the lemma is proved. □

So we consider the Riemann - Hilbert problem with $\theta(\lambda)$ as above for $0 \le x \le x_c(t)$,

$$\begin{aligned} H\psi^* &= \theta'(\lambda) \ \ on \ \ I_1 = (\alpha(x,t), \beta(x,t)) \\ \psi^* &= 0 \ \ on \ \ I_2 = \mathbf{R} \backslash [\alpha(x,t), \beta(x,t)] , \end{aligned} \tag{233}$$

(c.f. (160)). This problem has the familiar solution

$$\psi^*(\lambda) = Re \lim_{\epsilon \downarrow 0} f(\lambda + i\epsilon) , \tag{234}$$

where

$$f(z) = R(z)^{1/2} \frac{1}{\pi i} \int_\alpha^\beta \frac{i(\theta'(\lambda))}{R(\lambda)^{1/2}} \frac{d\lambda}{\lambda - z} , \tag{235}$$

and the conditions $\psi^* + iH\psi^* \in H^p$, along with $\int \psi^* = \pi x$, yield the two equations,

$$T(\alpha, \beta) \equiv \int_\alpha^\beta \frac{i}{[(\lambda - \alpha)(\lambda - \beta)]^{1/2}} \theta' d\lambda = \pi t , \tag{236}$$

$$X(\alpha, \beta) \equiv \int_\alpha^\beta i \frac{\lambda - m}{[(\lambda - \alpha)(\lambda - \beta)]^{1/2}} \theta' d\lambda = \pi x . \tag{237}$$

As our data falls into the same general category as that described in the beginning of Chapter 5, we have the following **facts**.

- 1. For $\alpha^0(\beta) < \alpha < 0$ and $\beta > 0$,

$$(238)\qquad T(\alpha,\beta)=\int_{x_-(\alpha)}^{x_-(\beta)}\int_{\beta^0(y)}^{\beta}\frac{i}{[R(\lambda)R_0(\lambda,y)]^{1/2}}\,d\lambda dy$$

$$(239)\qquad X(\alpha,\beta)=\int_{x_-(\alpha)}^{x_-(\beta)}\int_{\beta^0(y)}^{\beta}i\frac{\lambda-m}{[R(\lambda)R_0(\lambda,y)]^{1/2}}\,d\lambda dy\ +\pi x_-(\alpha)\ ,$$

$$(240)\quad T_\alpha=\frac{\pi\partial_\alpha x_-(\alpha)}{[(\alpha-\beta)(\alpha-\beta^0(x_-(\alpha)))]^{1/2}}+$$
$$+\int_{x_-(\alpha)}^{x_-(\beta)}\int_{\beta^0(y)}^{\beta}\frac{i}{[R(\lambda)R_0(\lambda,y)]^{1/2}}\frac{\frac{1}{2}}{\lambda-\alpha}\,d\lambda dy\quad>\ 0\ ,$$

$$(241)\quad T_\beta=\frac{\pi\partial_\beta x_-(\beta)}{[(\beta-\alpha^0(x_-(\beta)))(\beta-\alpha)]^{1/2}}+$$
$$-\int_{x_-(\alpha)}^{x_-(\beta)}\int_{\alpha^0(y)}^{\alpha}\frac{i}{[R(\lambda)R_0(\lambda,y)]^{1/2}}\frac{1/2}{\lambda-\beta}d\lambda dy\ .$$

• 2. For $\alpha<\alpha^0(\beta)$ and $\beta>0$,

$$(242)\quad T(\alpha,\beta)=-\int_{x_-(\beta)}^{x_-(\alpha)}\int_{\beta}^{\beta^0(y)}\frac{i}{[R(\lambda)R_0(\lambda,y)]^{1/2}}\,d\lambda dy$$

$$(243)\quad X(\alpha,\beta)=-\int_{x_-(\beta)}^{x_-(\alpha)}\int_{\beta}^{\beta^0(y)}i\frac{\lambda-m}{[R(\lambda)R_0(\lambda,y)]^{1/2}}\,d\lambda dy\ +\pi x_-(\alpha)\ ,$$

$$(244)\ T_\alpha=\frac{\pi\partial_\alpha x_-(\alpha)}{[(\alpha-\beta)(\alpha-\beta^0(x_-(\alpha)))]^{1/2}}+$$
$$-\int_{x_-(\beta)}^{x_-(\alpha)}\int_{\beta}^{\beta^0(y)}\frac{i}{[R(\lambda)R_0(\lambda,y)]^{1/2}}\frac{\frac{1}{2}}{\lambda-\alpha}\,d\lambda dy\quad>\ 0\ ,$$

$$(245)\ T_\beta=\frac{\pi\partial_\beta x_-(\beta)}{[(\beta-\alpha^0(x_-(\beta)))(\beta-\alpha)]^{1/2}}+$$
$$+\int_{x_-(\beta)}^{x_-(\alpha)}\int_{\alpha}^{\alpha^0(y)}\frac{i}{[R(\lambda)R_0(\lambda,y)]^{1/2}}\frac{1/2}{\lambda-\beta}d\lambda dy\ .$$

• 3. To solve (236) , (237) at $t=0$, the regions of integration in (240 - 241) must vanish, and hence we reconstruct the initial data.
• 4. From (240), we have that for $\alpha^0(\beta)<\alpha<0$ and $\beta>0$,

$$T_\alpha>0\ ,$$

• 5. For $0\le t\le\pi t_m\equiv\min_{\beta\in[0,1]}T(0,\beta)$ and $\beta>0$, there exists a unique solution $\alpha(\beta,t)$ to

$$T(\alpha,\beta)\ =\ \pi t\ .$$

• 6. Again substituting $\alpha(\beta,t)$ into $X(\alpha,\beta)$, we define

$$X(\beta,t) \equiv X(\alpha(\beta,t),\beta)\ ,$$

and also

$$\begin{aligned}\partial_\beta X(\beta,t) &= (X_\alpha)\,\partial_\beta\alpha(\beta,t) \;+\; X_\beta \\ &= \left(-\frac{\beta-\alpha}{2}T_\alpha\right)\left(-\frac{T_\alpha}{T_\beta}\right) \;+\; \frac{\beta-\alpha}{2}T_\beta \\ &= (\beta-\alpha)T_\beta\ . \qquad (246)\end{aligned}$$

• 7. The function $X(\beta,t)$ is a monotone decreasing function of β for $\beta\in(0,1)$ at $t=0$, and hence for small $t>0$.
• 8. We define, as usual,

$$x_L(t) \;=\; \lim_{\beta\uparrow 1}\frac{1}{\pi}X(\beta,t)\ ,$$

and the analog for the right region (see Chapter 5, equations (151 - 159)),

$$x_L(t) \;=\; \lim_{\beta\uparrow 1}\frac{1}{\pi}X_R(\beta,t)\ ,$$

and we have

$$x_L(t) = x_R(t) = x_c(t)\ .$$

• 9. For small $t>0$, and for some $x_c(t)\in(0,1)$, equations (236),(237) and their analogs on the "right" (see Chapter 5, equations (151 - 159)) uniquely determine two continuously differentiable functions $\alpha(x,t)$, $\beta(x,t)$ for $x\in[0,x_c(t)]$ and $x\in[x_c(t),1]$ respectively:
† The functions $\alpha(x,t),\beta(x,t)$ and their first derivatives match at $x=x_c(t)$.
† $\alpha(\cdot,t)$ is monotone decreasing on $[0,1]$.
† $\beta(\cdot,t)$ is monotone increasing on $[0,x_c(t)]$ and monotone decreasing on $[x_c(t),1]$. Hence
$\beta_x(x,t)\to\beta_x(x_c(t),t)=0$ as $x\to x_c(t)$.
• 10. For small $t\geq 0$, the functions ψ_L^* and ψ_R^* combine to solve the maximization problem for $x\in(0,1)$.

Now we will show that the function $\beta(x,t)$ develops a shock in finite time at a point x in the interior of the interval $(0,1)$.

LEMMA 6.2. *The infimum of* $T(0,\beta)$, *for* $0<\beta\leq 1$ *is achieved at some* $\bar{\beta}>0$.

Proof. From the representation (219) for $T(0,\beta)$,

$$T(0,\beta) \;=\; \int_{-1}^{1}\frac{1}{\sqrt{1-\mu^2}}\theta'\left(\frac{\beta}{2}\mu+\frac{\beta}{2}\right)d\mu$$

we see that

$$(247) \qquad T_\beta(0,\beta) \;=\; \int_{-1}^{1} \frac{\frac{1+\mu}{2}}{\sqrt{1-\mu^2}}\theta''\left(\frac{\beta}{2}\mu + \frac{\beta}{2}\right) d\mu \;<\; 0\,,$$

where the last inequality follows from the fact that $\theta'' < 0$. Thus we see that the continuous function $T(0,\beta)$ on $(0,1]$ must achieve its minimum at some $\bar{\beta} > 0$. □

If we approximate the data by straight lines near $x = 0$, then we expect a "boundary shock" in the sense of Chapter 5 to develop at time

$$t_b \;=\; \lim_{\beta \downarrow 0} \frac{1}{\pi} T(0,\beta)\,.$$

The above theorem tells us that

$$t_m < t_b\,.$$

We will show that a shock develops at some time $t < t_m$, in the interior of $(0,1)$. The above inequality then means that a shock develops in the interior before any singularities can form at the boundary $x = 0$.

THEOREM 6.1. *At $t = t_m$, the function $X(\beta,t)$ is* not *monotone in β.*

Proof. As $T(0,\bar{\beta}) = \pi t_m$, we must have $\alpha(\bar{\beta}) = 0$. Furthermore, we still have

$$\partial_\beta X(\beta,t_m) \;=\; (\beta - \alpha) T_\beta\,,$$

and using (247), we see that for $\beta = \bar{\beta}$,

$$\partial_\beta X(\beta,t_m) < 0\,.$$

On the other hand, for β near zero, the straight line calculations from Chapter 5 can be applied, and so, as $t_m < t_b$, $X(\beta,t_m)$ is increasing in β for β near zero (c.f. proof of Lemma 5.5). Thus $X(\beta,t_m)$ is neither monotone increasing nor monotone decreasing on all of $(0,x_c(t))$. □

The facts stated above, along with Lemma 6.1, establish the formation of a shock at some time $t < t_m$, and at some $x \in (0,x_c(t))$. Indeed, from the facts stated above, we know that for t near zero, $X(\cdot,t)$ is monotone increasing for $\beta \in (0,1)$. But by Lemma 6.1, $X(\cdot,t_m)$ is not monotone. Hence, as t evolves, $X_\beta(\beta,t)$ must develop a root for some $\bar{\beta} \in (0,1]$. From the relationship

$$\beta_x \;=\; \frac{\pi}{(\beta-\alpha)T_\beta} = \frac{\pi}{X_\beta}\,,$$

we see that $\beta_x(x,\bar{t}) \to \infty$ as $x \to \frac{1}{\pi}X(\bar{\beta},\bar{t})$.

Evolution of the zero gap ansatz past shock time.

Now we consider initial data which has a shock in the upper Riemann invariant (see Figure 1).

We will take the shock location to be at x_s, and the "height" of the shock to be β_s. We assume that the inverse function of the initial data $g(\mu) = \beta^{-1}(\mu)$ satisfies $g'''(\mu) > 0$ for all μ near β_s. Thus $\beta' \to \infty$ as

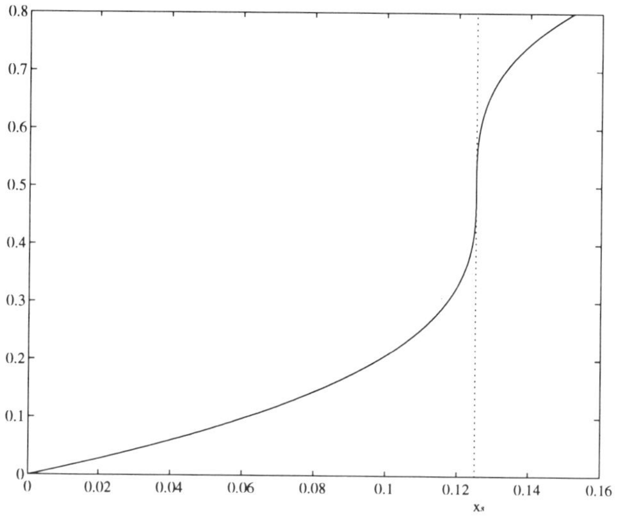

FIGURE 1. The upper Riemann invariant β, with a shock at x_s.

$x \to x_s$. The further assumptions on the initial data are as in the beginning of Chapter 5:

- $\beta^0(x)$ possesses one nondegenerate critical point, a maximum.
- $\alpha^0(x)$ is negative and strictly monotone decreasing, i.e. $\alpha^0_x < 0$ for all $x \in [0,1]$.
- For definiteness, $\alpha^0(0) = \beta^0(0) = 0, \quad \alpha^0(1) = \beta^0(1) = -1, \quad$, and $\beta^0(x_c) = \max \beta^0(x) = 1$.

Again, it will be clear from the analysis that follows, that the qualitative behavior of the continuum limit does not depend on the specific choices of constants, which are made only for convenience. Note that we no longer assume the symmetry $-\alpha^0(x) = \beta^0(x)$, and, of course, we no longer require $(\beta^0)'' > 0$.

Question: Can the above data at time $t = 0$ indeed arise from data at $t < 0$? That is, for $t < 0$ can we solve the zero gap ansatz (229 - 232) to obtain a solution $\alpha(x,t), \beta(x,t)$ for all $x \in (0,1)$, with the correct monotonicity properties, and such that $(\alpha(x,0-), \beta(x,0-)) = (\alpha^0(x), \beta^0(x))$, our given initial data (see Figure 1 for $\beta^0(x)$)? As we now show, the answer is 'yes'.

As before, the Riemann - Hilbert problem (233) is solved by (234- 235), and the associated equations for α and β are given by (236 - 237). At $t = 0$ these conditions reconstruct the initial data. However, T_β is not strictly positive at $t = 0$. Indeed, we have the formula (see 241 of Fact 1 , noting that at $t = 0$ the region of integration vanishes)

$$T_\beta = \frac{\pi \partial_\beta x_-(\beta)}{[(\beta - \alpha^0(x_-(\beta)))(\beta - \alpha)]^{1/2}} ,$$

and so at the location of the shock, $\beta = \beta_s$, $T_\beta = 0$, as $\partial_\beta x_-(\beta) = 0$ at β_s. However, for $\alpha < 0$, the derivation of

$$T_\alpha = \frac{\pi \partial_\alpha x_-(\alpha)}{[(\alpha - \beta)(\alpha - \beta^0(x_-(\alpha)))]^{1/2}} +$$

$$(248) \qquad + \int_{x_-(\alpha)}^{x_-(\beta)} \int_{\beta^0(y)}^{\beta} \frac{i}{[R(\lambda)R_0(\lambda, y)]^{1/2}} \frac{\frac{1}{2}}{\lambda - \alpha} \, d\lambda dy > 0 ,$$

(249)

still holds, and so Fact 5 remains true. Thus we can find $\alpha(\beta, t) \in (-1, \alpha^0(\beta))$ such that

$$T(\alpha(\beta, t), \beta) = \pi t ,$$

not only for all $0 \leq t \leq t_m$, with t_m defined, as before, by

$$t_m = \min_{\beta \in [0,1]} T(0, \beta) ,$$

but also for (small) $t < 0$. Thus

$$X(\beta, t) = X(\alpha(\beta, t), \beta)$$

is defined for $\hat{t} \leq t \leq t_m$, for some (small) $\hat{t} < 0$

THEOREM 6.2. *For small $t < 0$, we have $\partial_\beta X(\beta, t) > 0$ for all $\beta \in (0, 1)$.*

Proof.
Of course, we need only consider β near β_s. As $\alpha(\beta, t) < \alpha^0(\beta)$, our expression (242) for T can be written, using a contour integration, as

$$T(\alpha, \beta) = \int_{x_-(\beta)}^{x_-(\alpha)} \int_{\alpha}^{\alpha^0(y)} \frac{i}{[R(\lambda)R_0(\lambda, y)]^{1/2}} \, d\lambda dy ,$$

and hence

$$T_\beta = \frac{\pi \partial_\beta x_-(\beta)}{[(\beta - \alpha^0(x_-(\beta)))(\beta - \alpha)]^{1/2}} +$$

$$+ \int_{x_-(\beta)}^{x_-(\alpha)} \int_{\alpha}^{\alpha^0(y)} \frac{i}{[R(\lambda)R_0(\lambda, y)]^{1/2}} \frac{1/2}{\lambda - \beta} d\lambda dy > 0 .$$

Thus, as

$$\partial_\beta X(\beta, t_m) = (\beta - \alpha) T_\beta ,$$

the Theorem is proved. □

We leave it to the reader to verify that this immediately implies that the conditions (236 - 237), along with the associated conditions obtained from the "right" ansatz, uniquely determine functions $\alpha(x, t), \beta(x, t)$ for small $t < 0$ and for all $x \in (0, 1)$, and that these functions fall into the same category as the initial data described at the beginning of Chapter 5. This then shows, as advertised above, that the initial data under consideration, with a shock at $t = 0$, can be obtained from initial data in the general class described in the beginning of Chapter 5 by rolling t backwards, and starting at some (small) $t < 0$.

Our next goal is to show that for $t > 0$ there exists 3 solutions to the Riemann - Hilbert problem in a vicinity of the shock.

THEOREM 6.3. *For $0 < t \leq \gamma$, where γ is some small positive number, $\partial_\beta X(\beta, t)$ has exactly two roots, occuring for β near β_s.*

Proof. We first show that at least two roots exist.

We recall that

$$\partial_\beta X(\beta, t) \quad = \quad (\beta - \alpha) T_\beta \ ,$$

and we re-write (241),

$$(250) \qquad T_\beta \ = \ \frac{\pi \partial_\beta x_-(\beta)}{[(\beta - \alpha^0(x_-(\beta)))(\beta - \alpha)]^{1/2}} + \\ - \int_{x_-(\alpha)}^{x_-(\beta)} \int_{\alpha^0(y)}^{\alpha} \frac{i}{[R(\lambda) R_0(\lambda, y)]^{1/2}} \frac{1/2}{\lambda - \beta} d\lambda dy \ .$$

Since we have, for all λ in the region of integration of the second term in (250),

$$\left| \frac{1}{\lambda - \beta} \right| \leq \frac{1}{\beta - \alpha} \ ,$$

we can bound the integral by

$$\left| - \int_{x_-(\alpha)}^{x_-(\beta)} \int_{\alpha^0(y)}^{\alpha} \frac{i}{[R(\lambda) R_0(\lambda, y)]^{1/2}} \frac{1/2}{\lambda - \beta} d\lambda dy \right| \\ \leq - \frac{1}{2(\beta - \alpha)} \int_{x_-(\alpha)}^{x_-(\beta)} \int_{\alpha^0(y)}^{\alpha} \frac{i}{[R(\lambda) R_0(\lambda, y)]^{1/2}} d\lambda dy \ = \ \frac{1}{2(\beta - \alpha)} \pi t \ .$$

Thus, the second term in (250) is small for t small, and if we take β bounded away from β_s,

$$|\beta - \beta_s| \geq \delta > 0 \ ,$$

for some fixed $\delta > 0$, then the first term in (250) is bounded away from zero.

Hence, for small $t > 0$ and $|\beta - \beta_s| > \delta$,

$$\partial_\beta X(\beta, t) \ > \ 0 \ .$$

Now we evaluate $\partial_\beta X(\beta, t)$ at $\beta = \beta_s$, the height of the shock. Using the fact that $\partial_\beta x_-(\beta)\big|_{\beta = \beta_s} = 0$, we have from (250)

$$T_\beta \ = \ - \int_{x_-(\alpha)}^{x_-(\beta)} \int_{\alpha^0(y)}^{\alpha} \frac{i}{[R(\lambda) R_0(\lambda, y)]^{1/2}} \frac{1/2}{\lambda - \beta} d\lambda dy \ < \ 0 \ .$$

Thus $\partial_\beta X(\beta, t)$ changes signs at least twice.

To show that the sign of $\partial_\beta X(\beta,t)$ changes exactly twice, it is enough to show that

$$\partial_{\beta\beta\beta} X(\beta,t) > 0$$

for small $t > 0$. But this follows by differentiating the relations

$$\begin{aligned} \frac{\partial}{\partial\beta} X(\alpha,\beta) &= \frac{\beta-\alpha}{2} T_\beta \, , \\ \frac{\partial}{\partial\alpha} X(\alpha,\beta) &= -\frac{\beta-\alpha}{2} T_\alpha \, , \end{aligned}$$

and using the assumption $\beta'''(x,0) > 0$ near x_s. We leave the details to the reader. □

Now we define $\beta_1(t)$ and $\beta_2(t)$ to be the roots of $\partial_\beta X(\beta,t)$, $\beta_1(t) > \beta_2(t)$ for $t > 0$, and set

$$\begin{aligned} x_1(t) &= \frac{1}{\pi} X(\alpha(\beta_1(t),t),\beta_1(t)) \, , \\ x_2(t) &= \frac{1}{\pi} X(\alpha(\beta_2(t),t),\beta_2(t)) \, . \end{aligned}$$

As $\partial_\beta X(\beta,t) > 0$ for β near 0, we must have $x_1(t) < x_2(t)$.

Now note that for β in the two exterior regions, $(0,\beta_1(t))$ and $(\beta_2(t),1)$, $X(\beta,t)$ is increasing in β, and in the interior region, $(\beta_1(t),\beta_2(t))$, $X(\beta,t)$ is decreasing in β. This yields three sheets for the function $\beta(x,t)$, and hence for $\alpha(x,t) = \alpha(\beta(x,t),t)$ as well, and we see that out of our initial data with a shock at $t = 0$, we obtain for small positive t a time - dependent spatial interval $(x_1(t),x_2(t))$ on which the conditions (236 - 237) have three solutions. We label them $\{\alpha_0(x,t),\beta_0(x,t)\}$, $\{\alpha_1(x,t),\beta_1(x,t)\}$, and $\{\alpha_2(x,t),\beta_2(x,t)\}$. On the intervals $(0,x_1(t))$ and $(x_2(t),1)$, equations (229 - 232) continue to have a unique solution. Our notation is that $\{\alpha_0(x,t),\beta_0(x,t)\}$ is defined on $(0,x_2(t))$, $\{\alpha_1(x,t),\beta_1(x,t)\}$ is defined on $(x_1(t),x_2(t))$, and $\{\alpha_2(x,t),\beta_2(x,t)\}$ is defined on $(x_2(t),x_c(t))$.

The above calculations and formulae should be understood as providing the evolution of the continuous Toda PDE system past shock time by characteristics, or, equivalently, by inverting the Hodograph transform of the solution. Thus we have the explicit generalization to this PDE system of the classical folding phenomenon associated with the inviscid Burger's equation. Here each one of the pairs $(\alpha_j(x,t),\beta_j(x,t))$, $0 \le j \le 2$, is a solution of the continuous Toda equations $\alpha_t = -\frac{\beta-\alpha}{2}\alpha_x$, $\beta_t = \frac{\beta-\alpha}{2}\beta_x$.

To complete this section, we describe the Riemann invariants $\alpha_j(x,t)$ beyond shock time. Our specific goal is to establish the general shape of the curves $\alpha_j(x,t)$as portrayed in Figure 2 below. We note that the curves in Figure 2 are actual numerical soltions of $T(\alpha,\beta) = \pi t$, $X(\alpha,\beta) = \pi x$, with initial data $\alpha^0(x) = -\frac{4}{3}x$, $\beta^0(x) = \frac{1}{2} + (x-\frac{1}{8})^{1/3}$.

We begin by recalling the general relationship for $\alpha = \alpha_j(x,t)$

$$\alpha_x = -\frac{\pi}{(\beta-\alpha)T_\alpha} \, , \tag{251}$$

We note that the derivation of this formula only requires that

$$T_\alpha(\alpha(x,t),\beta(x,t)) \neq 0 \ ,$$

and

$$T_\beta(\alpha(x,t),\beta(x,t)) \neq 0 \ .$$

Thus, using (248) we see that for small $t > 0$, and for all $x \in [0,1]$, $x \neq x_1(t)$, $x \neq x_2(t)$,

$$\alpha_x(x,t) \quad < \quad 0 \ .$$

Further, note from formula (251) that α_x is continuous in α, β. Thus the individual sheets $\{\alpha_j(x,t)\}$ have continuous, and hence finite, derivatives on their closed intervals of definition, i.e.

- $\alpha_{0x}(x,t)$ is continuous on $[0, x_2(t)]$,
- $\alpha_{1x}(x,t)$ is continuous on $[x_1(t), x_2(t)]$,
- $\alpha_{2x}(x,t)$ is continuous on $[x_1(t), x_c(t)]$,

Note that $\beta_2(x,t)$ achieves its maximum at $x_c(t)$. Also, from (251), we must have

$$\alpha_{0x}(x_2(t),t) = \alpha_{1x}(x_2(t),t) \ , \quad \alpha_{1x}(x_1(t),t) = \alpha_{2x}(x_1(t),t) \ .$$

We need the following result.

LEMMA 6.3. *For small $t > 0$, and $\beta > 0$, $\alpha(\beta, t)$ is*

- *decreasing for $\beta \in (0, \beta_2(t))$*
- *increasing for $\beta \in (\beta_2(t), \beta_1(t))$*
- *decreasing for $\beta \in (\beta_1(t), 1)$*

Proof. This follows trivially from

$$\alpha_\beta \quad = \quad -\frac{T_\beta}{T_\alpha}$$

as $T_\alpha > 0$, and $T_\beta(\alpha(\beta,t),\beta)$ changes sign exactly at $\beta_1(t)$ and $\beta_2(t)$,and $T_\beta > 0$ for β near 0.

□

Combining Lemma 6.3 with the facts that $X(\beta, t)$ is

- increasing for $\beta \in (0, \beta_2(t))$
- decreasing for $\beta \in (\beta_2(t), \beta_1(t))$
- increasing for $\beta \in (\beta_1(t), 1)$

we see that $\{\alpha_j(x,t)\}$ are three decreasing functions of x, which match continuously at $x_1(t)$ and $x_2(t)$,

$$\begin{aligned} \alpha_0(x_2(t),t) &= \alpha_1(x_2(t),t) \ , \\ \alpha_1(x_1(t),t) &= \alpha_2(x_1(t),t) \ . \end{aligned}$$

To complete the picture, we consider the slope $\alpha_x(x,t) = \alpha_{jx}(x,t)$ as a function of β, which is quite natural, given the representation (251). First, we prove the following preparatory lemma (c.f. [**19**], page xx).

LEMMA 6.4. *The function $T(\alpha,\beta)$ satisfies the following identity,*

$$2(\beta-\alpha)T_{\alpha\beta} = T_\beta - T_\alpha \ . \tag{252}$$

Proof. We set

$$h = -\frac{\beta-\alpha}{2}T - X = \int_\alpha^\beta \left(\frac{\lambda-\beta}{\lambda-\alpha}\right)^{1/2} i\theta'(\lambda)d\lambda \ ,$$

and calculate

$$\begin{aligned}\partial_\beta h &= \frac{1}{2}\int_\alpha^\beta \frac{i\theta'(\lambda)}{[(\lambda-\alpha)(\lambda-\beta)]^{1/2}}d\lambda \\ &= \frac{1}{2}T \ .\end{aligned}$$

Thus, differentiating this formula with respect to α, we find

$$h_{\beta\alpha} = \frac{1}{2}T_\alpha \ . \tag{253}$$

Now we differentiate h with respect to α first, and obtain

$$\partial_\alpha h = -\frac{1}{2}T + \frac{\beta-\alpha}{2}T_\alpha + \frac{\beta-\alpha}{2}T_\alpha \ ,$$

and hence

$$h_{\alpha\beta} = -\frac{1}{2}T_\beta + T_\alpha + (\beta-\alpha)T_{\alpha\beta} \ . \tag{254}$$

Comparing equations (253) and (254), we must have, as $h(\alpha,\beta)$ is smooth,

$$2(\beta-\alpha)T_{\alpha\beta} = T_\beta - T_\alpha \ .$$

□

THEOREM 6.4. *For small $t>0$, and β near β_s, we have*

$$\partial_\beta\alpha_x(\beta,t) > 0 \ .$$

Proof. From (251), and using Lemma 6.4, we compute

$$\begin{aligned}\partial_\beta\alpha_x &= \pi\left((\beta-\alpha)T_\alpha\right)^{-2}\left[(1-\alpha_\beta)T_\alpha + (\beta-\alpha)\left(T_{\alpha\alpha}\alpha_\beta + T_{\alpha\beta}\right)\right] \\ &= \pi\left((\beta-\alpha)T_\alpha\right)^{-2}\left[\frac{1}{2}T_\alpha + \frac{3}{2}T_\beta - (\beta-\alpha)T_{\alpha\alpha}\frac{T_\beta}{T_\alpha}\right] \ .\end{aligned}$$

Now as we are close to β_s, and t is small, we are guaranteed that T_β is also small. Thus we see that the dominant behavior comes from

$$\pi\left((\beta-\alpha)T_\alpha\right)^{-2}\frac{1}{2}T_\alpha \ ,$$

which is positive. □

Hence, as we increase β, the slope of the curve $\alpha(x,t)$ increases. This shows that for small t the curve must have the shape shown in Figure 2 for x in the three sheeted region.

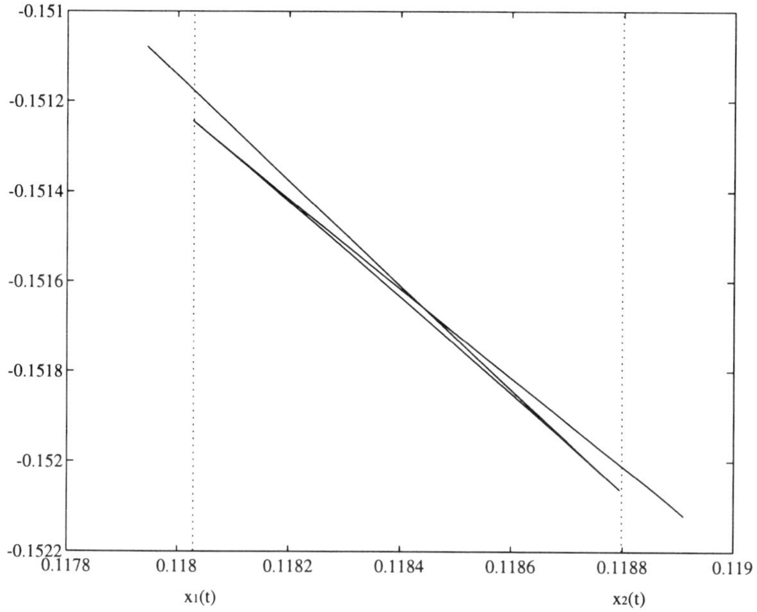

FIGURE 2. The continuation of the lower Riemann invariants $\alpha_j(x,t)$ past shock time, in a vicinity of the shock location.

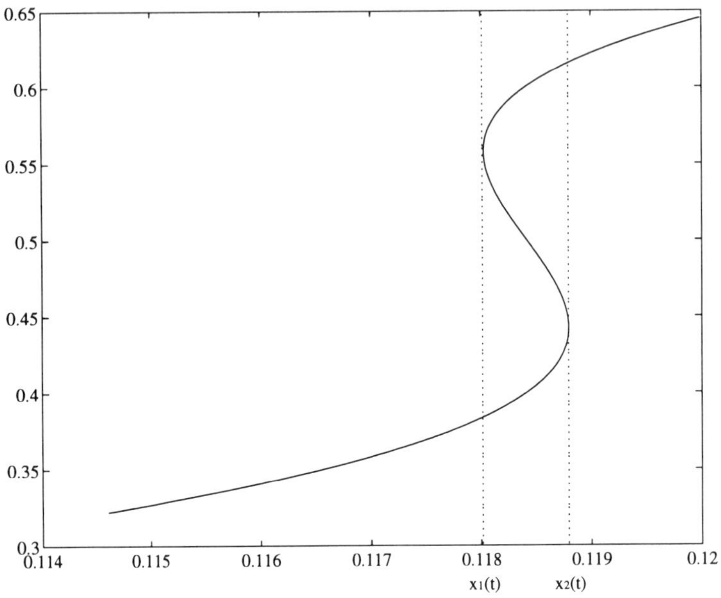

FIGURE 3. The continuation of the upper Riemann invariants $\beta_j(x,t)$ past shock time, in a vicinity of the shock location.

CHAPTER 7

Shock 3

In this section we will show, for the data considered at the end of the previous chapter and for $t > 0$, the existence of an interval $(x_L(t), x_R(t))$ on which the zero gap ansatz fails to solve the maximization problem on (x_L, x_R). That is, we will prove the existence of points of phase transition x_L and x_R such that the zero gap ansatz solves the maximization problem on $[0, x_L] \cup [x_R, 1]$, but not on $(x_L(t), x_R(t))$. We will show that these points of phase transition are the leading and trailing edge of a 1 - gap region, where the maximization problem is solved by a new ansatz for the Hilbert problem - one in which $H\psi = \theta' - t$ is to be satisfied on two bands, separated by a gap. Furthermore, as indicated in the introduction to the previous Chapter, we will show that $x_1(t) < x_L(t) < x_2(t) < x_R(t)$.

As in Chapter 6, we consider the R-H problem

$$(255) \qquad \begin{aligned} H\psi^* &= \theta'(\lambda) - t \quad on \quad I_1 = (\alpha(x,t), \beta(x,t)) \\ \psi^* &= 0 \quad on \quad I_2 = \mathbf{R} \backslash [\alpha(x,t), \beta(x,t)] \ , \end{aligned}$$

which has the solution

$$(256) \qquad \psi^*(\lambda) = Re \lim_{\epsilon \downarrow 0} f(\lambda + i\epsilon) \ ,$$

where

$$(257) \qquad f(z) = R(z)^{1/2} \frac{1}{\pi i} \int_\alpha^\beta \frac{i(\theta'(\lambda) - t)}{R(\lambda)^{1/2}} \frac{d\lambda}{\lambda - z} \ ,$$

and the conditions $\psi^* + iH\psi^* \in H^p$, along with $\int \psi^* = \pi x$, yield the two conditions,

$$(258) \qquad T(\alpha, \beta) \equiv \int_\alpha^\beta \frac{i}{[(\lambda - \alpha)(\lambda - \beta)]^{1/2}} \theta' d\lambda = \pi t \ ,$$

$$(259) \qquad X(\alpha, \beta) \equiv \int_\alpha^\beta i \frac{\lambda - m}{[(\lambda - \alpha)(\lambda - \beta)]^{1/2}} \theta' d\lambda = \pi x \ .$$

We recall that we have shown in Chapter 6 that for small $t > 0$ and for $x \in (x_1(t), x_2(t))$, there exists 3 solutions of (258), (259), and hence the R-H problem does not uniquely determine a candidate for the maximizer.

Recall from Chapter 6, that the sheets of $\alpha(x,t), \beta(x,t)$ are labeled as follows (see Figure 1):
$\alpha_0(x,t), \beta_0(x,t)$ on $(0, x_2(t))$,
$\alpha_1(x,t), \beta_1(x,t)$ on $(x_1(t), x_2(t))$,

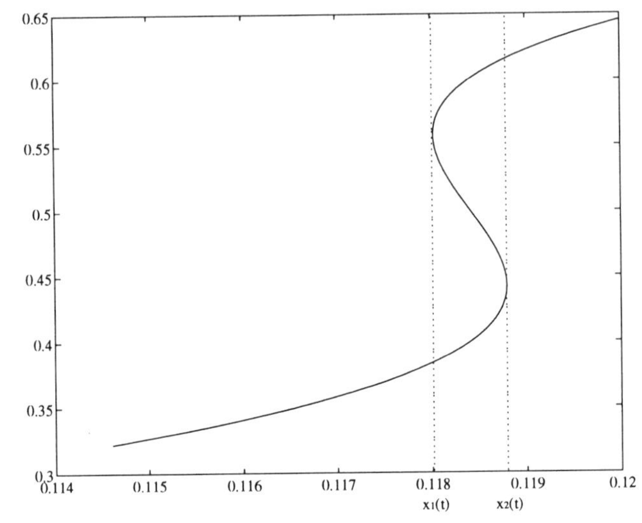

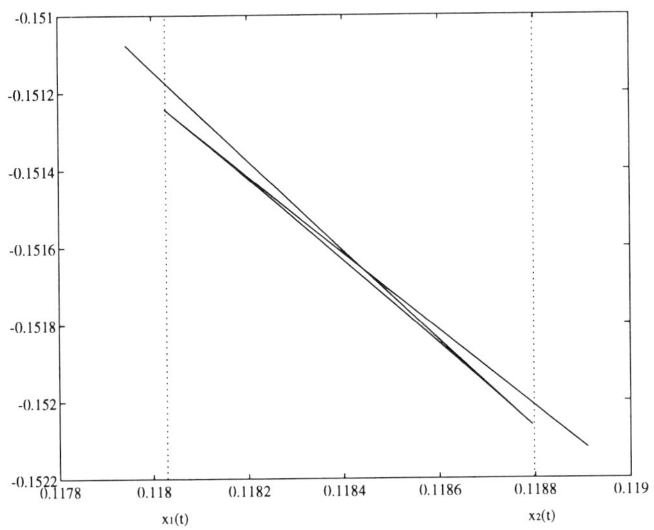

FIGURE 1. Plot, for small $t > 0$ of the Riemann invariant quantities $\alpha_j(x,t), \beta_j(x,t)$.

$\alpha_3(x,t), \beta_3(x,t)$ on $(x_1(t), x_c(t))$.
On the other hand, away from the three sheeted region, the zero gap problem uniquely determines $\alpha(x,t), \beta(x,t)$.

Notation: We label the function $\psi^*(\lambda; x, t)$ corresponding to each of $\{\alpha_j(x,t), \beta_j(x,t)\}$ for $j = 0, 1, 2$, by $\psi_j^*(\lambda; x, t)$. Also we set $\theta(\lambda, t) \equiv \theta(\lambda) - t\lambda$ (see Chapters 4 and 8 for an interpretation of $\theta(\lambda, t)$ as a (generalized) norming constant).

Recall from Theorem 5.1 in Chapter 5 that a solution $(\alpha(x,t),\beta(x,t))$ of (258), (259) gives rise to a maximizer ψ provided that

$$\text{(260)} \quad \int_{\beta(x,t)}^{\lambda} H\psi^*(\mu;\ x,t) - \Big(\theta'(\mu) - t\Big)\ d\mu < 0\ ,\quad \lambda > \beta(x,t)$$

$$\text{(261)} \quad \int_{\alpha(x,t)}^{\lambda} H\psi^*(\mu;\ x,t) - \Big(\theta'(\mu) - t\Big)\ d\mu < 0\ ,\quad \lambda < \alpha(x,t)$$

$$\text{(262)} \qquad 0 \le \psi^*(\mu;\ x,t) \le \phi\ .$$

Recall further that in Chapters 3, 4 and 5, the first two conditions were valid because the stronger inequalities

$$\text{(263)} \qquad H\psi^*(\lambda;\ x,t) - \big(\theta' - t\big) < 0 \ \ for\ \lambda > \beta(x,t)\ ,$$

$$\text{(264)} \qquad H\psi^*(\lambda;\ x,t) - \big(\theta' - t\big) > 0 \ \ for\ \lambda < \alpha(x,t)\ ,$$

were true. As we now show, for the present data, (263) is no longer true.

LEMMA 7.1. *For small $t > 0$, $x \in (x_1(t), x_2(t))$, and $\lambda \in \big(\beta_1(x,t)\ ,\ \beta_1(x_1(t),t)\big)$, we have*

$$H\psi_0^*(\lambda;\ x,t) - \theta'(\lambda,t) \ \ > \ \ 0\ .$$

Proof.
To prove this lemma, we will use the differentiated quantity $H\psi_x^*$. Observe once again from the identities

$$\alpha_x = -\frac{\pi}{(\beta-\alpha)T_\alpha}\ ,$$
$$\beta_x = \frac{\pi}{(\beta-\alpha)T_\beta}\ ,$$

which are valid provided $x \ne x_1(t)$, $x \ne x_2(t)$, that $\{\alpha_j(x,t),\beta_j(x,t)\}$ are continuously differentiable quantities, for $x \ne x_1(t)$, $x \ne x_2(t)$. As remarked in Chapter 4, this fact and the explicit formula (257) enable one to establish that ψ_{jx}^* and $H\psi_{jx}^*$ are given by

$$\text{(265)} \qquad \psi_{jx}^*(\lambda;x,t) = Re\left[\frac{i}{[(\lambda-\alpha_j(x,t))(\lambda-\beta_j(x,t))]^{1/2}}\right]\ ,$$

$$\text{(266)} \quad H\psi_{jx}^*(\lambda;x,t) = Im\left[\frac{i}{[(\lambda-\alpha_j(x,t))(\lambda-\beta_j(x,t))]^{1/2}}\right]\ .$$

We prove the Lemma by integrating (266).

Observe first that at $x = x_2(t)$, we have

$$H\psi_0^*(\lambda;\ x_2(t),t) \ = \ H\psi_1^*(\lambda;\ x_2(t),t)\ ,$$

as $\{\alpha_0(x_2(t),t),\beta_0(x_2(t),t)\} = \{\alpha_1(x_2(t),t),\beta_1(x_2(t),t)\}$. Thus, for $\lambda \in (\beta_1(x,t)\ ,\ \beta_1(x_1(t),t))$, we write

$$(267)\quad H\psi_0^*(\lambda;\ x,t) = H\psi_0^*(\lambda;\ x_2(t),t) + \\ + \int_{x_2(t)}^{x} \frac{1}{[(\lambda-\alpha_0(y,t))(\lambda-\beta_0(y,t))]^{1/2}}dy$$

$$(268)\quad = H\psi_1^*(\lambda;\ x_2(t),t) + \\ + \int_{x_2(t)}^{x} \frac{1}{[(\lambda-\alpha_0(y,t))(\lambda-\beta_0(y,t))]^{1/2}}dy$$

$$(269)\quad = H\psi_1^*(\lambda;\ x_-^1(\lambda,t),t) + \int_{x_-^1(\lambda,t)}^{x_2(t)} \frac{1}{[(\lambda-\alpha_1(y,t))(\lambda-\beta_1(y,t))]^{1/2}}dy \\ + \int_{x_2(t)}^{x} \frac{1}{[(\lambda-\alpha_0(y,t))(\lambda-\beta_0(y,t))]^{1/2}}dy\ ,$$

where in (267) and (269) we have integrated the quantities $H\psi_{0x}^*$ and $H\psi_{1x}^*$, respectively. Here $x_-^1(\cdot,t)$ is the inverse function of $\beta_1(\cdot,t)$,

$$\beta_1(x_-^1(\lambda,t),t) \ = \ \lambda\ ,$$

and hence, from the R-H problem (255),

$$H\psi_1^*(\lambda;\ x_-^1(\lambda,t),t) \ = \ \theta'(\lambda,t)\ .$$

Thus we have

$$H\psi_0^*(\lambda;\ x,t) \ = \ \theta'(\lambda,t) \ + \ \int_{x_-^1(\lambda)}^{x_2(t)} \frac{1}{[(\lambda-\alpha_1(y,t))(\lambda-\beta_1(y,t))]^{1/2}}dy \\ + \int_{x_2(t)}^{x} \frac{1}{[(\lambda-\alpha_0(y,t))(\lambda-\beta_0(y,t))]^{1/2}}dy\ ,$$

which implies

$$H\psi_0^*(\lambda;\ x,t) - \theta'(\lambda,t) = \int_{x_-^1(\lambda)}^{x} \frac{1}{[(\lambda-\alpha_1(y,t))(\lambda-\beta_1(y,t))]^{1/2}}dy \\ + \int_{x}^{x_2(t)} \frac{1}{[(\lambda-\alpha_1(y,t))(\lambda-\beta_1(y,t))]^{1/2}} - \frac{1}{[(\lambda-\alpha_0(y,t))(\lambda-\beta_0(y,t))]^{1/2}}dy\ ,$$

and, as we always take

$$[(\lambda-\alpha)(\lambda-\beta)]^{1/2} \ > \ 0\ for\ \lambda > \beta_1(y,t)\ ,$$

the proof will be complete upon establishing

$$\frac{1}{(\lambda-\alpha_1(y,t))(\lambda-\beta_1(y,t))} \geq \frac{1}{(\lambda-\alpha_0(y,t))(\lambda-\beta_0(y,t))}\ ,$$

for $\lambda > \beta_1(y,t)$. But this is equivalent to

$$\frac{\beta_1(y,t)-\beta_0(y,t)}{\lambda-\beta_0(y,t)} \geq \frac{\alpha_0(y,t)-\alpha_1(y,t)}{\lambda-\alpha_1(y,t)}\ ,$$

which we will now show is evident from the shapes of the curves $\alpha_j(x,t)$, $\beta_j(x,t)$. Indeed, it is clear that

$$\lambda - \beta_0(y,t) > \lambda - \beta_1(x,t) \ ,$$

and so it is sufficient to establish

$$\beta_1(y,t) - \beta_0(y,t) \geq \alpha_0(y,t) - \alpha_1(y,t) \ . \tag{270}$$

But, from the work of the previous Chapter, we know that the quantities $\beta_0(y,t)$ and $\beta_1(y,t)$ both have an infinite derivative at $y = x_2(t)$, and we have also established that the inverse function of the curve $\beta_j(y,t)$, $X(\beta,t)$ satisfies $\partial_\beta^3 X(\beta,t)$. Thus we surmise that $|\beta_1(y,t)-\beta_0(y,t)| \sim |y-x_2(t)|^{1/3}$. Further, as the quantities $\alpha_j(y,t)$ are displaying cusp - like behavior, i.e. the curves $\alpha_0(y,t)$ and $\alpha_1(y,t)$ have the *same slope and value* at $x_2(t)$, we have that $|\alpha_0(y,t) - \alpha_1(y,t)| = O(|y - x_2(t)|$. Thus, for small t, we have established (270), and hence completed the proof. □

Remark: By substituting $x = x_2(t)$ in Lemma 7.1, we see

$$H\psi_0^*(\lambda;\ x_2(t),t) - \theta'(\lambda,t) \ > \ 0$$

for $\lambda \in (\beta_1(x_2(t),t), \beta_1(x_1(t),t))$, and hence as $\beta_1(x_2(t),t) = \beta_0(x_2(t),t)$,

$$\int_{\beta_0(x_2(t),t)}^{\lambda} H\psi_0^*(\mu;\ x_2(t),t) - \theta'(\mu,t)d\mu > 0 \ ,$$

for $\lambda \in (\beta_0(x_2(t),t), \beta_1(x_1(t),t))$, i.e. $\psi_0^*(\lambda;\ x_2(t),t)$ fails to satisfy the side constraint (260).

THEOREM 7.1. *For small $t > 0$, the function $\psi_0^*(\lambda;\ x,t)$ solves the maximization problem for $x \in [0, x_1(t) + \gamma_1]$, for some (small) $\gamma_1 > 0$.*

Proof.

To show this, we take $x = x_1(t)$, and first we consider $\lambda \in (\beta_0(x_1(t),t)$, $\beta_0(x_2(t),t))$. We have

$$H\psi_0^*(\lambda;\ x_1(t),t) = H\psi_0^*(\lambda;\ x_-^0(\lambda,t),t) + \int_{x_-^0(\lambda,t)}^{x_1(t)} \frac{1}{[(\lambda-\alpha_0(y,t))(\lambda-\beta_0(y,t))]^{1/2}}dy \ . \tag{271}$$

As $x_-^0(\lambda,t)$ is defined by

$$\beta_0(x_-^0(\lambda,t),t) \ = \ \lambda \ ,$$

we see from the R-H problem (255) that $H\psi_0^*(\lambda;\ x_-^0(\lambda,t),t) = \theta'(\lambda,t)$. Thus we obtain

$$H\psi_0^*(\lambda;\ x_1(t),t) \ = \ \theta'(\lambda,t) - \int_{x_1(t)}^{x_-^0(\lambda,t)} \frac{1}{[(\lambda-\alpha_0(y,t))(\lambda-\beta_0(y,t))]^{1/2}}dy \ ,$$

which clearly shows

$$H\psi_0^*(\lambda;\ x_1(t),t) \ - \ \theta'(\lambda,t) < 0 \ ,$$

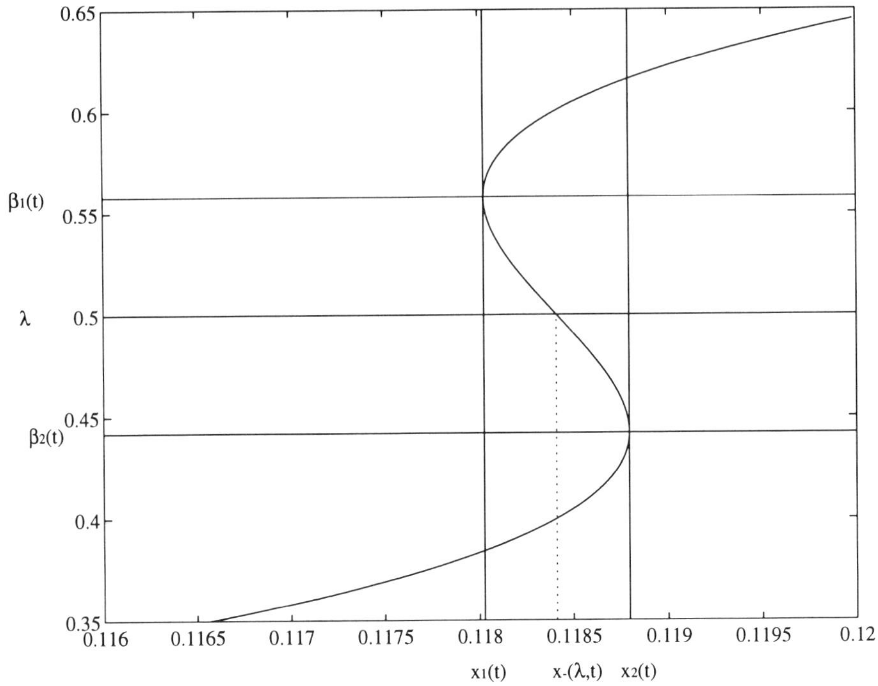

FIGURE 2. The Riemann invariant quantities $\beta_j(x,t)$ for small $t > 0$, along with λ, and $x_-^1(\lambda,t)$, used in the proof of Theorem 40.

and hence, for $\lambda \in (\beta_0(x_1(t),t), \beta_0(x_2(t),t))$,

$$\int_{\beta_0(x_1(t),t)}^{\lambda} H\psi_0^*(\mu;\ x_1(t),t) - \theta'(\mu,t)d\mu < 0\ .$$

Next we consider $\lambda \in (\beta_1(x_2(t),t), \beta_1(x_1(t),t)) = (\beta_0(x_2(t),t), \beta_1(x_1(t),t))$, and we refer the reader to Figure (2). A straightforward technical calculation, which we leave to Appendix II, shows that for any $x \in [x_1(t), x_2(t)]$,

$$(272)\int_{\beta_0(x,t)}^{\lambda} H\psi_0^*(\mu;\ x,t) - \theta'(\mu,t)d\mu = -\int_{x}^{x_2(t)} \cosh^{-1}\left(\frac{\lambda - \frac{\beta_0+\alpha_0}{2}}{\frac{\beta_0-\alpha_0}{2}}\right) dy$$
$$+ \int_{x_-^1(\lambda)}^{x_2(t)} \cosh^{-1}\left(\frac{\lambda - \frac{\beta_1+\alpha_1}{2}}{\frac{\beta_1-\alpha_1}{2}}\right) dy\ ,$$

where $\alpha_j = \alpha_j(y,t)$, $\beta_j = \beta_j(y,t)$, $j = 0, 1$.

Now we substitute $x = x_1(t) < x^1_-(\lambda, t)$ into (272), and obtain

$$\int_{\beta_0(x_1(t),t)}^{\lambda} H\psi_0^*(\mu;\ x_1(t),t) - \theta'(\mu,t)d\mu$$
$$= -\int_{x_1(t)}^{x^1_-(\lambda,t)} \cosh^{-1}\left(\frac{\lambda - \frac{\beta_0+\alpha_0}{2}}{\frac{\beta_0-\alpha_0}{2}}\right) dy$$
$$-\int_{x^1_-(\lambda,t)}^{x_2(t)} \left[\cosh^{-1}\left(\frac{\lambda - \frac{\beta_0+\alpha_0}{2}}{\frac{\beta_0-\alpha_0}{2}}\right) - \cosh^{-1}\left(\frac{\lambda - \frac{\beta_1+\alpha_1}{2}}{\frac{\beta_1-\alpha_1}{2}}\right)\right] dy \ .$$

Thus, to show $\int_{\beta_0(x_1(t),t)}^{\lambda} H\psi_0^*(\mu;\ x_1(t),t) - \theta'(\mu,t)d\mu < 0$, it is enough to establish

$$\cosh^{-1}\left(\frac{\lambda - \frac{\beta_0+\alpha_0}{2}}{\frac{\beta_0-\alpha_0}{2}}\right) > \cosh^{-1}\left(\frac{\lambda - \frac{\beta_1+\alpha_1}{2}}{\frac{\beta_1-\alpha_1}{2}}\right) \ , \tag{273}$$

which is implied if we can establish

$$\frac{\lambda - \frac{\beta_0+\alpha_0}{2}}{\frac{\beta_0-\alpha_0}{2}} > \frac{\lambda - \frac{\beta_1+\alpha_1}{2}}{\frac{\beta_1-\alpha_1}{2}} \ , \tag{274}$$

Or, equivalently,

$$\frac{\lambda - \beta_0}{\beta_0 - \alpha_0} > \frac{\lambda - \beta_1}{\beta_1 - \alpha_1} \ . \tag{275}$$

But this is true by inspection of Figure 1.

Thus, we have shown that for $\lambda \in (\ \beta_0(x_1(t),t)\ \ , \beta_1(x_1(t),t)\ \ )$,

$$\int_{\beta_0(x_1(t),t)}^{\lambda} H\psi(\mu;\ x_1(t),t) - \theta'(\mu,t)d\mu \ < \ 0 \ .$$

Finally, we consider $\lambda > \beta_2(x_1(t),t) = \beta_1(x_1(t),t)$.

The extension of formula (272) to this region yields

$$\int_{\beta_0(x,t)}^{\lambda} H\psi^*(\mu;\ x,t) - \theta'(\mu,t)d\mu = -\int_{x}^{x_2(t)} \cosh^{-1}\left(\frac{\lambda - \frac{\beta_0+\alpha_0}{2}}{\frac{\beta_0-\alpha_0}{2}}\right) dy$$
$$+\int_{x_1(t)}^{x_2(t)} \cosh^{-1}\left(\frac{\lambda - \frac{\beta_1+\alpha_1}{2}}{\frac{\beta_1-\alpha_1}{2}}\right) dy$$
$$-\int_{x_1(t)}^{x^2_-(\lambda,t)} \cosh^{-1}\left(\frac{\lambda - \frac{\beta_2+\alpha_2}{2}}{\frac{\beta_2-\alpha_2}{2}}\right) dy \ , \tag{276}$$

where again $\alpha_j = \alpha_j(y,t)$, $\beta_j = \beta_j(y,t)$, $j = 1, 2$.

Rearranging terms and evaluating at $x = x_1(t)$, we obtain

$$\int_{\beta_0(x_1(t),t)}^{\lambda} H\psi_0^*(\mu;\ x_1(t),t) - \theta'(\mu,t)d\mu$$
$$= -\int_{x_1(t)}^{x_-^2(\lambda,t)} \cosh^{-1}\left(\frac{\lambda - \frac{\beta_2+\alpha_2}{2}}{\frac{\beta_2-\alpha_2}{2}}\right) dy$$
$$-\int_{x_1(t)}^{x_2(t)} \left[\cosh^{-1}\left(\frac{\lambda - \frac{\beta_0+\alpha_0}{2}}{\frac{\beta_0-\alpha_0}{2}}\right) - \cosh^{-1}\left(\frac{\lambda - \frac{\beta_1+\alpha_1}{2}}{\frac{\beta_1-\alpha_1}{2}}\right)\right] dy\ .$$

Now we have already established (see (273 - 275)) that for $\lambda > \beta_1(y,t)$,

$$\cosh^{-1}\left(\frac{\lambda - \frac{\beta_0+\alpha_0}{2}}{\frac{\beta_0-\alpha_0}{2}}\right) > \cosh^{-1}\left(\frac{\lambda - \frac{\beta_1+\alpha_1}{2}}{\frac{\beta_1-\alpha_1}{2}}\right)\ .$$

Hence we conclude

$$\int_{\beta_0(x_1(t),t)}^{\lambda} H\psi_0^*(\mu;\ x_1(t),t) - \theta'(\mu,t)d\mu \ < \ 0\ ,\ \textit{for all}\ \lambda > \beta_0(x_1(t),t)\ ,$$

which verifies (260) for ψ_0^* at $(x_1(t),t)$.

The next step is to show that for $\lambda \in (\alpha_0(x_1(t),t), \beta_0(x_1(t),t))$, we have

$$0 \ < \ \psi_0^*(x_1(t),t)\ .$$

By integrating (265), we obtain

$$\psi_0^*(x_1(t),t) = \psi_0^*(x_-^0(\lambda,t),t) + \int_{x_-^0(\lambda,t)}^{x_1(t)} \frac{i}{[(\lambda-\alpha_0(y,t))(\lambda-\beta_0(y,t))]^{1/2}}dy\ .$$

Now as $x_-^0(\lambda,t)$ is defined by

$$\begin{aligned}\beta_0(x_-^0(\lambda,t),t) &= \lambda\ ,\ \lambda > 0\ ,\\ \alpha_0(x_-^0(\lambda,t),t) &= \lambda\ ,\ \lambda < 0\ ,\end{aligned}$$

we see from (255) that

$$\psi_0^*(\lambda;\ x_-^0(\lambda,t),t) \ = \ 0\ .$$

Hence we conclude that

$$\psi_0^*(\lambda;\ x_1(t),t) \ = \ \int_{x_-^0(\lambda,t)}^{x_1(t)} \frac{i}{[(\lambda-\alpha_0(y,t))(\lambda-\beta_0(y,t))]^{1/2}}dy \ > \ 0$$

for $\lambda \in (\alpha_0(x_1(t),t), \beta_0(x_1(t),t))$, and hence for all λ.

Next we show $\psi_0^*(\lambda;\ x_1(t),t) \leq \phi$, for $\lambda \in (\alpha_0(x_1(t),t), \beta_0(x_1(t),t))$. But this follows by continuity. Indeed, at $t = 0$ we have for $\lambda \in (\alpha^0(x), \beta^0(x))$,

$$\psi_0^*(\lambda;\ x,0) \ = \ \int_{x_-(\lambda,0)}^{x} \frac{i}{[(\lambda-\alpha^0(y))(\lambda-\beta^0(y))]^{1/2}}dy\ ,$$

and

$$\phi(\lambda) \;=\; \int_{x_-(\lambda,0)}^{x_+(\lambda,0)} \frac{i}{[(\lambda-\alpha^0(y))(\lambda-\beta^0(y))]^{1/2}} dy \ ,$$

and hence for x near x_s, the shock location, and so away from x_c, we have, at $t=0$,

$$\psi_0^*(\lambda;\ x,0) \le \phi(\lambda) - c \ < \ \phi(\lambda) \ ,$$

for some fixed $c > 0$.

Thus we have established (260) and (262) at $x = x_1(t)$.

Now we show that

$$(277) \quad \int_{\alpha_0(x_1(t),t)}^{\lambda} H\psi_0^*(\lambda\ ;\ x_1(t),t) - \theta'(\mu,t)d\mu \ < \ 0 \ , \ \lambda < \alpha_0(x_1(t),t) \ ,$$

which is the missing piece in proving that $\psi_0^*(\lambda;\ x_1(t),t)$ solves the side constraints (260 - 262). We show that (277) does not fail, a fortiori, for *any* x in $(0,1)$.

We use the relation

$$H\psi_{0t}^* \;=\; Im\, i\left(\frac{\lambda - m_0}{R_0(\mu;\ x,t)^{1/2}} - 1\right) \ , \ .$$

where $m_0 = m_0(t) = \frac{\alpha_0(x,t)+\beta_0(x,t)}{2}$, and $R_0(\mu;\ x,t) = (\mu - \alpha_0(x,t))(\mu - \beta_0(x,t))$. This relation is obtained by considering the R-H problem for the quantity ψ_{0t}^*. Now

$$\begin{aligned} \frac{d}{dt}\int_{\alpha_0(x,t)}^{\lambda} H\psi_0^*(\lambda\ ;\ x,t) - \theta'(\mu,t)d\mu \;&=\; \int_{\alpha_0(x,t)}^{\lambda} H\psi_{0t}^* + 1 d\mu \\ &=\; \int_{\alpha_0(x,t)}^{\lambda} \frac{\mu - m}{R_0(\mu;x,t)^{1/2}}\, d\mu < 0 \ . \end{aligned}$$

Thus, for a fixed λ, as we increase t, this integral decreases. We conclude that this condition cannot fail, as it is true at $t=0$.

Thus we have established (260 - 262), for $x = x_1(t)$. Now the above arguments work for all $(\alpha_0(x,t),\beta_0(x,t))$, provided x is to the *left* of the folding region $x \le x_1(t)$; all one has to do is check that the above formulae, e.g. (271), contain an additional integral from x to $x_1(t)$, and that this additional integral has the correct sign, etc.

Finally, it is easy to see, using continuity, that the above proof goes through a little ways into the folding region. This shows that the maximization problem is solved with the zero gap ansatz $(\alpha_0(x,t),\beta_0(x,t))$ for $x \in [0, x_1(t)+\gamma_1]$, for some $\gamma_1 > 0$.

□

For high oscillations to develop in the continuum limit as $\epsilon \downarrow 0$, a gap must appear in the associated R-H problem. The above calculations show us, in particular, the intriguing fact that high oscillations do not develop at $x_1(t)$, the point at which the characteristics start to fold over. High

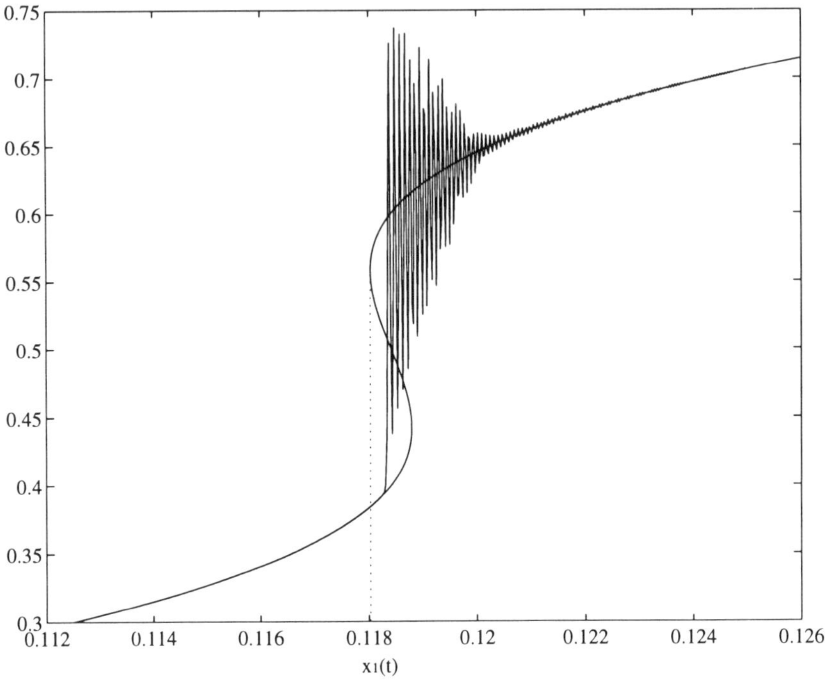

FIGURE 3. Plot, for $t = 0.02$ of the finite Toda lattice with $\epsilon^{-1} = 40,000$, and initial data $\beta^0(x) = (x - 1/8)^{1/3} + 1/2$, $\alpha^0 = (-4/3)x$. The figure contains two curves. (i.) The values of $\beta_{[x/\epsilon]}(t) = a_{[x/\epsilon]}(t) + 2b_{[x/\epsilon]}(t)$, $0 \leq x \leq 1$, obtained by solving the Toda equations (1.6 - 1.9). (ii.) The Riemann invariants $\beta_0(x,t)$, $\beta_1(x,t)$, $\beta_2(x,t)$ obtained by inverting $T(\alpha, \beta) = \pi t$, $X(\alpha, \beta) = \pi x$. The dashed line indicates the location of $x_1(t)$. Observe that the two curves coincide for $x \leq x_L(t) \approx .1184$, which is clearly greater than $x_1(t) \approx 0.118$. Similarly, the curves coincide for x far to the right. Furthermore, the value $x_L(t) = 0.1184$, the point where the oscilllations sets in, solves the equations (6.24), (6.25).

oscillations begin to develop only at $x = x_L(t) > x_1(t)$. This phenomenon is illustrated in Figure 3.

We have shown that the zero-gap ansatz solves the maximization problem at $x_1(t)$, and that it does not do so at $x_2(t)$ (see Remark following Lemma 7.1). Define

$$x_L(t) = \inf\{x_1(t) \leq x \leq x_2(t) \ : \ \textit{zero gap ansatz fails at } x\} \ .$$

Clearly $x_1(t) < x_1(t) + \gamma_1 \leq x_L(t) \leq x_2(t)$.

At $x = x_L(t)$, the previous calculations show that (261) and (262) are still satisfied. We now show that for some $\lambda > \beta_0(x_L(t),t)$, we must have

$$(278)\qquad \int_{\beta_0(x_L(t),t)}^{\lambda} H\psi_0^*(\mu;\ x_L(t),t) - \theta'(\mu,t)d\mu \ = 0\ ,$$

$$(279)\qquad H\psi_0^*(\lambda\ ;\ x_L(t),t) - \theta'(\lambda,t)\ =\ 0\ .$$

THEOREM 7.2. *For small $t > 0$, and $x \in (x_1(t), x_2(t))$, there exists a unique root $\lambda(x) \in (\beta_2(x_1(t),t), \beta_2(x_2(t),t))$ of*

$$H\psi_0^*(\lambda\ ;\ x,t) - \theta'(\lambda,t)\ .$$

Proof. Fix $x \in (x_1(t), x_2(t))$. We first note from Lemma (7.1) that, for $\lambda \in (\beta_1(x,t), \beta_1(x_1(t),t))$, we have

$$(280)\qquad H\psi_0^*(\lambda;\ x,t) - \theta'(\lambda,t)\ >\ 0\ .$$

Now observe that for $\lambda \geq \beta_2(x_2(t),t)$, we have as before,

$$\begin{aligned}
&H\psi_0^*(\lambda\ ;\ x,t)\\
&= \int_{x_2(t)}^{x} \frac{1}{[(\lambda-\alpha_0(y))(\lambda-\beta_0(y))]^{1/2}}dy\ +\\
&\quad + \int_{x_1(t)}^{x_2(t)} \frac{1}{[(\lambda-\alpha_1(y))(\lambda-\beta_1(y))]^{1/2}}dy\\
&\quad + \int_{x_-^2(\lambda,t)}^{x_1(t)} \frac{1}{[(\lambda-\alpha_2(y))(\lambda-\beta_2(y))]^{1/2}}dy\ +\ \theta'(\lambda,t)\\
&= \int_{x_1(t)}^{x_2(t)} \frac{1}{[(\lambda-\alpha_1(y))(\lambda-\beta_1(y))]^{1/2}} - \frac{1}{[(\lambda-\alpha_2(y))(\lambda-\beta_2(y))]^{1/2}}dy\\
&\quad + \int_{x_2(t)}^{x} \frac{1}{[(\lambda-\alpha_0(y))(\lambda-\beta_0(y))]^{1/2}}dy\\
&\quad + \int_{x_-^2(\lambda,t)}^{x_2(t)} \frac{1}{[(\lambda-\alpha_2(y))(\lambda-\beta_2(y))]^{1/2}}dy\ +\ \theta'(\lambda,t)\ ,
\end{aligned}$$

and as in the proof of Lemma 7.1,

$$\frac{1}{[(\lambda-\alpha_2(y,t))(\lambda-\beta_2(y,t))]^{1/2}}\ >\ \frac{1}{[(\lambda-\alpha_1(y,t))(\lambda-\beta_1(y,t))]^{1/2}}\ ,$$

for $\lambda > \beta_2(y,t)$. Thus we see that

$$(281)\qquad H\psi_0^*(\lambda\ ;\ x,t) - \theta'(\lambda,t)\ <\ 0\ ,\ \textit{for}\ \lambda \geq \beta_2(x_2(t),t)\ .$$

Thus from (280) and (281), we have, for $x \in (x_1(t), x_2(t))$, a root $\lambda(x) \in (\beta_2(x_1(t),t)\ ,\ \beta_2(x_2(t),t))$ of the function $H\psi_0^*(\lambda\ ;\ x,t) - \theta'(\lambda,t)$. Uniqueness follows from

$$\lambda\ \in\ (\beta_2(x_1(t),t)\ ,\ \beta_2(x_2(t),t)\ )\ \Rightarrow\ \frac{d}{d\lambda}\left(H\psi_0^*(\lambda\ ;\ x,t) - \theta'(\lambda,t)\right) < 0\ ,$$

established in Appendix II.

□

LEMMA 7.2. *For small $t > 0$, there exists a unique root $\hat{x}(t) \in (x_1(t), x_2(t))$ of the function*

$$\int_{\beta_0(x,t)}^{\lambda(x)} H\psi_0^*(\mu\ ;\ x,t) - \theta'(\mu,t)\ d\mu\ . \tag{282}$$

Proof.
We begin by showing

$$\int_{\beta_0(x_2(t),t)}^{\lambda(x_2(t))} H\psi_0^*(\mu\ ;\ x_2(t),t) - \theta'(\mu,t)\ d\mu\ >\ 0\ . \tag{283}$$

Indeed, from the Remark following Lemma 7.1, we see that

$$\int_{\beta_0(x_2(t),t)}^{\lambda} H\psi_0^*(\mu\ ;\ x_2(t),t) - \theta'(\mu,t)\ d\mu\ >\ 0$$

for $\lambda \in (\beta_0(x_2(t),t), \beta_1(x_1(t),t))$. Now

$$\int_{\beta_0(x_2(t),t)}^{\lambda(x_2(t))} H\psi_0^*(\mu\ ;\ x_2(t),t) - \theta'(\mu,t)\ d\mu\ =$$

$$\int_{\beta_0(x_2(t),t)}^{\lambda} H\psi_0^*(\mu\ ;\ x_2(t),t) - \theta'(\mu,t)\ d\mu\ + \tag{284}$$

$$+\int_{\beta_1(x_1(t),t)}^{\lambda(x_2(t))} H\psi_0^*(\mu\ ;\ x_2(t),t) - \theta'(\mu,t)\ d\mu\ , \tag{285}$$

and since $H\psi_0^*(\mu\ ;\ x_2(t),t) - \theta'(\mu,t) > 0$ for $\mu \in (\beta_2(x_1(t),t), \lambda(x_2(t))) = (\beta_1(x_1(t),t), \lambda(x_2(t)))$, we see that both terms in (284), (285) are positive, and hence we have established (283).

On the other hand, at $x = x_1(t)$, (282) is negative, so there is a root $\hat{x}(t)$ in the interior.

However, differentiating (282), we obtain

$$\frac{d}{dx}\int_{\beta_0(x,t)}^{\lambda(x)} H\psi - \theta'(\mu,t)d\mu \tag{286}$$

$$= \int_{\beta_0(x,t)}^{\lambda} \frac{1}{[(\mu-\alpha_0(x,t))(\mu-\beta_0(x,t))]^{1/2}} d\mu\ \ > 0\ .$$

Thus the root $\hat{x}(t)$ unique. □

A priori it is possible that there exists $\tilde{x}(t) \neq \hat{x}(t)$, and $\tilde{\lambda} > \beta_0(\tilde{x}(t),t)$, $\tilde{\lambda} \notin (\beta_2(x_1(t),t), \beta_2(x_2(t),t))$, such that

$$\int_{\beta_0(\tilde{x}(t),t)}^{\tilde{\lambda}} H\psi_0^*(\mu;\ \tilde{x}(t),t) - \theta'(\mu,t)d\mu\ = 0\ \ ,$$

$$H\psi_0^*(\lambda\ ;\ \tilde{x}(t),t) - \theta'(\lambda,t)\ =\ 0\ \ .$$

We now show that this does not occur.

We know from Lemma 7.1 that

$$\lambda \in (\beta_1(x,t)\ ,\ \beta_1(x_1(t),t)) \Rightarrow H\psi_0^*(\lambda\ ;\ x,t) - \theta'(\lambda,t) > 0\ .$$

Thus, the only root of $H\psi_0^*(\lambda;\ x,t) - \theta'(\lambda,t)$ for $\lambda\ \in\ (\beta_1(x,t)\ ,\ \beta_2(x_2(t),t))$ is the previously determined $\lambda(x)$.

Now for $\beta_1(x_2(t),t) < \lambda < \beta_1(x,t)$, we have from (272)

$$\begin{aligned}\int_{\beta_0(x,t)}^{\lambda} H\psi_0^*(\mu;\ x,t) - \theta'(\mu,t)d\mu \ &=\ - \int_x^{x_-^1(\lambda,t)} \cosh^{-1}\left(\frac{\lambda - \frac{\beta_0+\alpha_0}{2}}{\frac{\beta_0-\alpha_0}{2}}\right) dy \\ + \int_{x_-^1(\lambda,t)}^{x_2(t)} &\left[\cosh^{-1}\left(\frac{\lambda - \frac{\beta_1+\alpha_1}{2}}{\frac{\beta_1-\alpha_1}{2}}\right) - \cosh^{-1}\left(\frac{\lambda - \frac{\beta_0+\alpha_0}{2}}{\frac{\beta_0-\alpha_0}{2}}\right)\right] dy\ .\end{aligned}$$

As in (273 - 275), the second integral is negative, and hence

$$\int_{\beta_0(x,t)}^{\lambda} H\psi_0^*(\mu;\ x,t) - \theta'(\mu,t)d\mu\ <\ 0\ , \tag{287}$$

for $\beta_1(x_2(t),t) \le \lambda \le \beta_1(x,t)$.

Further, for $\lambda \in (\beta_0(x,t)\ ,\ \beta_0(x_2(t),t)]$, (272) becomes

$$\begin{aligned}&\int_{\beta_0(x,t)}^{\lambda} H\psi_0^*(\lambda;\ x,t) - \theta'(\mu,t)d\mu \\ &= - \int_x^{x_-^0(\lambda,t)} \cosh^{-1}\left(\frac{\lambda - \frac{\beta_0+\alpha_0}{2}}{\frac{\beta_0-\alpha_0}{2}}\right) dy\ < 0\ ,\end{aligned} \tag{288}$$

thus the *only* solution (λ, x) of

$$\begin{aligned}\int_{\beta_0(x,t)}^{\lambda} H\psi_0^*(\mu;\ x,t) - \theta'(\mu,t)d\mu\ &=\ 0 \\ H\psi_0^*(\lambda;\ x,t) - \theta'(\lambda,t)\ &=\ 0\end{aligned}$$

is the previously determined root.

LEMMA 7.3. *For $j = 1, 2$, the equations*

$$\begin{aligned}\int_{\beta_j(x,t)}^{\lambda} H\psi_j^*(\mu;\ x,t) - \theta'(\mu,t)d\mu = 0\ \ , \\ H\psi_j^*(\lambda;\ x,t) - \theta'(\lambda,t) = 0\end{aligned}$$

have <u>*no*</u> *solutions (x, λ) with $x \in [x_1(t), x_2(t)]$, $\lambda > \beta_j(x,t)$.*

Proof.

Fix $x \in (x_1(t), x_2(t))$. We will prove this Lemma by showing the following three identities,

$$H\psi_2^*(\lambda;\ x,t) - \theta'(\lambda,t) < 0 \quad , \ for\ \lambda > \beta_2(x,t)\ , \tag{289}$$

$$H\psi_1^*(\lambda;\ x,t) - \theta'(\lambda,t) < 0 \quad , for\ \lambda \geq \beta_2(x,t), \tag{290}$$

$$\int_{\beta_j(x,t)}^{\lambda} H\psi_j^*(\mu;\ x,t) - \theta'(\mu,t)d\mu > 0\ , \quad for\ \lambda \in (\beta_1(x,t), \beta_2(x,t)]\ . \tag{291}$$

To establish (289), we use the formula (266) for the quantity $H\psi_{2x}^*$, and the Riemann - Hilbert problem (255) to write, for $\lambda > \beta_2(x,t)$

$$H\psi_2^*(\lambda;\ x,t) - \theta'(\lambda,t) = -\int_x^{x_-^2(\lambda,t)} \frac{1}{R_2(\lambda,y,t)^{1/2}}dy\ . \tag{292}$$

Formula (292) establishes (289), as $R_2(\lambda,y,t)^{1/2} > 0$ for $\lambda > \beta_2(y,t)$, and $x < x_-^2(\lambda,t)$.

To show (290) for $\lambda \geq \beta_2(x,t)$, we use the following formula, obtained in similar fashion,

$$H\psi_1^*(\lambda;\ x,t) - \theta'(\lambda,t) = -\int_x^{x_1(t)} \frac{1}{R_1(\lambda,y,t)^{1/2}}dy - \int_{x_1(t)}^{x_-^2(\lambda,t)} \frac{1}{R_2(\lambda,y,t)^{1/2}}dy$$

$$= -\int_x^{x_1(t)} \left(\frac{1}{R_1(\lambda,y,t)^{1/2}} - \frac{1}{R_2(\lambda,y,t)^{1/2}}\right) dy \tag{293}$$

$$- \int_x^{x_-^2(\lambda,t)} \frac{1}{R_2(\lambda,y,t)^{1/2}}dy\ . \tag{294}$$

Now (294) above is negative, and following familiar arguments using the shape of the curves $\{\alpha_j(x,t), \beta_j(x,t)\}$, one can show that the integrand in (293) is negative, and thus we have established (290).

To establish (291), we note first that for $\lambda \in (\beta_1(x,t), \beta_1(x_1(t),t))$, we have

$$H\psi_1^*(\lambda;\ x,t) - \theta'(\lambda,t) = -\int_x^{x_-^1(\lambda,t)} \frac{1}{R_1(\lambda,y,t)^{1/2}}dy \ < \ 0\ ,$$

and hence we conclude that

$$\int_{\beta_1(x,t)}^{\lambda} H\psi_1^*(\mu;\ x,t) - \theta'(\mu,t)d\mu \ < \ 0\ ,$$

for $\lambda \in (\beta_1(x,t), \beta_1(x_1(t),t))$.

Finally, to show (291) in the case $\lambda \in (\beta_1(x_1(t),t), \beta_2(x,t))$, we use the formula analogous to (272), for $\lambda \in (\beta_1(x_1(t),t), \beta_2(x,t)]$,

$$\int_{\beta_1(x,t)}^{\lambda} H\psi_1^*(\mu;\ x,t) - \theta'(\mu,t)d\mu = -\int_x^{x_1(t)} \cosh^{-1}\left(\frac{\lambda - \frac{\beta_1+\alpha_1}{2}}{\frac{\beta_1-\alpha_1}{2}}\right) dy$$

$$-\int_{x_1(t)}^{x_-^2(\lambda,t)} \cosh^{-1}\left(\frac{\lambda - \frac{\beta_2+\alpha_2}{2}}{\frac{\beta_2-\alpha_2}{2}}\right) dy$$

$$= -\int_x^{x_-^2(\lambda,t)} \cosh^{-1}\left(\frac{\lambda - \frac{\beta_1+\alpha_1}{2}}{\frac{\beta_1-\alpha_1}{2}}\right) dy$$

$$-\int_{x_-^2(\lambda,t)}^{x_1(t)} \left(\cosh^{-1}\left(\frac{\lambda - \frac{\beta_1+\alpha_1}{2}}{\frac{\beta_1-\alpha_1}{2}}\right) - \cosh^{-1}\left(\frac{\lambda - \frac{\beta_2+\alpha_2}{2}}{\frac{\beta_2-\alpha_2}{2}}\right)\right) dy ,$$

and in the above the first term is clearly negative. The second term is shown to be negative by establishing

$$\left(\frac{\lambda - \frac{\beta_1+\alpha_1}{2}}{\frac{\beta_1-\alpha_1}{2}}\right) > \left(\frac{\lambda - \frac{\beta_2+\alpha_2}{2}}{\frac{\beta_2-\alpha_2}{2}}\right) ,$$

which follows, as before, from the shape of the curves. This completes the proof of the Lemma. □

THEOREM 7.3. *For small $t > 0$, and $x < \hat{x}(t)$, the function $\psi^*(\lambda;\ x,t)$ solves the maximization conditions (260 - 262).*

Proof.

As observed in the proof of Theorem 7.1, the condition (261) is true for all $x \in (x_1(t), x_2(t))$. Also as shown in Theorem 7.1, the condition $\psi_0^*(\lambda;\ x,t) \le \phi(\lambda)$, is true for small t by continuity, as we have

$$\psi_0^*(\lambda;\ x,t=0) \le \phi(\lambda) - c \ <\ \phi(\lambda) .$$

The condition $\psi_0^*(\lambda;\ x,t) \ge 0$ for $\lambda \in (\alpha_0(x,t), \beta_0(x,t))$ follows from the identity

$$\psi_0^*(\lambda;\ x,t) = \psi_0^*(\lambda;\ x_-^0(\lambda,t),t) + \int_{x_-^0(\lambda,t)}^{x} \frac{i}{R_0(\lambda,y)^{1/2}} dy ,$$

as the above integral is positive, and $x_-^0(\lambda,t)$ is chosen so that $\beta_0(x_-^0(\lambda,t),t) = \lambda$, and hence $\psi_0^*(\lambda;\ x_-^0(\lambda,t),t) = 0$. Of course $\psi_0^*(\lambda;\ x,t) \equiv 0$ for $\lambda \notin (\alpha_0(x,t), \beta_0(x,t))$.

We now complete the proof of the Theorem by establishing

$$\int_{\beta_0(x,t)}^{\lambda} H\psi_0^*(\mu;\ x,t) - \theta'(\mu,t)d\mu < 0 \tag{295}$$

for all $\lambda > \beta_0(x,t)$.

First, we observe from (286) that

$$\frac{d}{dx}\int_{\beta_0(x,t)}^{\lambda(x)} H\psi_0^*(\mu;\ x,t) - \theta'(\mu,t)d\mu$$
$$= \int_{\beta_0(x,t)}^{\lambda(x)} \frac{1}{[(\mu-\alpha_0(x,t))(\mu-\beta_0(x,t))]^{1/2}}d\mu \quad > 0$$

for all $x \in (x_1(t), x_2(t))$. This implies that for $x < \hat{x}(t)$, we have

$$\int_{\beta_0(x,t)}^{\lambda(x)} H\psi_0^*(\mu;\ x,t) - \theta'(\mu,t)d\mu \ < \ 0\ .$$

Now we will consider λ in four intervals,

- $(\beta_0(x,t), \beta_0(x_2(t),t)]$,
- $[\beta_1(x_2(t),t), \beta_1(x,t)]$,
- $(\beta_1(x,t), \lambda(x)]$,
- $(\lambda(x), 1]$.

First, observe from (288) that (295) is true for $\lambda \in (\beta_0(x,t), \beta_0(x_2(t),t)]$. Secondly, observe from (287) that (295) is true for $\lambda \in [\beta_1(x_2,t), \beta_1(x,t)]$.

Third, we consider $\lambda \in (\lambda(x), 1]$. from Theorem 7.2, and (281), we see that $\lambda(x)$ is the only root of $H\psi_0^*(\lambda;\ x,t) - \theta'(\lambda,t)$ for $\lambda \in (\beta_2(x_1(t),t), 1]$. Hence, we have

$$\int_{\beta_0(x,t)}^{\lambda} H\psi_0^*(\mu;\ x,t) - \theta'(\mu,t)d\mu = \int_{\beta_0(x,t)}^{\lambda(x)} H\psi_0^*(\mu;\ x,t) - \theta'(\mu,t)d\mu$$
$$+ \int_{\lambda(x)}^{\lambda} H\psi_0^*(\mu;\ x,t) - \theta'(\mu,t)d\mu\ ,$$

and, as $H\psi_0^*(\mu;\ x,t) - \theta'(\mu,t) < 0$ for $\lambda \in (\lambda(x), 1]$, and the first term above is negative for $x < \hat{x}(t)$, we see that (295) is true for $\lambda \in (\lambda(x), 1]$.

Finally we consider $\lambda \in (\beta_1(x,t), \lambda(x)]$. From Lemma 7.1, we recall that for $\lambda \in (\beta_1(x,t), \beta_1(x_1(t),t)]$, we have

$$H\psi_0^*(\lambda;\ x,t) - \theta'(\lambda,t) > 0\ ,$$

and, from Theorem 7.2, we see that

$$H\psi_0^*(\lambda;\ x,t) - \theta'(\lambda,t) > 0\ ,$$

for $\lambda \in (\beta_1(x_1(t),t), \lambda(x))$.

Using these two inequalities, we have, for $\lambda \in (\beta_1(x,t), \lambda(x)]$,

$$\int_{\beta_0(x,t)}^{\lambda} H\psi_0^*(\mu;\ x,t) - \theta'(\mu,t)\ d\mu = \int_{\beta_0(x,t)}^{\lambda(x)} H\psi_0^*(\mu;\ x,t) - \theta'(\mu,t)\ d\mu +$$
$$+ \int_{\lambda(x)}^{\lambda} H\psi_0^*(\mu;\ x,t) - \theta'(\mu,t)\ d\mu$$
$$< \int_{\beta_0(x,t)}^{\lambda(x)} H\psi_0^*(\mu;\ x,t) - \theta'(\mu,t)\ d\mu\ ,$$

which is negative as $x < \hat{x}(t)$.

Thus the condition (260) is true for any $x < \hat{x}(t)$. This completes the proof of the Theorem. □

Therefore, for all $x < \hat{x}(t)$, the function $\psi_0^*(\lambda;\ x, t)$ solves the maximization conditions (260 - 262), and for $\hat{x}(t) \leq x < x_2(t)$, it does not. Therefore we conclude that

$$x_L(t) = \hat{x}(t) \ ,$$

and we have established that $x_L(t)$ is determined by the unique solution of (278), (279).

Of course, for x well to the right of the region $(x_1(t), x_2(t)$, the zero gap ansatz will again solve the maximization problem. There is therefore a second point of phase transition, for which the zero gap ansatz governs the continuum limit to the *right*, and fails to the left. One may check that as we approach from the right, the condition which fails is $0 \leq \psi_2^*$, which can be seen by integrating the differentiated Hilbert problem directly, as described above.

Thus, we see that there exists two points of phase transition, defining a spatial region $(x_L(t)\ ,\ x_R(t))$, in which the nature of the continuum limit has changed. These points are determined solely by the zero gap ansatz, and suggest strongly the birth of a new gap in the Riemann - Hilbert problem.

CHAPTER 8

Shock 4

In this Chapter we will consider the initial data introduced in Chapter 6 and studied further in Chapter 7 (see description and Figure (1) on pages 85-86 of Chapter 6).

We recall that in Chapter 7 we have proven, for this class of data, and for small $t > 0$, the existence of an interval $(x_1(t), x_2(t))$ such that, for $x \in (x_1(t), x_2(t))$, the zero gap ansatz does not uniquely determine a candidate $\psi^*(\lambda;\ x,t)$ for the maximizer. Rather, the conditions $T = \pi t$, $X = \pi x$ determine, for each $x \in (x_1(t), x_2(t))$, three pairs $\{\alpha_j(x,t), \beta_j(x,t)\}$, $0 \leq j \leq 2$. The functions $\alpha_1(x,t)$ and $\beta_1(x,t)$ are defined only for $x \in (x_1(t), x_2(t))$, while $\{\alpha_j(x,t), \beta_j(x,t)\}$ for $j = 0, 2$ are defined on the larger domains $(0, x_2(t))$, $(x_1(t), x_c(t))$, respectively. Next, we showed that the candidate $\psi_0^*(\lambda;\ x,t)$ solves the maximization problem on $[0, x_1(t)]$, but fails for $x \geq x_L(t)$. The point $x_L(t)$ is given as the unique solution of the pair of equations

$$\int_{\beta_0(x,t)}^{\lambda} H\psi_0^*(\lambda;\ x,t) - \theta'(\mu,t)d\mu = 0 \ , \tag{296}$$

$$H\psi_0^*(\lambda\ ;\ x,t) - \theta'(\lambda,t) = 0 \ , \tag{297}$$

for some $\lambda > \beta_0(x,t)$. Thus we are lead intuitively to a new ansatz for the (integrated version of the) Riemann - Hilbert problem,

$$-\frac{2}{\pi}\theta(\lambda,t) + \frac{1}{\pi^2}L\psi - l = 0 \quad , \quad \lambda \in I_1 = (\alpha\ ,\ \beta^{(1)}) \cup (\beta^{(2)}\ ,\ \beta^{(3)}) \ , \tag{298}$$

$$\psi = 0 \quad , \quad \lambda \in I_2 = \mathbf{R}\backslash[\alpha\ ,\ \beta^{(1)}] \cup [\beta^{(2)}\ ,\ \beta^{(3)}] \ , \tag{299}$$

where $\alpha < \beta^{(1)} \leq \beta^{(2)} \leq \beta^{(3)}$.

Remark: Of course, if we differentiate (298), (299) with respect to λ, we obtain the differentiated Riemann - Hilbert problem

$$H\psi = \theta'(\lambda,t) \quad , \quad \lambda \in I_1 = (\alpha\ ,\ \beta^{(1)}) \cup (\beta^{(2)}\ ,\ \beta^{(3)}) \ , \tag{300}$$

$$\psi = 0 \quad , \quad \lambda \in I_2 = \mathbf{R}\backslash[\alpha\ ,\ \beta^{(1)}] \cup [\beta^{(2)}\ ,\ \beta^{(3)}] \ . \tag{301}$$

The additional information contained in (298), (299) over and above (300), (301), is that the constant of integration l is the <u>same</u> in <u>both</u> intervals comprising I_1. If only one band is prescribed, as in the previous Chapters, then clearly (298), (299) and (300), (301) are equivalent.

Our goal in this Chapter is to show, for small t, how to use the Riemann - Hilbert problem (300),(301) to construct a solution of the maximization problem for x in a (small) interval to the right of $x_L(t)$. The argument in this chapter consists of the following steps.

•1 Solution of the Riemann - Hilbert problem (300),(301) and derivation of conditions governing the endpoints $\{\alpha, \beta^{(j)}\}_{j=1}^{3}$ of the set I_1.

•2 Verification that if the new band $(\beta^{(2)}, \beta^{(3)})$ collapses to a point, the conditions collapse to the zero gap conditions.

•3 Proof that for x in a neighborhood to the right of x_L we can solve for $\alpha, \beta^{(j)}(x,t)$.

•4 Derivation of PDE's satisfied by $\alpha, \beta^{(j)}$.

•5 Explicit evaluation of the differentiated Riemann - Hilbert problem.

•6 Proof that the one gap ansatz solves the maximization problem near the leading edge.

•1 We recall that the zero gap ansatz considered in the preceding chapters leads to two conditions

$$T(\alpha,\beta) = \pi t \ ,$$
$$X(\alpha,\beta) = \pi x \ ,$$

which, prior to shock formation, uniquely determine the two endpoints of the band I_1. We now need 4 conditions for $\alpha, \beta^{(j)}$. As in the case of the zero gap ansatz, the (differentiated) Riemann - Hilbert problem (300 - 301) also has an explicit solution,

$$\psi \quad = \quad Re \lim_{\epsilon \downarrow 0} f(\lambda + i\epsilon) \ , \tag{302}$$

where

$$f(z) = R(z)^{1/2} \frac{1}{\pi i} \left[\int_{\alpha}^{\beta^{(1)}} + \int_{\beta^{(2)}}^{\beta^{(3)}} \right] \frac{i\theta'(\lambda,t)}{R(\lambda)^{1/2}} \frac{1}{\lambda - z} \, d\lambda \ , \tag{303}$$

and where $R(z)$ is now given by

$$R(z) = (z-\alpha)(z-\beta^{(1)})(z-\beta^{(2)})(z-\beta^{(3)}) \ ,$$
$$R(\lambda)^{1/2} > 0 \ \ for \ \lambda > \beta^{(3)} \ .$$

From the requirement that $f \in H^p$, for some $p \in [1,2)$, we obtain the conditions

$$\left[\int_{\alpha}^{\beta^{(1)}} + \int_{\beta^{(2)}}^{\beta^{(3)}} \right] \frac{i\theta'(\lambda,t)}{R(\lambda)^{1/2}} \, d\lambda \quad = \quad 0 \ ,$$
$$\left[\int_{\alpha}^{\beta^{(1)}} + \int_{\beta^{(2)}}^{\beta^{(3)}} \right] \frac{i\theta'(\lambda,t)}{R(\lambda)^{1/2}} \, \lambda \, d\lambda \quad = \quad 0 \ .$$

From the condition that $\int \psi = \pi x$, we obtain, as in (78 - 82) of Chapter 3,

$$\left[\int_{\alpha}^{\beta^{(1)}} + \int_{\beta^{(2)}}^{\beta^{(3)}}\right] \frac{i\theta'(\lambda,t)}{R(\lambda)^{1/2}} \lambda^2 \, d\lambda = \pi x \ .$$

We obtain the fourth condition from the integrated Riemann - Hilbert problem (298),(299), as follows.

For $\lambda \in (\beta^{(1)}, \beta^{(2)})$, we have from (298) and the general representation of the operator L (125),

$$\text{(304)} \qquad \begin{aligned} \frac{1}{\pi^2} L\psi - l &= \frac{1}{\pi^2}\left[L\psi(\beta^{(1)}) + 2\pi \int_{\beta^{(1)}}^{\lambda} H\psi \, d\mu\right] - l \\ &= \frac{2}{\pi}\, \theta(\beta^{(1)},t) + \frac{2}{\pi} \int_{\beta^{(1)}}^{\lambda} H\psi \, d\mu \ , \end{aligned}$$

and from 298 and 304 we see that at $\lambda = \beta^{(2)}$ we must have

$$\theta(\beta^{(2)},t) = \theta(\beta^{(1)},t) + \int_{\beta^{(1)}}^{\beta^{(2)}} H\psi(\mu) \, d\mu$$

$$\Longleftrightarrow \int_{\beta^{(1)}}^{\beta^{(2)}} H\psi(\mu) - \theta'(\mu) \, d\mu = 0 \ .$$

Thus we obtain four conditions for $\alpha < \beta^{(1)} < \beta^{(2)} < \beta^{(3)}$,

$$\text{(305)} \qquad T_0 \equiv \left[\int_{\alpha}^{\beta^{(1)}} + \int_{\beta^{(2)}}^{\beta^{(3)}}\right] \frac{i\theta'(\lambda,t)}{R(\lambda)^{1/2}} \, d\lambda = 0 \ ,$$

$$\text{(306)} \qquad T_1 \equiv \left[\int_{\alpha}^{\beta^{(1)}} + \int_{\beta^{(2)}}^{\beta^{(3)}}\right] \frac{i\theta'(\lambda,t)}{R(\lambda)^{1/2}} \lambda \, d\lambda = 0 \ ,$$

$$\text{(307)} \qquad T_2 \equiv \left[\int_{\alpha}^{\beta^{(1)}} + \int_{\beta^{(2)}}^{\beta^{(3)}}\right] \frac{i\theta'(\lambda,t)}{R(\lambda)^{1/2}} \lambda^2 \, d\lambda = \pi x \ ,$$

$$\text{(308)} \qquad N \equiv \int_{\beta^{(1)}}^{\beta^{(2)}} H\psi(\mu) - \theta'(\mu,t) \, d\mu = 0 \ .$$

Any solution of (305 - 308) produces a solution of the Riemann - Hilbert problem (300), (301) through (303).

•2

Observe that if we denote $(T_0, T_1, T_2, N)^T$ by Y, then (305 - 308) have the form

$$\text{(309)} \qquad Y(\alpha, \beta^{(1)}, \beta^{(2)}, \beta^{(3)}) = \pi x e_3 \ ,$$

where $e_3 = (0,0,1,0)^T$ is the standard unit vector.

The function Y is defined for $\alpha < \beta^{(1)} < \beta^{(2)} < \beta^{(3)}$. It turns out that Y extends smoothly to certain boundary points of its region of definition. In particular we have the following.

LEMMA 8.1. *$Y(\alpha,\beta^{(1)},\beta^{(2)},\beta^{(3)})$ has an extension as a C^1 function to a neighborhood of $(\alpha,\beta^{(1)},\tilde{\beta},\tilde{\beta})$, for any triple $-1<\alpha<\beta^{(1)}<\tilde{\beta}<1$, and also to a neighborhood of $(\alpha,\beta_*,\beta_*,\beta^{(3)})$ for any triple $-1<\alpha<\beta_*<\beta^{(3)}<1$.*

Proof.
Observe that, by a change of variables,

$$T_0(\alpha,\beta^{(1)},\beta^{(2)},\beta^{(3)}) = \int_\alpha^{\beta^{(1)}} \frac{i\theta'(\lambda,t)\,d\lambda}{\sqrt{(\lambda-\alpha)(\lambda-\beta^{(1)})(\lambda-\beta^{(2)})(\lambda-\beta^{(3)})}}$$
$$+\int_{-1}^{1} \frac{i\theta'\left(\frac{\beta^{(3)}-\beta^{(2)}}{2}\mu+\frac{\beta^{(3)}+\beta^{(2)}}{2}\right)}{\sqrt{(\frac{\beta^{(3)}-\beta^{(2)}}{2}\mu+\frac{\beta^{(3)}+\beta^{(2)}}{2}-\alpha)(\frac{\beta^{(3)}-\beta^{(2)}}{2}\mu+\frac{\beta^{(3)}+\beta^{(2)}}{2}-\beta^{(1)})}}\frac{d\mu}{\sqrt{\mu^2-1}}\ .$$

But the right hand side clearly extends to a smooth function of $\alpha,\beta^{(1)},\beta^{(2)},\beta^{(3)}$ in a neighborhood of $\alpha<\beta^{(1)}<\beta^{(2)}=\beta^{(3)}=\tilde{\beta}$, etc. The argument is similar for T_1,T_2.

On the other hand, we have

$$(310)\quad f(z)=R(z)^{1/2}\frac{1}{\pi i}\int_\alpha^{\beta^{(1)}}\frac{i\theta'(\lambda,t)}{R(\lambda)^{1/2}}\frac{1}{\lambda-z}\,d\lambda$$
$$+\frac{R(z)^{1/2}}{\pi i}\int_{-1}^{1}\frac{i\theta'\left(\frac{\beta^{(3)}-\beta^{(2)}}{2}\mu+\frac{\beta^{(3)}+\beta^{(2)}}{2}\right)}{\sqrt{(\frac{\beta^{(3)}-\beta^{(2)}}{2}\mu+\frac{\beta^{(3)}+\beta^{(2)}}{2}-\alpha)(\frac{\beta^{(3)}-\beta^{(2)}}{2}\mu+\frac{\beta^{(3)}+\beta^{(2)}}{2}-\beta^{(1)})}}\cdot$$
$$\cdot\frac{1}{\sqrt{\mu^2-1}}\frac{d\mu}{\frac{\beta^{(3)}-\beta^{(2)}}{2}\mu+\frac{\beta^{(3)}+\beta^{(2)}}{2}-z}\ ,$$

and for $z\in(\beta^{(1)},\beta^{(2)})$, $H\psi=f$, and hence

$$iN=\int_{\beta^{(1)}}^{\beta^{(2)}}R(z)^{1/2}\frac{1}{\pi i}\int_\alpha^{\beta^{(1)}}\frac{i\theta'(\lambda,t)}{R(\lambda)^{1/2}}\frac{1}{\lambda-z}\,d\lambda dz$$
$$+\int_{\beta^{(1)}}^{\beta^{(2)}}R(z)^{1/2}\Bigg($$
$$\frac{1}{\pi i}\int_{-1}^{1}\frac{i\theta'\left(\frac{\beta^{(3)}-\beta^{(2)}}{2}\mu+\frac{\beta^{(3)}+\beta^{(2)}}{2}\right)}{\sqrt{(\frac{\beta^{(3)}-\beta^{(2)}}{2}\mu+\frac{\beta^{(3)}+\beta^{(2)}}{2}-\alpha)(\frac{\beta^{(3)}-\beta^{(2)}}{2}\mu+\frac{\beta^{(3)}+\beta^{(2)}}{2}-\beta^{(1)})}}\cdot$$
$$\frac{1}{\sqrt{\mu^2-1}}\frac{d\mu}{\frac{\beta^{(3)}-\beta^{(2)}}{2}\mu+\frac{\beta^{(3)}+\beta^{(2)}}{2}-z}$$
$$-i\theta'(z,t)\Bigg)\ dz\ ,$$

which shows that N is C^1 as well. □

For $t > 0$ and for $x = x_L(t)$, the function $\int_{\beta_0(x,t)}^{\lambda} H\psi_0^*(\mu;\ x,t) - \theta'(\mu)d\mu$ has a zero at $\lambda = \lambda(x_L(t)) > \beta_0(x,t)$. As $x \downarrow x_L(t)$, we expect

$$\begin{aligned} \alpha(x,t) &\to \alpha_0(x_L(t),t)\ , \\ \beta^{(1)}(x,t) &\to \beta_0(x_L(t),t)\ , \\ \beta^{(2)}(x,t),\ \beta^{(3)}(x,t) &\to \lambda(x_L(t))\ . \end{aligned} \tag{311}$$

The first step towards proving this is the following result.

THEOREM 8.1. *Suppose $t > 0$ is given and small. A triple $\alpha < \beta^{(1)} < \tilde{\beta}$, solves the equation $Y(\alpha, \beta^{(1)}, \tilde{\beta}, \tilde{\beta}) = \pi x e_3$, if and only if $\alpha = \alpha_0(x_L(t), t)$, $\beta^{(1)} = \beta_0(x_L(t), t)$, $x = x_L(t)$, and $\tilde{\beta} = \lambda(x_L(t))$.*

Proof. First we note that if we allow $\beta^{(2)}, \beta^{(3)} \to \tilde{\beta}$ for some $\tilde{\beta}$, we obtain

$$\begin{aligned} \left(T_1 - \tilde{\beta}T_0\right)(\alpha, \beta^{(1)}, \tilde{\beta}, \tilde{\beta}) &= \lim_{\beta^{(2)},\beta^{(3)} \to \tilde{\beta}} \left(T_1 - \tilde{\beta}T_0\right) \\ &= \lim_{\beta^{(2)},\beta^{(3)} \to \tilde{\beta}} \left[\int_{\alpha}^{\beta^{(1)}} + \int_{\beta^{(2)}}^{\beta^{(3)}}\right] \frac{i\theta'(\lambda,t)}{R(\lambda)^{1/2}}(\lambda - \tilde{\beta})\, d\lambda \\ &= \int_{\alpha}^{\beta^{(1)}} \frac{i\theta'(\lambda, t=0)}{\left[(\lambda-\alpha)(\lambda-\beta^{(1)})\right]^{1/2}}\, d\lambda \ - \ \pi t \\ &= T(\alpha, \beta^{(1)}) - \pi t\ , \end{aligned} \tag{312}$$

and

$$\begin{aligned} \left(T_2 - \tilde{\beta}T_1\right)(\alpha, \beta^{(1)}, \tilde{\beta}, \tilde{\beta}) &= \lim_{\beta^{(2)},\beta^{(3)} \to \tilde{\beta}} \left(T_2 - \tilde{\beta}T_1\right) \\ &= \lim_{\beta^{(2)},\beta^{(3)} \to \tilde{\beta}} \left[\int_{\alpha}^{\beta^{(1)}} + \int_{\beta^{(2)}}^{\beta^{(3)}}\right] \frac{i\theta'(\lambda,t)}{R(\lambda)^{1/2}}\lambda(\lambda - \tilde{\beta})\, d\lambda \\ &= \int_{\alpha}^{\beta^{(1)}} \frac{i\theta'(\lambda, t=0)}{\left[(\lambda-\alpha)(\lambda-\beta^{(1)})\right]^{1/2}}(\lambda - \frac{\beta^{(1)}+\alpha}{2})\, d\lambda\ + \\ &\quad + \frac{\beta^{(1)}+\alpha}{2}\int_{\alpha}^{\beta^{(1)}} \frac{i\theta'(\lambda, t=0)}{\left[(\lambda-\alpha)(\lambda-\beta^{(1)})\right]^{1/2}} \ - \ \frac{\beta^{(1)}+\alpha}{2}\pi t \\ &= X(\alpha, \beta^{(1)}) + \frac{\beta^{(1)}+\alpha}{2}(T - \pi t)\ . \end{aligned}$$

We see that to satisfy $Y(\alpha, \beta^{(1)}, \tilde{\beta}, \tilde{\beta}) = \pi x e_3$, we must have

$$\alpha = \alpha_j(x,t)\ , \tag{313}$$

$$\beta^{(1)} = \beta_j(x,t)\ , \tag{314}$$

that is, that α and $\beta^{(1)}$ must solve the one band Riemann - Hilbert problem.

Second, we consider the quantity T_0. In the limit as $\beta^{(2)}, \beta^{(3)} \to \tilde{\beta}$, we have

$$T_0 \to \int_\alpha^{\beta^{(1)}} \frac{i\theta'(\lambda,t)}{[(\lambda-\alpha)(\lambda-\beta^{(1)})]^{1/2}} \frac{1}{\lambda-\tilde{\beta}}\, d\lambda + \pi \frac{\theta'(\tilde{\beta},t)}{\left[(\tilde{\beta}-\alpha)(\tilde{\beta}-\beta^{(1)})\right]^{1/2}} .$$

Thus we see that to satisfy $Y(\alpha, \beta^{(1)}, \tilde{\beta}, \tilde{\beta}) = \pi x e_3$, we must have (313), (314), and the following,

$$\begin{aligned} [(\tilde{\beta}-\alpha)(\tilde{\beta}-\beta^{(1)})]^{1/2} \frac{1}{\pi i} \int_\alpha^{\beta^{(1)}} & \frac{i\theta'(\lambda,t)}{[(\lambda-\alpha)(\lambda-\beta^{(1)})]^{1/2}} \frac{1}{\lambda-\tilde{\beta}}\, d\lambda \\ & -i\theta'(\tilde{\beta},t) = 0 , \end{aligned} \tag{315}$$

that is, using the notation of the previous Chapters (256), (257),

$$H\psi_j^*(\tilde{\beta};\ x,t) - \theta'(\tilde{\beta},t) = 0 . \tag{316}$$

Finally, we consider the quantity N. To evaluate this, we note that in the limit, $f(z)$, defined in (303), satisfies

$$\begin{aligned} \lim_{\beta^{(2)},\beta^{(3)}\to\tilde{\beta}} f(z) = \tilde{f} = & \\ \frac{[(z-\alpha)(z-\beta^{(1)})]^{1/2}(z-\tilde{\beta})}{\pi i} & \int_\alpha^{\beta^{(1)}} \frac{i\theta'(\lambda,t)}{[(\lambda-\alpha)(\lambda-\beta^{(1)})]^{1/2}} \frac{d\lambda}{(\lambda-\tilde{\beta})(\lambda-z)} \\ + \frac{[(z-\alpha)(z-\beta^{(1)})]^{1/2}}{[(\tilde{\beta}-\alpha)(\tilde{\beta}-\beta^{(1)})]^{1/2}} i\theta'(\tilde{\beta},t) . & \end{aligned} \tag{317}$$

We leave it to the reader to verify that $\tilde{f}$ satisfies the Riemann - Hilbert problem

$$Im\ \tilde{f} = H\psi = \theta'(\lambda,t) \quad , \quad \lambda \in (\alpha\ ,\ \beta^{(1)}) , \tag{318}$$

$$Re\ \tilde{f} = \psi = 0 \quad , \quad \lambda \in \mathbf{R}\backslash[\alpha\ ,\ \beta^{(1)}] . \tag{319}$$

Now suppose that $\alpha = \alpha_j(x,t)$, $\beta^{(1)} = \beta_j(x,t)$, and that $\tilde{\beta}$ is such that

$$H\psi_j^*(\tilde{\beta};\ x,t) - \theta'(\tilde{\beta},t) = 0 , \tag{320}$$

as in (316) above, and we compute the asymptotics for $\tilde{f}(z)$ as $z \to \infty$. We have, from (317),

$$\tilde{f} = [(z-\alpha)(z-\beta^{(1)})]^{1/2}\left[-\frac{1}{\pi i}\int_{\alpha}^{\beta^{(1)}}\frac{i\theta'(\lambda,t)}{[(\lambda-\alpha)(\lambda-\beta^{(1)})]^{1/2}}\frac{1}{\lambda-\tilde{\beta}}\frac{1}{1-\frac{\lambda-\tilde{\beta}}{z-\tilde{\beta}}}d\lambda\right.$$

$$\left. + \frac{i\theta'(\tilde{\beta},t)}{[(\tilde{\beta}-\alpha)(\tilde{\beta}-\beta^{(1)})]^{1/2}}\right]$$

$$= [(z-\alpha)(z-\beta^{(1)})]^{1/2}\left[-\frac{1}{\pi i}\int_{\alpha}^{\beta^{(1)}}\frac{i\theta'(\lambda,t)}{[(\lambda-\alpha)(\lambda-\beta^{(1)})]^{1/2}}\frac{1}{\lambda-\tilde{\beta}}\cdot\right.$$

$$\cdot\left(1+\frac{\lambda-\tilde{\beta}}{z-\tilde{\beta}}+\left(\frac{\lambda-\tilde{\beta}}{z-\tilde{\beta}}\right)^2+\cdots\right)d\lambda$$

$$\left.+\frac{i\theta'(\tilde{\beta},t)}{[(\tilde{\beta}-\alpha)(\tilde{\beta}-\beta^{(1)})]^{1/2}}\right] .$$

Now the leading term is precisely

$$[(z-\alpha)(z-\beta^{(1)})]^{1/2}\left[-\frac{1}{\pi i}\int_{\alpha}^{\beta^{(1)}}\frac{i\theta'(\lambda,t)}{[(\lambda-\alpha)(\lambda-\beta^{(1)})]^{1/2}}\frac{1}{\lambda-\tilde{\beta}}d\lambda+\right.$$

$$\left.\frac{i\theta'(\tilde{\beta},t)}{[(\tilde{\beta}-\alpha)(\tilde{\beta}-\beta^{(1)})]^{1/2}}\right]$$

$$= \frac{[(z-\alpha)(z-\beta^{(1)})]^{1/2}}{[(\tilde{\beta}-\alpha)(\tilde{\beta}-\beta^{(1)})]^{1/2}}\left(H\psi_j^*(\tilde{\beta};x,t)-\theta'(\tilde{\beta},t)\right) = 0 ,$$

from (320). Moreover, the next term in the expansion for $\tilde{f}$ is

$$[(z-\alpha)(z-\beta^{(1)})]^{1/2}\left[-\frac{1}{\pi i}\int_{\alpha}^{\beta^{(1)}}\frac{i\theta'(\lambda,t)}{[(\lambda-\alpha)(\lambda-\beta^{(1)})]^{1/2}}\frac{1}{z-\tilde{\beta}}d\lambda\right]$$

$$= -\left[\frac{z-\alpha}{z-\beta^{(1)}}\right]^{1/2}\left(T(\alpha,\beta^{(1)})-\pi t\right) ,$$

which is zero as well, because we have chosen α and $\beta^{(1)}$ to solve the one band problem. Hence $\tilde{f}$ solves the zero gap Riemann - Hilbert problem (318 - 319), and lies in L^p, for some $1 \le p < 2$. That is, $\tilde{f} = f_j$, where f_j is the zero gap quantity corresponding the $\alpha_j(x,t)$, $\beta_j(x,t)$, and so the condition $N = 0$ becomes

$$\int_{\beta_j(x,t)}^{\tilde{\beta}} H\psi_j^*(\mu;\ x,t) - \theta'(\mu,t)d\mu = 0 .$$

Thus we see that we have obtained the conditions for the zero gap ansatz, along with the additional conditions (278), (279) for $x_L(t)$, and $\lambda(x_L(t))$ (see Proposition 7.3, and also (278), (279)). Hence $x = x_L(t)$, and $\tilde{\beta} = \lambda(x_L(t))$ □

As remarked at the end of Chapter 7, the "trailing edge" $x_R(t)$ of the region where the zero gap ansatz fails to satisfy the side constraints is determined from the zero gap ansatz by the failure, as x decreases from the right, of the maximization condition

$$\psi_2^* \geq 0 \; , \tag{321}$$

i.e. that there exists a unique pair $x_R(t), \beta_*$ such that

$$\begin{aligned} \beta_* &< \beta_2(x_R(t), t) \; , \\ \psi_2^*(\beta_* \; ; \; x_R(t), t) &= 0 \; , \\ \frac{\partial}{\partial \lambda}\psi_2^*(\beta_* \; ; \; x_R(t), t) &= 0 \; , \end{aligned}$$

and if we evaluate $\psi_2^*(\lambda \; ; \; x, t)$, for $x > x_R(t)$, and *any* $\lambda \in (\alpha_2(x,t), \beta_2(x,t))$, the condition 321 is true, while it fails for $x < x_R(t)$, for λ in some subset of the interval $(\alpha_2(x,t), \beta_2(x,t))$.

The proof of the following theorem follows from similar arguments used to establish Theorem 8.1.

THEOREM 8.2. *Suppose $t > 0$ is given and small. A triple $\alpha < \beta_* < \beta^{(3)}$, solves the equation $Y(\alpha, \beta_*, \beta_*, \beta^{(3)}) = \pi x e_3$, if and only if $\alpha = \alpha_0(x_R(t), t)$, $\beta^{(3)} = \beta_2(x_R(t), t)$, $x = x_R(t)$, and $\beta_* = \lambda(x_R(t))$.*

Thus, the existence of a pair $(\beta_*, x_R(t))$ such that the zero gap ansatz $\psi_2^*(\lambda; \; x_R(t), t)$ satisfies

$$\begin{aligned} \psi_2^*(\beta_*; \; x_R(t), t) &= 0 \; , \\ \partial_\lambda \psi_2^*(\beta_*; \; x_R(t), t) &= 0 \; , \end{aligned}$$

is equivalent to the existence of a solution to the one gap ansatz in the limiting case $\beta^{(1)}, \beta^{(2)} \to \beta_* < \beta^{(3)}$.

• 3 Now we would like to see that in a neighborhood of the leading edge $x_L(t)$, determined in Chapter 7, we can continue to solve the system (305 - 308).

We seek (α , $\beta^{(1)}$, $\beta^{(3)}$) in terms of ($\beta^{(2)}$, x , t), to solve $T_0 = 0$, $T_1 = 0$, $N = 0$. As we know we can solve these three equations at $x_L(t)$,

$$\begin{aligned} \alpha &= \alpha_0(x_L(t), t) \; , \\ \beta^{(1)} &= \beta_0(x_L(t), t) \; , \\ \beta^{(2)} = \beta^{(3)} &= \lambda(x_L(t)) \; , \end{aligned}$$

we need only show (recall Lemma 8.1) that the Jacobian does not vanish at this point $P_L(t) \equiv (\alpha_0(x_L(t),t), \beta_0(x_L(t),t), \lambda(x_L(t)), \lambda(x_L(t)))$.

$$\frac{\partial\,(T_1\ ,\ T_2\ ,\ N\ )}{\partial\,(\alpha\ ,\ \beta^{(1)}\ ,\ \beta^{(3)}\ )} \neq 0\ .$$

Towards this end, we show the following identities.

LEMMA 8.2. *Define, for $k \geq 0$,* [1]

$$T_k = \left[\int_\alpha^{\beta^{(1)}} + \int_{\beta^{(2)}}^{\beta^{(3)}}\right] \frac{i\theta'(\lambda,t)}{R(\lambda)^{1/2}}\ \lambda^k\ d\lambda\ .$$

Then for each $k \geq 0$, T_k satisfies

$$\frac{\partial}{\partial\beta^{(j)}}\ T_{k+1}\ -\ \beta^{(j)}\ \frac{\partial}{\partial\beta^{(j)}}\ T_k\ -\ \frac{1}{2}T_k\ =\ 0\ .$$

Remark: in the proof of this Lemma, and in that of the succeeding Lemma, we will use the notation $\alpha = \beta^{(0)}$.

Proof.

We have

$$\frac{\partial}{\partial\beta^{(j)}}\ T_{k+1} - \beta^{(j)}\ \frac{\partial}{\partial\beta^{(j)}}\ T_k\ -\ \frac{1}{2}T_k = \frac{\partial}{\partial\beta^{(j)}}\left(T_{k+1}\ -\ \beta^{(j)}T_k\right) + \frac{1}{2}\ T_k$$

$$= \frac{\partial}{\partial\beta^{(j)}}\left[\int_\alpha^{\beta^{(1)}} + \int_{\beta^{(2)}}^{\beta^{(3)}}\right] \frac{i\theta'(\lambda,t)}{R(\lambda)^{1/2}}\ \lambda^k\ \left(\lambda - \beta^{(j)}\right)\ d\lambda\ +\ \frac{1}{2}\ T_k$$

$$=\ -\frac{1}{2}T_k\ +\ \frac{1}{2}T_k\ =\ 0\ .$$

□

REMARK 8.1.

If we are at a solution $\alpha = \beta^{(0)}, \beta^{(1)}, \beta^{(2)}, \beta^{(3)}$ to (305 - 308), then we have the simplification

$$\frac{\partial}{\partial\beta^{(j)}}\ T_1\ =\ \beta^{(j)}\ \frac{\partial}{\partial\beta^{(j)}}\ T_0\ , \tag{322}$$

$$\begin{aligned}\frac{\partial}{\partial\beta^{(j)}}\ T_2\ &=\ \beta^{(j)}\ \frac{\partial}{\partial\beta^{(j)}}\ T_1 \\ &=\ (\beta^{(j)})^2\ \frac{\partial}{\partial\beta^{(j)}}\ T_0\ ,\end{aligned} \tag{323}$$

LEMMA 8.3. *Take N as defined in (308). Then*

$$\frac{\partial}{\partial\beta^{(j)}}\ N\ =\ g_j\frac{\partial}{\partial\beta^{(j)}}\ T_0\ ,$$

---

[1] As in Lemma 8.1 the functions $T_k(\alpha, \beta^{(1)}, \beta^{(2)}, \beta^{(3)})$, defined a priori only for $\alpha < \beta^{(1)} < \beta^{(2)} < \beta^{(3)}$, have C^1 continuations to a neighborhood of $(\alpha, \beta^{(1)}, \tilde\beta, \tilde\beta)$ for $\alpha < \beta^{(1)} < \tilde\beta$, etc.

where

$$g_j = \frac{1}{\pi}\int_{\beta^{(1)}}^{\beta^{(2)}} R(z)^{1/2}\frac{1}{z-\beta^{(j)}}\, dz\ .$$

Proof.
As

$$N = \int_{\beta^{(1)}}^{\beta^{(2)}} \frac{1}{i}\, f(z) - \theta'(z,t)dz\ ,$$

and $\frac{1}{i}f(\lambda) = Im\ f = H\psi = \theta'(\lambda,t)$ for $\lambda = \beta^{(1)}$, $\lambda = \beta^{(2)}$, we need only establish

$$\frac{\partial}{\partial\beta^{(j)}}\, f(z) = R(z)^{1/2}\frac{1}{\pi i}\left(\frac{-\frac{\partial}{\partial\beta^{(j)}}\, T_0}{z-\beta^{(j)}}\right)\ . \tag{324}$$

Expand $\frac{1}{\lambda - z}$ in a geometric series in the formula (303) for $f(z)$, to obtain

$$\begin{aligned} f(z) &= R(z)^{1/2}\frac{1}{\pi i}\left[\int_{\alpha}^{\beta^{(1)}} + \int_{\beta^{(2)}}^{\beta^{(3)}}\right]\frac{i\theta'(\lambda,t)}{R(\lambda)^{1/2}}\,\frac{1}{\lambda - z}\, d\lambda \\ &= R(z)^{1/2}\frac{1}{\pi i}\,\frac{-1}{z}\left[\int_{\alpha}^{\beta^{(1)}} + \int_{\beta^{(2)}}^{\beta^{(3)}}\right]\frac{i\theta'(\lambda,t)}{R(\lambda)^{1/2}}\sum_{k=0}^{\infty}\left(\frac{\lambda}{z}\right)^k d\lambda \\ &= R(z)^{1/2}\frac{1}{\pi i}\,\frac{-1}{z}\sum_{k=0}^{\infty}\frac{T_k}{z^k}\ , \end{aligned} \tag{325}$$

where T_k is defined in Lemma (8.2) above.

Differentiating (325) with respect to $\beta^{(j)}$, we obtain

$$\begin{aligned} \frac{\partial}{\partial\beta^{(j)}}\, f(z) &= R(z)^{1/2}\frac{-\frac{1}{2}}{z-\beta^{(j)}}\frac{1}{\pi i}\,\frac{-1}{z}\sum_{k=0}^{\infty}\frac{T_k}{z^k} + \\ &\qquad R(z)^{1/2}\frac{1}{\pi i}\,\frac{-1}{z}\sum_{k=0}^{\infty}\frac{\frac{\partial}{\partial\beta^{(j)}}T_k}{z^k} \\ &= R(z)^{1/2}\frac{1}{\pi i}\,\frac{-1}{z}\,\frac{1}{z-\beta^{(j)}}\sum_{k=0}^{\infty}\frac{(z-\beta^{(j)})\frac{\partial}{\partial\beta^{(j)}}T_k - \frac{1}{2}T_k}{z^k} \\ &= R(z)^{1/2}\frac{1}{\pi i}\,\frac{-1}{z}\,\frac{1}{z-\beta^{(j)}}\Bigg[z\,\frac{\partial}{\partial\beta^{(j)}}\, T_0 + \\ &\qquad \sum_{k=0}^{\infty}\frac{\frac{\partial}{\partial\beta^{(j)}}\, T_{k+1} - \beta^{(j)}\,\frac{\partial}{\partial\beta^{(j)}}\, T_k - \frac{1}{2}T_k}{z^k}\Bigg] \\ &= R(z)^{1/2}\frac{1}{\pi i}\left(\frac{-\frac{\partial}{\partial\beta^{(j)}}\, T_0}{z-\beta^{(j)}}\right)\ , \end{aligned} \tag{326}$$

where in (326) we have used Lemma (8.2). This completes the proof of the lemma. □

We are now in a position to analyze the Jacobian at the particular solution $P_L(t)$ of (305 - 308) (see Theorem 8.1). Using Remark 8.1 and Lemma 8.3, we have,

(327)
$$\begin{vmatrix} \frac{\partial}{\partial\alpha}T_0 & \frac{\partial}{\partial\beta^{(1)}}T_0 & \frac{\partial}{\partial\beta^{(3)}}T_0 \\ \frac{\partial}{\partial\alpha}T_1 & \frac{\partial}{\partial\beta^{(1)}}T_1 & \frac{\partial}{\partial\beta^{(3)}}T_1 \\ \frac{\partial}{\partial\alpha}N & \frac{\partial}{\partial\beta^{(1)}}N & \frac{\partial}{\partial\beta^{(3)}}N \end{vmatrix}$$
$$= \frac{\partial}{\partial\alpha}T_0 \; \frac{\partial}{\partial\beta^{(1)}}T_0 \; \frac{\partial}{\partial\beta^{(3)}}T_0 \begin{vmatrix} 1 & 1 & 1 \\ \alpha & \beta^{(1)} & \beta^{(3)} \\ g_0 & g_1 & g_3 \end{vmatrix}$$
$$= \frac{1}{\pi}\frac{\partial}{\partial\alpha}T_0 \; \frac{\partial}{\partial\beta^{(1)}}T_0 \; \frac{\partial}{\partial\beta^{(3)}}T_0 \times$$
$$\int_{\beta^{(1)}}^{\beta^{(2)}} dz\; R(z)^{1/2}\frac{(\beta^{(3)}-\alpha)(\beta^{(3)}-\beta^{(1)})(\beta^{(1)}-\alpha)}{(z-\alpha)(z-\beta^{(1)})(z-\beta^{(3)})} .$$

To show the Jacobian is nonzero, we must only establish

$$\frac{\partial}{\partial\beta^{(j)}}T_0 \;\neq\; 0 \;,$$

for $j = 0,\ 1,\ 3$.

THEOREM 8.3. *We have*

$$\frac{\partial}{\partial\alpha}T_0(P_L(t)) < 0 \;.$$

Proof.
The following formula generalizes (141) for times t beyond shock time.

(328)
$$\theta(\lambda,t) = \begin{cases} \int_0^{x_-^0(\lambda,t)} \cosh^{-1}\left(\frac{\lambda-\frac{\beta_0+\alpha_0}{2}}{\frac{\beta_0-\alpha_0}{2}}\right) dy, \; \lambda \in (0\;,\;\beta_0(x_2(t),t)) \\[2ex] \int_0^{x_2(t)} \cosh^{-1}\left(\frac{\lambda-\frac{\beta_0+\alpha_0}{2}}{\frac{\beta_0-\alpha_0}{2}}\right) dy \;-\; \int_{x_-^1(\lambda,t)}^{x_2(t)} \cosh^{-1}\left(\frac{\lambda-\frac{\beta_1+\alpha_1}{2}}{\frac{\beta_1-\alpha_1}{2}}\right) dy\;, \\ \qquad\qquad \lambda \in (\beta_0(x_2(t),t)\;,\;\beta^1(x_1(t),t)) \\[2ex] \int_0^{x_2(t)} \cosh^{-1}\left(\frac{\lambda-\frac{\beta_0+\alpha_0}{2}}{\frac{\beta_0-\alpha_0}{2}}\right) dy \;-\; \int_{x_1(t)}^{x_2(t)} \cosh^{-1}\left(\frac{\lambda-\frac{\beta_1+\alpha_1}{2}}{\frac{\beta_1-\alpha_1}{2}}\right) dy \\ \quad + \int_{x_1(t)}^{x_-^2(\lambda,t)} \cosh^{-1}\left(\frac{\lambda-\frac{\beta_2+\alpha_2}{2}}{\frac{\beta_2-\alpha_2}{2}}\right) dy\;,\;\; \lambda > \beta_2(x_1(t),t) \end{cases}$$

The proof of this formula is again straightforward. One shows directly that $\frac{\partial}{\partial t}\theta(\lambda,t) = -\lambda$, and at $t=0$, one sees easily that $\theta(\lambda,0)=\theta(\lambda)$.

Now we return to the definition (305), and, differentiating the representation (328) of the function $\theta(\lambda,t)$, we interchange orders of integration to obtain

$$
\begin{aligned}
T_0 \;=\;& \int_0^{x_-^0(\alpha,t)} \left\{ \int_\alpha^{\alpha_0(y,t)} + \int_{\beta_0(y,t)}^{\beta^{(1)}} + \int_{\beta^{(2)}}^{\beta^{(3)}} \right\} \frac{i}{[R(\lambda)R_0(\lambda,y)]^{1/2}}\, d\lambda dy \,+ \\
&+ \int_{x_-^0(\alpha,t)}^{x_-^0(\beta^{(1)},t)} dy \int_{\beta_0(y,t)}^{\beta^{(1)}} d\lambda\, \frac{i}{[R(\lambda)R_0(\lambda,y)]^{1/2}} \\
&+ \int_{x_-^0(\alpha,t)}^{x_2(t)} dy \int_{\beta^{(2)}}^{\beta^{(3)}} d\lambda\, \frac{i}{[R(\lambda)R_0(\lambda,y)]^{1/2}} + \\
&+ \int_{x_2(t)}^{x_1(t)} dy \int_{\beta^{(2)}}^{\beta^{(3)}} d\lambda\, \frac{i}{[R(\lambda)R_1(\lambda,y)]^{1/2}} + \\
&\int_{x_1(t)}^{x_-^2(\beta^{(2)},t)} dy \int_{\beta^{(2)}}^{\beta^{(3)}} d\lambda\, \frac{i}{[R(\lambda)R_2(\lambda,y)]^{1/2}} + \\
&+ \int_{x_-^2(\beta^{(2)},t)}^{x_-^2(\beta^{(3)},t)} dy \int_{\beta_2(y,t)}^{\beta^{(3)}} d\lambda\, \frac{i}{[R(\lambda)R_0(\lambda,y)]^{1/2}} \\
=\;& \int_{\beta_0(x_-^0(\alpha,t),t)}^{\beta^{(1)}} \int_{x_-^0(\alpha,t)}^{x_-^0(\lambda,t))} \frac{i}{[R(\lambda)R_0(\lambda,y)]^{1/2}}\, d\lambda dy \,+ \\
&+ \int_{\beta^{(2)}}^{\beta^{(3)}} d\lambda \int_{x_-^0(\alpha,t)}^{x_2(t)} dy \frac{i}{[R(\lambda)R_0(\lambda,y)]^{1/2}} + \\
&+ \int_{\beta^{(2)}}^{\beta^{(3)}} d\lambda \int_{x_2(t)}^{x_1(t)} dy \frac{i}{[R(\lambda)R_1(\lambda,y)]^{1/2}} + \\
&+ \int_{\beta^{(2)}}^{\beta^{(3)}} d\lambda \int_{x_1(t)}^{x_-^2(\lambda,t)} dy \frac{i}{[R(\lambda)R_2(\lambda,y)]^{1/2}} \;.
\end{aligned}
\tag{329}
$$

Now we use the integral representation of $H\psi_0^*(\lambda;\ x,t) - \theta'(\lambda,t)$ defined on the multi-sheeted zero gap curves $\alpha_j(x,t), \beta_j(x,t)$, used in Chapter 7 and which we re-write, for example, in the case $\beta_0(x,t) < \lambda < \beta_0(x_2(t),t)$,

$$
H\psi_0^*(\lambda;\ x,t) - \theta'(\lambda,t) = -\int_x^{x_-^0(\lambda,t)} \frac{1}{R_0(\lambda,y)^{1/2}} dy \;,
$$

and in the case $\lambda > \beta_2(x_1(t), t)$,

$$H\psi_0^*(\lambda;\ x, t) - \theta'(\lambda, t) = -\int_x^{x_2(t)} \frac{1}{R_0(\lambda, y)^{1/2}} dy$$
$$- \int_{x_2(t)}^{x_1(t)} \frac{1}{R_1(\lambda, y)^{1/2}} dy$$
$$- \int_{x_1(t)}^{x_-^2(\lambda,t)} \frac{1}{R_2(\lambda, y)^{1/2}} dy \ ,$$

to obtain, from (329),

$$(330)\quad T_0 = \int_{\beta_0(x_-^0(\alpha,t),t)}^{\beta^{(1)}} \frac{i}{R(\lambda)^{1/2}} \left(\theta'(\lambda, t) - H\psi_0^*(\lambda\ ; x_-^0(\alpha, t), t)\right)\, d\lambda \ +$$
$$+ \int_{\beta^{(2)}}^{\beta^{(3)}} \frac{i}{R(\lambda)^{1/2}} \left(\theta'(\lambda, t) - H\psi_0^*(\lambda\ ; x_-^0(\alpha, t), t)\right)\, d\lambda \ .$$

Using the expression (266) of the zero gap quantity $H\psi_{0,x}^*$, we differentiate (330),

$$\frac{\partial}{\partial\alpha} T_0 \ = -\frac{\partial}{\partial\alpha} x_-^0(\alpha, t)) \left\{ \int_{\beta_0(x_-^0(\alpha,t),t)}^{\beta^{(1)}} + \int_{\beta^{(2)}}^{\beta^{(3)}} \right\} \frac{i}{\tilde{R}(\lambda)^{1/2}} \frac{d\lambda}{\lambda - \alpha}$$
$$+ \left\{ \int_{\beta_0(x_-^0(\alpha),t)}^{\beta^{(1)}} + \int_{\beta^{(2)}}^{\beta^{(3)}} \right\} \frac{i}{R(\lambda)^{1/2}} \left(\theta'(\lambda, t) - H\psi_0^*(\lambda\ ; x_-(\alpha, t), t)\right) \frac{d\lambda/2}{\lambda - \alpha} \ ,$$

where

$$\tilde{R}(\lambda) = (\lambda - \beta_0(x_-^0(\alpha, t), t))(\lambda - \beta^{(1)})(\lambda - \beta^{(2)})(\lambda - \beta^{(3)}) \ .$$

Evaluating the first integral explicitly, we obtain

$$(331) T_{0\alpha} = \frac{\pi \frac{\partial}{\partial\alpha} x_-^0(\alpha, t)}{\left[(\alpha - \beta_0(x_-^0(\alpha, t), t))(\alpha - \beta^{(1)})(\alpha - \beta^{(2)})(\alpha - \beta^{(3)})\right]^{1/2}} \ +$$
$$+ \left\{ \int_{\beta_0(x_-^0(\alpha,t),t)}^{\beta^{(1)}} + \int_{\beta^{(2)}}^{\beta^{(3)}} \right\} \frac{i}{R(\lambda)^{1/2}} \left(\theta'(\lambda, t) - H\psi_0^*(\lambda\ ; x_-(\alpha, t), t)\right) \frac{d\lambda/2}{\lambda - \alpha} \ .$$

The first term in (331) is negative. At $P_L(t)$, the first integral vanishes as $\beta^{(1)} = \beta_0(x_L(t), t)$, $\alpha = \alpha_0(x_L(t), t)$. Moreover, as $\lambda = \lambda(x_L(t))$ is a root of $H\psi_0^*(\lambda; x_L(t), t) - \theta'(\lambda, t)$, we see that the second integral also vanishes at $\beta^{(2)} = \beta^{(3)} = \lambda(x_L(t))$. Hence, as the first term in (331) is negative, the proof is complete.

□

LEMMA 8.4. *At the leading edge, $x = x_L(t)$, we have*

$$\frac{\partial}{\partial \beta^{(1)}} T_0 < 0 \,,$$
$$\frac{\partial}{\partial \beta^{(2)}} T_0 = \frac{\partial}{\partial \beta^{(3)}} T_0 \quad > \quad 0 \,.$$

Proof. We have the representation for T_0, in analogy with (330),

$$(332)\ T_0 = \int_{\alpha_0(x_-^0(\beta^{(1)},t),t)}^{\alpha} \frac{i}{R(\lambda)^{1/2}} \left(\theta'(\lambda,t) - H\psi_0^*(\lambda\,; x_-(\beta^{(1)},t),t)\right) \, d\lambda$$
$$+ \int_{\beta^{(2)}}^{\beta^{(3)}} \frac{i}{R(\lambda)^{1/2}} \left(\theta'(\lambda,t) - H\psi_0^*(\lambda\,; x_-(\beta^{(1)},t),t)\right) \, d\lambda \,.$$

This formula is obtained in a similar way to (330), except with a slightly different break up of the region of x integration. We leave the details to the interested reader. Differentiating this with respect to $\beta^{(1)}$, we obtain the analog of (331),

$$\frac{\partial}{\partial \beta^{(1)}} T_0 = \frac{\pi \frac{\partial}{\partial \beta^{(1)}} x_-^0(\beta^{(1)},t)}{\left[(\beta^{(1)} - \alpha(x_-^0(\beta^{(1)},t),t))(\beta^{(1)} - \alpha)(\beta^{(1)} - \beta^{(2)})(\beta^{(1)} - \beta^{(3)})\right]^{1/2}}$$
$$+ \int_{\alpha_0(x_-^0(\beta^{(1)},t),t)}^{\alpha} \frac{i}{R(\lambda)^{1/2}} \left(\theta'(\lambda,t) - H\psi_0^*(\lambda\,; x_-(\beta^{(1)},t),t)\right) \frac{\frac{d\lambda}{2}}{\lambda - \beta^{(1)}} +$$
$$+ \int_{\beta^{(2)}}^{\beta^{(3)}} \frac{i}{R(\lambda)^{1/2}} \left(\theta'(\lambda,t) - H\psi_0^*(\lambda\,; x_-(\beta^{(1)},t),t)\right) \frac{\frac{1}{2}}{\lambda - \beta^{(1)}} \, d\lambda \,.$$

At $P_L(t)$, the first integral is zero because $\alpha_0(x_-^0(\beta^{(1)}),t) = \alpha_0(x_L(t),t) = \alpha$. The second term is zero because $\beta^{(2)} = \beta^{(3)} = \lambda(x_L(t))$ is a root of $H\psi_0^*(\lambda\,; x_L(t),t) - \theta'(\lambda,t)$. Thus we have

$$\frac{\partial T_0(P_L(t))}{\partial \beta^{(1)}} =$$
$$\frac{\pi \frac{\partial x_-^0(\beta^{(1)})}{\partial \beta^{(1)}}}{\left[(\beta^{(1)} - \alpha(x_L(t),t))(\beta^{(1)}),t))(\beta^{(1)} - \alpha)(\beta^{(1)} - \beta^{(2)})(\beta^{(1)} - \beta^{(3)})\right]^{1/2}} < 0 \,.$$

Now we consider $\frac{\partial T_0}{\partial \beta^{(3)}}$ at the leading edge. Using 332, we have

$$\frac{\partial}{\partial\beta^{(3)}}T_0 =$$

$$\int_{\alpha_0(x_-^0(\beta^{(1)},t),t)}^{\alpha} \frac{i}{R(\lambda)^{1/2}}\left(\theta'(\lambda,t) - H\psi_0^*(\lambda\,; x_-(\beta^{(1)},t),t)\right)\frac{\frac{d\lambda}{2}}{\lambda-\beta^{(3)}}$$

$$+\frac{\partial}{\partial\beta^{(3)}}\int_{\beta^{(2)}}^{\beta^{(3)}} \frac{i}{R(\lambda)^{1/2}}\left(\theta'(\lambda,t) - H\psi_0^*(\lambda\,; x_-(\beta^{(1)},t),t)\right)\, d\lambda\ .$$

The first term vanishes at $P_L(t)$ as before, so we only need to consider the limit as $\beta^{(3)},\beta^{(2)} \to \lambda(x_L(t))$ of

$$\frac{\partial}{\partial\beta^{(3)}}\int_{\beta^{(2)}}^{\beta^{(3)}} \frac{i}{R(\lambda)^{1/2}}\left(\theta'(\lambda,t) - H\psi_0^*(\lambda\,; x_-(\beta^{(1)},t),t)\right)\, d\lambda =$$

$$\frac{\partial}{\partial\beta^{(3)}}\int_{-1}^{1} \frac{\theta'(\lambda(\mu)) - H\psi_0^*(\lambda(\mu)\,; x_-(\beta^{(1)},t),t)}{\left[(\lambda(\mu)-\alpha)(\lambda(\mu)-\beta^{(1)})\right]^{1/2}}\frac{1}{\sqrt{1-\mu^2}}d\mu\ ,$$

where we have introduced

$$\lambda(\mu) = \frac{\beta^{(3)}-\beta^{(2)}}{2}\mu + \frac{\beta^{(3)}+\beta^{(2)}}{2}\ .$$

This gives us

$$\frac{\partial}{\partial\beta^{(3)}}\int_{\beta^{(2)}}^{\beta^{(3)}} \frac{i}{R(\lambda)^{1/2}}\left(\theta'(\lambda,t) - H\psi_0^*(\lambda\,; x_-(\beta^{(1)},t),t)\right)\, d\lambda =$$

$$\int_{-1}^{1}\left(\frac{\partial}{\partial(\lambda(\mu))}\left(\frac{\theta'(\lambda(\mu)) - H\psi_0^*(\lambda(\mu)\,; x_-(\beta^{(1)},t),t)}{\left[(\lambda(\mu)-\alpha)(\lambda(\mu)-\beta^{(1)})\right]^{1/2}}\right)\right)\frac{\frac{1+\mu}{2}}{\sqrt{1-\mu^2}}d\mu\ ,$$

and evaluating at $PL(t)$, we obtain

$$\frac{\partial}{\partial\beta^{(3)}}T_0 = \pi\frac{\frac{\partial}{\partial\lambda}\left(\theta'(\lambda,\ t) - H\psi_0^*(\lambda;\ x_L(t),t)\right)}{\left[(\lambda-\alpha)(\lambda-\beta^{(1)})\right]^{1/2}}\bigg|_{P_L(t)} > 0\ ,$$

where the last inequality uses the fact that at the location of failure of the zero gap ansatz, we have $\frac{\partial}{\partial\lambda}\left(H\psi_0^*(\lambda\,; x_-(\beta^{(1)},t),t) - \theta'(\lambda,t)\right) < 0$, established in Appendix II, Claim 2. The calculations for $\frac{\partial T_0}{\partial\beta^{(2)}}$ are similar.

This completes the proof of the theorem. □

Thus we have established that the Jacobian (327) is nonzero at the leading edge, and hence we learn that for t small and positive, and for $\beta^{(2)}$ near $\lambda(x_L(t))$, we can determine $(\alpha,\beta^{(1)},\beta^{(3)})$ as C^1 functions of $(\beta^{(2)},t)$ which solve

(333) $$T_0 = 0\ ,$$

(334) $$T_1 = 0\ ,$$

(335) $$N = 0\ ,$$

and such that

$$\begin{aligned} \alpha(\beta^{(2)},t) &\to \alpha_0(x_L(t),t)\ , \\ \beta^{(1)}(\beta^{(2)},t) &\to \beta_0(x_L(t),t)\ , \\ \beta^{(3)}(\beta^{(2)},t) &\to \lambda(x_L(t))\ , \end{aligned}$$

as $\beta^{(2)} \to \lambda(x_L(t))$.

LEMMA 8.5. *Set*

$$T_2(\beta^{(2)},t) = T_2(\alpha(\beta^{(2)},t),\beta^{(1)}(\beta^{(2)},t),\beta^{(2)},\beta^{(3)}(\beta^{(2)},t))\ .$$

Then for $\beta^{(2)}$ near $\lambda(x_L(t))$, we have

$$\begin{aligned} &\frac{\partial}{\partial\beta^{(2)}}T_2(\beta^{(2)}) < 0\quad,\quad for\beta^{(2)} < \lambda(x_L(t))\ , \\ &\frac{\partial}{\partial\beta^{(2)}}T_2(\beta^{(2)}) > 0\quad,\quad for\beta^{(2)} > \lambda(x_L(t))\ . \end{aligned} \tag{336}$$

Proof. By Remark 8.1,

$$\begin{aligned} \frac{\partial}{\partial\beta^{(2)}}X(\beta^{(2)}) = \alpha^2\frac{\partial T_0}{\partial\alpha}\frac{\partial\alpha}{\partial\beta^{(2)}} \quad + \quad (\beta^{(1)})^2\frac{\partial T_0}{\partial\beta^{(1)}}\frac{\partial\beta^{(1)}}{\partial\beta^{(2)}} \quad + \\ (\beta^{(2)})^2\frac{\partial T_0}{\partial\beta^{(2)}} \quad + \quad (\beta^{(3)})^2\frac{\partial T_0}{\partial\beta^{(3)}}\frac{\partial\beta^{(3)}}{\partial\beta^{(2)}}\quad . \end{aligned} \tag{337}$$

Now because we are solving (333 - 335), we have

$$\begin{pmatrix} 1 & 1 & 1 \\ \alpha & \beta^{(1)} & \beta^{(3)} \\ g_0 & g_1 & g_3 \end{pmatrix} \begin{pmatrix} \frac{\partial T_0}{\partial\alpha} & 0 & 0 \\ 0 & \frac{\partial T_0}{\partial\beta^{(1)}} & 0 \\ 0 & 0 & \frac{\partial T_0}{\partial\beta^{(3)}} \end{pmatrix} \frac{\partial}{\partial\beta^{(2)}} \begin{pmatrix} \alpha \\ \beta^{(1)} \\ \beta^{(3)} \end{pmatrix} = -\frac{\partial T_0}{\partial\beta^{(2)}} \begin{pmatrix} 1 \\ \beta^{(2)} \\ g_2 \end{pmatrix} \tag{338}$$

And so, using Kramer's rule, and 337, we obtain

$$\frac{\partial}{\partial\beta^{(2)}}T_2(\beta^{(2)}) =$$

$$-\frac{\partial T_0}{\partial\beta^{(2)}}\frac{\left(\alpha^2\begin{vmatrix}1&1&1\\\beta^{(2)}&\beta^{(1)}&\beta^{(3)}\\g_2&g_1&g_3\end{vmatrix}+(\beta^{(1)})^2\begin{vmatrix}1&1&1\\\alpha&\beta^{(2)}&\beta^{(3)}\\g_0&g_2&g_3\end{vmatrix}-(\beta^{(2)})^2\begin{vmatrix}1&1&1\\\alpha&\beta^{(1)}&\beta^{(3)}\\g_0&g_1&g_3\end{vmatrix}+(\beta^{(3)})^2\begin{vmatrix}1&1&1\\\alpha&\beta^{(1)}&\beta^{(2)}\\g_0&g_1&g_2\end{vmatrix}\right)}{\begin{vmatrix}1&1&1\\\alpha&\beta^{(1)}&\beta^{(3)}\\g_2&g_1&g_3\end{vmatrix}}$$

$$= \frac{\partial T_0}{\partial\beta^{(2)}}\frac{\begin{vmatrix}1&1&1&1\\\alpha&\beta^{(1)}&\beta^{(2)}&\beta^{(3)}\\\alpha^2&(\beta^{(1)})^2&(\beta^{(2)})^2&(\beta^{(3)})^2\\g_0&g_1&g_2&g_3\end{vmatrix}}{\begin{vmatrix}1&1&1\\\alpha&\beta^{(1)}&\beta^{(3)}\\g_0&g_1&g_3\end{vmatrix}}$$

$$(339) = (\beta^{(3)}-\beta^{(2)})(\beta^{(2)}-\alpha)(\beta^{(2)}-\beta^{(1)})\frac{\partial}{\partial\beta^{(2)}}T_0 \times \frac{\int_{\beta^{(1)}}^{\beta^{(2)}} dz R(z)^{-1/2}}{\int_{\beta^{(1)}}^{\beta^{(2)}} dz R(z)^{1/2}\frac{1}{\prod_{j\neq 2}(z-\beta^{(j)})}},$$

which shows that $\frac{\partial}{\partial\beta^{(2)}}T_2(\beta^{(2)},t)$ has the opposite sign to $\beta^{(3)}-\beta^{(2)}$. On the other hand, solving (338) for $\frac{\partial\beta^{(3)}}{\partial\beta^{(2)}}$, at $\beta^{(2)}=\lambda(x_L(t))$, we find

$$\frac{\partial\beta^{(3)}}{\partial\beta^{(2)}} = -\left.\frac{\frac{\partial T_0}{\partial\beta^{(2)}}}{\frac{\partial T_0}{\partial\beta^{(3)}}}\right|_{P_L(t)} = -1$$

by Lemma 8.4. This proves the Lemma. □

Thus decreasing $\beta^{(2)}$ away from $\lambda(x_L(t))$ increases X, which implies that we can solve **(305 - 308)** for x slightly to the right of $x_L(t)$.

•4 Before beginning the proof that the function $\psi(\lambda;\, x,t)$ constructed through (302), (303) out of $\alpha(x,t),\beta^{(j)}(x,t)$ solves the maximization problem, we note that just as the shape of the functions $\beta(x,t),\alpha(x,t)$ was essential in the proof that the zero gap ansatz solved the maximization problem before shock time, the shape of the curves $\alpha(x,t),\beta^{(j)}(x,t)$ also plays an important role here. So at this point we present the following description of the curves $\alpha,\beta^{(j)}(x,t)$.

THEOREM 8.4. *For $x > x_L(t)$, but still close, we have*

$$\alpha_x < 0 , \tag{340}$$

$$\beta_x^{(1)} > 0 , \tag{341}$$

$$\beta_x^{(2)} < 0 , \tag{342}$$

$$\beta_x^{(3)} > 0 . \tag{343}$$

Proof.
Using (305 - 308), we obtain equations for the quantities $\alpha_x, \beta_x^{(j)}$,

$$\begin{pmatrix} 1 & 1 & 1 & 1 \\ \alpha & \beta^{(1)} & \beta^{(2)} & \beta^{(3)} \\ \alpha^2 & (\beta^{(1)})^2 & (\beta^{(2)})^2 & (\beta^{(3)})^2 \\ g_0 & g_1 & g_2 & g_3 \end{pmatrix} \times$$

$$\begin{pmatrix} \frac{\partial}{\partial\alpha}T_0 & 0 & 0 & 0 \\ 0 & \frac{\partial}{\partial\beta^{(1)}}T_0 & 0 & 0 \\ 0 & 0 & \frac{\partial}{\partial\beta^{(2)}}T_0 & 0 \\ 0 & 0 & 0 & \frac{\partial}{\partial\beta^{(3)}}T_0 \end{pmatrix} \frac{\partial}{\partial x} \begin{pmatrix} \alpha \\ \beta^{(1)} \\ \beta^{(2)} \\ \beta^{(3)} \end{pmatrix}$$

$$= \pi \begin{pmatrix} 0 \\ 0 \\ 1 \\ 0 \end{pmatrix} ,$$

The solution is,

$$\frac{\partial T_0}{\partial \alpha}\alpha_x = \pi \frac{\begin{vmatrix} 0 & 1 & 1 & 1 \\ 0 & \beta^{(1)} & \beta^{(2)} & \beta^{(3)} \\ 1 & (\beta^{(1)})^2 & (\beta^{(2)})^2 & (\beta^{(3)})^2 \\ 0 & g_1 & g_2 & g_3 \end{vmatrix}}{\begin{vmatrix} 1 & 1 & 1 & 1 \\ \alpha & \beta^{(1)} & \beta^{(2)} & \beta^{(3)} \\ \alpha^2 & (\beta^{(1)})^2 & (\beta^{(2)})^2 & (\beta^{(3)})^2 \\ g_0 & g_1 & g_2 & g_3 \end{vmatrix}}, \tag{344}$$

$$\frac{\partial T_0}{\partial \beta^{(1)}}\beta_x^{(1)} = \pi \frac{\begin{vmatrix} 1 & 0 & 1 & 1 \\ \alpha & 0 & \beta^{(2)} & \beta^{(3)} \\ \alpha^2 & 1 & (\beta^{(2)})^2 & (\beta^{(3)})^2 \\ g_0 & 0 & g_2 & g_3 \end{vmatrix}}{\begin{vmatrix} 1 & 1 & 1 & 1 \\ \alpha & \beta^{(1)} & \beta^{(2)} & \beta^{(3)} \\ \alpha^2 & (\beta^{(1)})^2 & (\beta^{(2)})^2 & (\beta^{(3)})^2 \\ g_0 & g_1 & g_2 & g_3 \end{vmatrix}}, \tag{345}$$

$$\frac{\partial T_0}{\partial \beta^{(2)}}\beta_x^{(2)} = \pi \frac{\begin{vmatrix} 1 & 1 & 0 & 1 \\ \alpha & \beta^{(1)} & 0 & \beta^{(3)} \\ \alpha^2 & (\beta^{(1)})^2 & 1 & (\beta^{(3)})^2 \\ g_0 & g_1 & 0 & g_3 \end{vmatrix}}{\begin{vmatrix} 1 & 1 & 1 & 1 \\ \alpha & \beta^{(1)} & \beta^{(2)} & \beta^{(3)} \\ \alpha^2 & (\beta^{(1)})^2 & (\beta^{(2)})^2 & (\beta^{(3)})^2 \\ g_0 & g_1 & g_2 & g_3 \end{vmatrix}}, \tag{346}$$

$$\frac{\partial T_0}{\partial \beta^{(3)}}\beta_x^{(3)} = \pi \frac{\begin{vmatrix} 1 & 1 & 1 & 0 \\ \alpha & \beta^{(1)} & \beta^{(2)} & 0 \\ \alpha^2 & (\beta^{(1)})^2 & (\beta^{(2)})^2 & 1 \\ g_0 & g_1 & g_2 & 0 \end{vmatrix}}{\begin{vmatrix} 1 & 1 & 1 & 1 \\ \alpha & \beta^{(1)} & \beta^{(2)} & \beta^{(3)} \\ \alpha^2 & (\beta^{(1)})^2 & (\beta^{(2)})^2 & (\beta^{(3)})^2 \\ g_0 & g_1 & g_2 & g_3 \end{vmatrix}}. \tag{347}$$

All of these determinants can be calculated explicitly, and we leave verification of the inequalities (340 - 343) to the reader. □

Formulae (344 - 347) immediately imply the following result.

COROLLARY 8.5. *As $x \downarrow x_L(t)$, α_x, $\beta_x^{(1)}$ converge, but $\beta_x^{(2)}$, $\beta_x^{(3)}$ blow up.*

The results in Theorem 8.4 and 8.5 are summarized in figure (1).

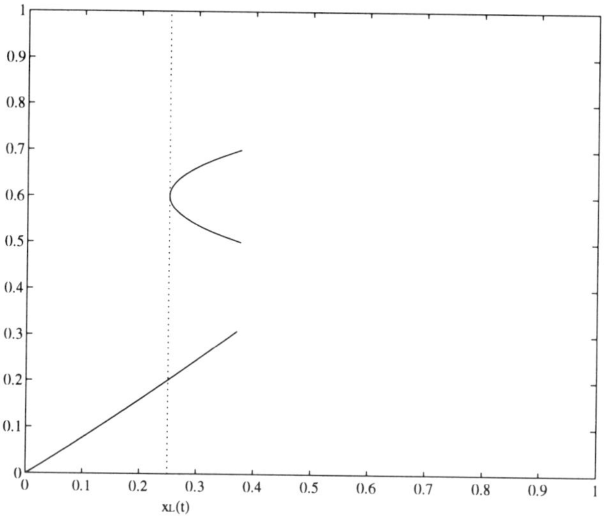

FIGURE 1. The leading edge $x_L(t)$, and the curves $\beta^{(j)}(x,t)$ in a neighborhood to the right of $x_L(t)$. To the left of $x_L(t)$, the curve is $\beta_0(x,t)$.

We can obtain similar conditions to (344-347) for the quantities $\frac{\partial}{\partial t}\alpha$, $\frac{\partial}{\partial t}\beta^{(j)}$. For example, we find

$$\frac{\partial T_0}{\partial \alpha}\alpha_t = \pi \frac{\begin{vmatrix} 0 & 1 & 1 & 1 \\ 1 & \beta^{(1)} & \beta^{(2)} & \beta^{(3)} \\ \frac{\sigma}{2} & (\beta^{(1)})^2 & (\beta^{(2)})^2 & (\beta^{(3)})^2 \\ 0 & g_1 & g_2 & g_3 \end{vmatrix}}{\begin{vmatrix} 1 & 1 & 1 & 1 \\ \alpha & \beta^{(1)} & \beta^{(2)} & \beta^{(3)} \\ \alpha^2 & (\beta^{(1)})^2 & (\beta^{(2)})^2 & (\beta^{(3)})^2 \\ g_0 & g_1 & g_2 & g_3 \end{vmatrix}} . \tag{348}$$

Equations (344) and (348) enable us to write a PDE for $\alpha(x,t)$ that does not contain the function T_0,

$$\alpha_t = \frac{\begin{vmatrix} 0 & 1 & 1 & 1 \\ 1 & \beta^{(1)} & \beta^{(2)} & \beta^{(3)} \\ \frac{\sigma}{2} & (\beta^{(1)})^2 & (\beta^{(2)})^2 & (\beta^{(3)})^2 \\ 0 & g_1 & g_2 & g_3 \end{vmatrix}}{\begin{vmatrix} 0 & 1 & 1 & 1 \\ 0 & \beta^{(1)} & \beta^{(2)} & \beta^{(3)} \\ 1 & (\beta^{(1)})^2 & (\beta^{(2)})^2 & (\beta^{(3)})^2 \\ 0 & g_1 & g_2 & g_3 \end{vmatrix}} \alpha_x .$$

The same procedure can be applied to all of the $\beta^{(j)}$, and further, the determinants can be calculated in terms of completely elliptic integrals, and one obtains,

$$\frac{\partial}{\partial t}\beta^{(j)} = \left[\frac{\int_{\beta^{(1)}}^{\beta^{(2)}} \frac{(\mu-\beta^{(j)})\,\mu}{R(\mu)^{1/2}} d\mu}{\int_{\beta^{(1)}}^{\beta^{(2)}} \frac{(\mu-\beta^{(j)})}{R(\mu)^{1/2}} d\mu} + \frac{1}{2}\left(\beta^{(j)} - \sum_{k\neq j} \beta_k\right)\right] \frac{\partial}{\partial x}\beta^{(j)} . \tag{349}$$

which are the Whitham averaged equations [**23**].

A second preliminary step that is needed before beginning the proof of (350-353) is the evaluation of $\psi_x, H\psi_x$.

Since $\alpha(x,t)$ and $\beta^{(j)}(x,t)$ are continuously differentiable functions of x, $x \neq x_L(t)$, we can compute the Riemann - Hilbert problem satisfied by the differentiated quantity $\psi_x^*(\lambda; x, t)$,

$$\begin{aligned} H\psi_x &= 0 \quad , \quad \lambda \in \ (\alpha, \beta^{(1)}) \cup (\beta^{(2)}, \beta^{(3)}) \ , \\ \psi_x &= 0 \quad , \quad \lambda \notin \ (\alpha, \beta^{(1)}) \cup (\beta^{(2)}, \beta^{(3)}) \ . \end{aligned}$$

We leave it to the reader to verify that the unique function which satisfies the differentiated Riemann - Hilbert problem and the differentiated constraints

$$\left(\int_{\alpha}^{\beta^{(1)}} + \int_{\beta^{(2)}}^{\beta^{(3)}}\right) \psi_x \, d\lambda = \pi \ ,$$

$$\int_{\beta^{(1)}}^{\beta^{(2)}} H\psi_x = 0 \ ,$$

and which has at worst a square root singularity at $\alpha, \beta^{(j)}$, is given by

$$f_x = \psi_x + i\, H\psi_x = i\frac{\lambda - d}{R(\lambda; y, t)^{1/2}} \ ,$$

$$R(\lambda; y, t) = (\lambda - \alpha(y,t)) \prod_{j=1}^{3} (\lambda - \beta^{(j)}) \ ,$$

where d is chosen such that

$$\int_{\beta^{(1)}}^{\beta^{(2)}} H\psi_x = 0 \ .$$

Clearly $d \in (\ \beta^{(1)}\ ,\ \beta^{(2)}\ )$.

Using the solution of the differentiated problem, and Theorem 8.4, we now prove the following result.

THEOREM 8.6. *For $x_L(t) \leq x$, but close to $x_L(t)$, and for small $t > 0$, the function $\psi(\lambda;\ x, t)$ defined by (302 - 303) solves the maximization problem (41).*

Proof.

The arguments of the previous chapters used to obtain the maximization conditions apply here, so we merely state that we must show

$$(350)\qquad \lambda > \beta^{(3)} \Rightarrow \int_{\beta^{(3)}}^{\lambda} H\psi - \theta' d\mu \;<\; 0 \;,$$

$$(351)\qquad \lambda \in (\beta^{(1)}\,,\,\beta^{(2)}) \Rightarrow \int_{\beta^{(1)}}^{\lambda} H\psi - \theta' d\mu \;<\; 0 \;,$$

$$(352)\qquad \lambda < \alpha \Rightarrow \int_{\alpha}^{\lambda} H\psi - \theta' d\mu \;<\; 0 \;,$$

$$(353)\qquad 0 \;\leq\; \psi \;\leq\; \phi \;.$$

We define $\tilde{x}(\lambda)$ to be the (unique) point of intersection ,if it exists, of the curves $\alpha(x,t), \beta^{(j)}(x,t)$ with λ.

Now fix $x > x_L(t)$, but close to $x_L(t)$. For λ close to (but bigger than) $\beta^{(3)}(x,t)$, it is clear that there is an intersection of $\beta^{(3)}(y,t)$ with the horizontal line of height λ. We thus can write

$$H\psi(\lambda\;;\;x,t) \;=\; H\psi(\lambda;\;\tilde{x}(\lambda),t) + \int_{\tilde{x}(\lambda)}^{x} \frac{\lambda - d}{R(\lambda;y,t)^{1/2}} dy \;,$$

But from the Riemann - Hilbert problem (300),

$$H\psi(\lambda;\;\tilde{x}(\lambda),t) = \theta'(\lambda,t)\;,$$

from which we see that for x close to $x_L(t)$,

$$(354)\qquad H\psi(\lambda\;;\;x,t) - \theta'(\lambda,t) \;=\; -\int_{x}^{\tilde{x}(\lambda)} \frac{\lambda - d}{R^{1/2}} dy \;,$$

which is negative, by the shape of the curves in Figure (1).

For $\lambda > \beta^{(3)}(x,t)$, but further away, the argument is as follows.

We write

$$H\psi(\lambda\;;\;x,t) \;=\; H\psi(\lambda;x_L(t),t) \;+\; \int_{x_L(t)}^{x} \frac{\lambda - d}{R^{1/2}} dy \;.$$

As $H\psi(\lambda;\;x_L(t),t) = H\psi_0^*(\lambda;\;x_L(t),t)$ (see proof of Theorem 8.1), we have

$$H\psi(\lambda\;;\;x,t) - \theta'(\lambda,t) \;=\; H\psi_0^*(\lambda\;;x_L(t),t) - \theta'(\lambda,t) + \int_{x_L(t)}^{x} \frac{\lambda - d}{R^{1/2}} dy \;.$$

Now we have shown in Chapter 7 that $H\psi_0^*(\lambda;x_L(t),t) - \theta'(\lambda,t)$ has no root for $\lambda \;>\; \lambda(x_L(t))$. We fix $\delta > 0$ such that $\beta^{(3)}(x,t)$ exists for $x_L(t) \leq x \leq x_L(t) + \delta$. We will consider all $x \in (x_L(t), x_L(t) + \eta)$, for some $\eta < \delta/2$. Clearly, from the above argument, we have that for each $x \in (x_L(t), x_L(t) + \eta)$,

$$H\psi(\lambda;\;x,t) - \theta'(\lambda;\;x,t) < 0\;,$$

for all $\lambda \in (\beta^{(3)}(x,t), \beta^{(3)}(x_L(t)+\delta,t))$. Furthermore, for $\lambda > \beta^{(3)}(x_L(t)+\delta,t)$, we have

$$H\psi_0^*(\lambda; x_L(t),t) - \theta'(\lambda,t) < c < 0 \ ,$$

and so, as $\int_{x_L(t)}^{x} \frac{\lambda-d}{R(\lambda;y,t)^{1/2}}dy$ is continuous in x, we see that for $x \in (x_L(t), x_L(t)+\eta)$,

$$H\psi(\lambda; xt) - \theta'(\lambda,t) < 0 \ \ ,$$

for all $\lambda > \beta^{(3)}(x,t)$, provided $0 < \eta < \frac{\delta}{2}$ is chosen suitably small. This proves (350). The inequality (352) follows from similar arguments.

From the Riemann - Hilbert problem (300), (301), we see that $\psi = 0$ for $\lambda \notin [\alpha, \beta^{(1)}] \cup [\beta^{(2)}, \beta^{(3)}]$, and so we must only show (353) for $\lambda \in (\alpha, \beta^{(1)}) \cup (\beta^{(2)}, \beta^{(3)})$. We consider $\lambda \in (\beta^{(2)}, \beta^{(3)})$. It is evident from Figure (1) that there is an intersection of λ with $\beta^{(2)}(x,t)$ or $\beta^{(3)}(x,t)$. Thus we can write

$$\psi(\lambda; x,t) = \psi(\lambda, \tilde{x}(\lambda), t) + \int_{\tilde{x}(\lambda)}^{x} i\frac{\lambda-d}{R(\lambda;y,t)^{1/2}}dy \ ,$$

and from the Riemann - Hilbert problem (300), (301), we have $\psi(\lambda; \tilde{x}(\lambda), t) = 0$, and hence

$$\psi(\lambda; x,t) = \int_{\tilde{x}(\lambda)}^{x} i\frac{\lambda-d}{R(\lambda;y,t)^{1/2}}dy \ > 0 \ ,$$

by the shape of the curves in Figure (1). Similar considerations verify (353) for $\lambda \in (\alpha(x,t), \beta^{(1)}(x,t))$.

We now consider (351).

By using the shape of the curves and the differentiated problem, we can show that

$$H\psi(\lambda; x,t) - \theta'(\lambda,t) \ = \ \int_{\tilde{x}(\lambda)}^{x} \frac{\lambda-d}{R^{1/2}}dy \ \ ,$$

provided λ is close to $\beta^{(2)}$, and $\beta^{(1)} < \lambda < \beta^{(2)}$. Thus, using (308),

$$\begin{aligned}\int_{\beta^{(1)}}^{\lambda} H\psi(\mu; x,t) - \theta'(\mu,t)\ d\mu &= -\int_{\lambda}^{\beta^{(2)}} H\psi(\mu; x,t) - \theta'(\mu,t)\ d\mu \\ &= -\int_{\lambda}^{\beta^{(2)}} \int_{\tilde{x}(\mu)}^{x} \frac{\mu-d}{R^{1/2}}\ dy\ d\mu \\ &= -\int_{\tilde{x}(\lambda)}^{x} \int_{\lambda}^{\beta^{(2)}(y,t)} \frac{\mu-d}{R^{1/2}}\ d\mu\ dy \ < \ 0 \ .\end{aligned}$$

Similarly, if λ is close to $\beta^{(1)}(x,t)$, we have

$$\int_{\beta^{(1)}}^{\lambda} H\psi(\mu;x,t)-\theta'(\mu,t)\,d\mu = \int_{\beta^{(1)}}^{\lambda}\int_{x_-(\mu)}^{x} \frac{\mu-d}{R^{1/2}}\,dy\,d\mu$$

$$= \int_{x_-(\lambda)}^{x}\int_{\beta^{(1)}(y,t)}^{\lambda} \frac{\mu-d}{R^{1/2}}\,d\mu\,dy \;<\; 0\;. \tag{355}$$

So condition (351) is not violated for λ in the interval $(\beta^{(1)}(x,t),\beta^{(2)}(x,t))$ and close to $\beta^{(1)}(x,t)$ or $\beta^{(2)}(x,t)$.

On the other hand, we have shown in Chapter 7 that at $x=x_L(t)$, there is precisely one root of

$$\int_{\beta^{(1)}}^{\lambda} H\psi_0^*(\mu\;;x_L(t),t)-\theta'(\mu,t)\;,$$

at $\lambda=\lambda(x_L(t))$. Thus, for δ small and positive and for x close to $x_L(t)$, we have for, $\lambda\in(\beta^{(1)}(x,t)+\delta\,,\,\beta^{(2)}(x,t)-\delta)\subset(\beta^{(1)}(x_L(t),t)\,,\,\beta^{(2)}(x_L(t),t))$,

$$\int_{\beta^{(1)}(x_L(t),t)}^{\lambda} H\psi(\mu;x_L(t),t)-\theta'(\mu,t)\;<\;\epsilon\;<0\;, \tag{356}$$

for some $\epsilon(\delta)>0$. But then by continuity

$$\int_{\beta^{(1)}(x,t)}^{\lambda} H\psi(\mu;x,t)-\theta'(\mu,t)d\mu<\epsilon/2$$

for x in a small neighborhood to the right of $x_L(t)$, and for $\lambda\in(\beta^{(1)}(x,t)+\delta,\beta^{(2)}(x,t)-\delta)$. Clearly (355) and (356) together prove (351).

This shows that for some small interval $(x_L(t),x_L(t)+\eta_1)$, the condition (351) does not fail.

□

We have thus shown that the continuum limit of the Toda lattice is governed by a "one gap Riemann - Hilbert problem" on an interval $(x_L(t),$ $x_L(t)+\eta_1)$ where $\eta_1=\eta_1(t)$ is small.

Thus we have, finally, established the following result.

THEOREM 8.7. *For $0<x<x_L(t)$, we have*

$$w-\lim_{\epsilon\to0}\hat{b}_\epsilon^2 \;=\; \frac{1}{4}Q_{tt} \;=\; \left(\frac{\beta(x,t)-\alpha(x,t)}{4}\right)^2,$$

and similarly,

$$w-\lim_{\epsilon\to0}\hat{a}_\epsilon \;=\; \frac{1}{2}Q_{xt} \;=\; \frac{\beta(x,t)+\alpha(x,t)}{2}\;,$$

For $x_L(t) < x < x_L(t) + \delta$, we have

$$
\begin{aligned}
w - \lim_{\epsilon \to 0} \hat{b}_\epsilon^2 &= \frac{1}{4} Q_{tt} \\
&= \frac{1}{2}\left[c - \frac{1}{8}\left(\alpha + \sum_{j=1}^{3} \beta^{(j)}\right)^2 + \frac{1}{4}\left(\alpha^2 + \sum_{j=1}^{3}\left(\beta^{(j)}\right)^2\right)\right] ,
\end{aligned}
$$

where c is defined by

$$
\int_{\beta^{(1)}}^{\beta^{(2)}} \frac{\lambda^2 - \frac{\alpha + \sum_{j=1}^{3} \beta^{(j)}}{2}\lambda + c}{R(\lambda)^{1/2}} d\lambda = 0 .
$$

Similarly,

$$
\begin{aligned}
w - \lim_{\epsilon \to 0} \hat{a}_\epsilon &= \frac{1}{2} Q_{xt} \\
&= \frac{1}{2}\left(\alpha + \sum_{j=1}^{3} \beta^{(j)}\right) - d ,
\end{aligned}
$$

where d is defined by

$$
\int_{\beta^{(1)}}^{\beta^{(2)}} \frac{\lambda - d}{R(\lambda)^{1/2}} d\lambda = 0 .
$$

The above formulae are derived following similar calculations to those presented on page 13, using the explicit knowledge of the quantities $\psi_x(\lambda;\, x, t)$ and $\psi_t(\lambda;\, x, t)$.

REMARK 8.2.

If the first point $x = x^{\#}$ at which the Jacobian vanishes is a point at which $\beta^{(1)}$ intersects $\beta^{(2)}$, then we can continue to solve (305 - 308) all the way from $x_L(t)$ to $x^{\#}$, which must then coincide with the "trailing edge" $x = x_R(t)$ of the region where the zero gap ansatz fails to solve the maximization problem . We now show that in this case, we can employ the differentiated Riemann - Hilbert problem and conclude that the 1-gap ansatz solves the maximization conditions all the way across the interval $(x_L(t), x_R(t))$.

Indeed, if the above is true, we can write

$$
H\psi(\lambda; x, t) - \theta'(\lambda, t) = \int_{\tilde{x}(\lambda)}^{x} \frac{\lambda - d}{R^{1/2}} dy ,
$$

for every $\lambda \in (\beta^{(1)} , \beta^{(2)})$, and so we have

$$
\int_{\beta^{(1)}}^{\lambda} H\psi(\mu; x, t) - \theta'(\mu, t) d\mu = \int_{\tilde{x}(\lambda)}^{x} \int_{\beta^{(1)}(y,t)}^{\lambda} \frac{\mu - d}{R^{1/2}} d\mu \, dy .
$$

The inner integral is positive. This follows because

$$\int_{\beta^{(1)}(y,t)}^{\lambda} \frac{\mu - d}{R^{1/2}} \, d\mu \Big|_{\lambda=\beta^{(2)}(y,t)} = 0 ,$$

$\frac{\mu-d}{R^{1/2}}$ changes sign precisely once in $(\beta^{(1)}(y,t), \beta^{(2)}(y,t))$, and

$$\int_{\beta^{(1)}(y,t)}^{\lambda} \frac{\mu - d}{R^{1/2}} \, d\mu$$

is clearly positive for λ near $\beta^{(1)}(y,t)$.

The condition

$$\int_{\beta^{(1)}}^{\lambda} H\psi(\mu; x, t) - \theta'(\mu, t) \;\, d\mu \;\; < \;\; 0 \;\; , \; for \; \lambda \in (\beta^{(1)}, \beta^{(2)})$$

is therefore satisfied, as $x < \tilde{x}(\lambda)$. The other conditions in Theorem 8.6 follow in similar fashion.

On the other hand if $\beta^{(1)}(x^{\#}, t) < \beta^{(2)}(x^{\#}, t)$, where $x^{\#}$ is the first point at which the Jacobian vanishes, then we expect another band to open up in the Riemann - Hilbert problem.

In the next Chapter we show that for a different, and more restrictive class of data $\beta^{(1)}(x^{\#}, t) = \beta^{(2)}(x^{\#}, t)$, as described above, and hence for $t > 0$, the continuum limit is completely desribed, for all $0 \leq x \leq 1$, by zero and one gap situations.

In Chapter 10 we prove that the same is true for the more general class of data considered in this chapter.

CHAPTER 9

Symmetric data

In this chapter we consider the initial $\alpha^0(x), \beta^0(x)$ with the same properties (see Remark 9.1 below) as in Chapter 8, but in addition we asume that

- In a neighborhood of x_s,

$$(\beta^0(x))^2 = \beta_s^2 + (x - x_s)^{1/3} . \tag{357}$$

- For some (small) $\delta > 0$, and for $x \in [0, x_s + \delta]$, the data is symmetric, i.e.

$$-\alpha^0(y) = \beta^0(y) .$$

Our goal in this section is to study the one gap R-H problem (300), (301), as described in Chapter 8, and show that for the restricted initial data the quantities

$$T_{0\beta^{(j)}}, \quad 0 \le j \le 3 ,$$

do not vanish on a solution of

$$T_0 = 0 , \tag{358}$$

$$T_1 = 0 , \tag{359}$$

$$T_2 = \pi x , \tag{360}$$

$$N = 0 , \tag{361}$$

where T_j and N are defined in (305 - 308), provided $\beta^{(j)} \neq \beta^{(k)}$.

This clearly has the implication (see formula (327)) that if we establish the existence of a solution to (358 - 361) at the leading edge, for example, and if we show that we can solve for $(\beta^{(0)}, \beta^{(1)}, \beta^{(3)})$ in terms of $\beta^{(2)}$ such that

$$T_0 = 0 ,$$
$$T_1 = 0 ,$$
$$N = 0 ,$$

for $\{\beta^{(j)}\}$ near the leading edge, then we can solve the full system for $\{\beta^{(j)}(x,t)\}$ for all $x \ge x_L(t)$, until an x is reached for which $\beta^{(2)}(x,t) = \beta^{(1)}(x,t)$. This then proves that, for the special symmetric data above, the continuum limit of the Toda equations is described by zero and one gap solutions of the Riemann - Hilbert problems (c.f. Theorem 8.6) for **all** $0 \le x \le 1$.

REMARK 9.1.

For the data in Chapter 8, $\alpha^0(x)$ was assumed to be smooth at x_s, but here $\alpha^0(x)$ blows up as $x \to x_s$. This leads to additional technical difficulties. Nevertheless, as the reader may check, the following facts remain true for the initial data described above.

- For $t = 0$, the zero gap ansatz (255), analyzed in Chapter 7 reconstructs the initial data.
- For small $t > 0$, there exists 3 solutions $\{\alpha_j(x,t), \beta_j(x,t)\}$ to the zero gap conditions (258), (259).
- The inverse function $X(\beta, t)$ of the curves $\beta_j(x,t)$ has two critical values $\beta_1(t), \beta_2(t)$, where $\partial_\beta X(\beta, t) = 0$, and hence the general shape of the $\beta_j(x,t)$ obtained in Chapter 6 does not change.
- The quantities $\alpha_j(x,t)$ remain monotone decreasing functions of x, as described in Chapter 6.
- For small $t > 0$, $\alpha_{jx}(x,t)$ are continuous and negative on their respective intervals of definition, and exhibit the same cusp behavior at

$$x_1(t) = \frac{1}{\pi} X(\beta_1(t), t) \ , \qquad x_2(t) = \frac{1}{\pi} X(\beta_2(t), t) \ ,$$

as before.

However, the fact that $\alpha_x^0 \to \infty$ as $x \to x_s$ invalidates the proof (see Theorem 6.4) that, for all β close to β_s, $\partial_\beta \alpha_x(\beta, t) > 0$. However, for β close to $\beta_1(t)$or $\beta_2(t)$, the proof that $\partial_\beta \alpha_x(\beta, t) > 0$ does go through and hence, for x approaching $x_1(t)$ or $x_2(t)$, the shapes established in Chapter 6 are correct. On the other hand, the knowledge of the ordering, in particular $\alpha_1(x,t) \le \alpha_0(x,t)$ (see Figure (3)) for **all** $x \in (x_1(t), x_2(t))$, is now lost.

The above properties are enough to show that all the results of Chapter 7 and 8 remain true for the above data. But, in addition, we have the following result.

THEOREM 9.1. *For small $t > 0$, and the initial data described above, if $-\alpha < \beta^{(1)} < \beta^{(2)} < \beta^{(3)}$ are close to β_s, then we have*

$$\frac{\partial}{\partial \beta^{(j)}} T_0 = 0 \ \Rightarrow \ N \neq 0 \ .$$

Proof.

Setting $t = 0$ in (330), we obtain

$$(362)\, T_0(\alpha, \beta^{(1)}, \beta^{(2)}, \beta^{(3)}) = \left(\int_{\beta^0(x_-^{x^0}(\alpha))}^{\beta^{(1)}} + \int_{\beta^{(2)}}^{\beta^{(3)}} \right) \frac{i}{R(\lambda)^{1/2}} \cdot$$
$$\cdot \left(\theta'(\lambda) - H\psi_0^*(\lambda \ ; \ x_-^0(\alpha), 0)\right) d\lambda \ ,$$

where

$$R(\lambda) = (\lambda - \alpha)(\lambda - \beta^{(1)})(\lambda - \beta^{(2)})(\lambda - \beta^{(3)}) \ ,$$
$$R(\lambda)^{1/2} > 0 \ \ for \ \ \lambda > \beta^{(3)} \ ,$$

and from the proof of Theorem 5.1, we have

$$(363)\ \theta'(\lambda) - H\psi_0^*(\lambda\ ;\ x_-^0(\alpha), 0) = \int_{x_-^0(\alpha)}^{x_-^0(\lambda)} \frac{dy}{R^0(\lambda, y)^{1/2}}\ ,\ \lambda > \beta^0(x_-^0(\alpha))\ ,$$
$$R^0(\lambda, y) = (\lambda - \alpha^0(y))(\lambda - \beta^0(y))\ ,$$
$$R^0(\lambda, y)^{1/2} > 0\ \ for\ \ \lambda > \beta^0(y)\ .$$

The first step in the proof is to simplify formula (363). We have, using $\alpha^0(y) = -\beta^0(y)$, and the explicit form (357) for $\beta^0(y)$,

$$\theta'(\lambda) - H\psi_0^*(\lambda\ ;\ x_-^0(\alpha), 0) = \int_{x_-(\alpha)}^{x_-(\lambda)} \frac{1}{(\lambda^2 - \beta^0(y)^2)^2}\, dy$$
$$= \int_{(\beta^0(x_-(\alpha)))^2}^{\lambda^2} \frac{\partial_\mu \left((\beta^0)^2 \right)^{-1}}{(\lambda^2 - \mu)^{1/2}}\, d\mu$$
$$= \int_{-1}^{1} \frac{\partial_\mu \left((\beta^0)^2 \right)^{-1} \circ \left(\lambda^2 \left(\frac{1+x}{2}\right) + \beta^0(x_-(\alpha))^2 \left(\frac{1-x}{2}\right)\right)}{\sqrt{1-x}} \times \left(\frac{(\lambda - \alpha)(\lambda - \beta^0(x_-(\alpha)))}{2} \right)^{1/2} dx$$
$$= \left(\frac{(\lambda - \alpha)(\lambda - \beta^0(x_-(\alpha)))}{2} \right)^{1/2} \times \int_{-1}^{1} \frac{3\left(\lambda^2 \left(\frac{1+x}{2}\right) + \beta^0(x_-(\alpha))^2 \left(\frac{1-x}{2}\right)\ -\ \beta_s^2\right)^2}{\sqrt{1-x}}\, dx$$
$$(364)\qquad = \left(\frac{(\lambda - \alpha)(\lambda - \beta^0(x_-(\alpha)))}{2} \right)^{1/2} q(\lambda\ ;\ \beta^0(x_-(\alpha))\ ,\ \beta_s)\ ,$$

where

$$q(\lambda\ ;\ \beta^0(x_-(\alpha))\ ,\ \beta_s) = \int_{-1}^{1} \frac{3\left(\lambda^2 \left(\frac{1+x}{2}\right) + \beta^0(x_-(\alpha))^2 \left(\frac{1-x}{2}\right)\ -\ \beta_s\right)^2}{\sqrt{1-x}}\, dx$$

is a quartic polynomial in λ, which can be evaluated explicitly,

$$(365)\quad q(\lambda) = q(\lambda, \alpha) = \frac{2\sqrt{2}}{5} \left\{ 8\lambda^4 + 4((\beta^0)^2 - 5\beta_s^2)\lambda^2 + 15\beta_s^4 - 10\beta_s^2(\beta^0)^2 + 3(\beta^0)^4 \right\}\ .$$

From these considerations, we can write

$$T_0 = \left(\int_{\beta^0(x_-(\alpha))}^{\beta^{(1)}} + \int_{\beta^{(2)}}^{\beta^{(3)}} \right) \frac{i}{\sqrt{2}} \left(\frac{\lambda - \beta^0(x_-(\alpha))}{(\lambda-\beta^{(1)})(\lambda-\beta^{(2)})(\lambda-\beta^{(3)})} \right)^{1/2} \cdot$$
$$\cdot q(\lambda \; ; \; \beta^0(x_-^0(\alpha)) \; , \; \beta_s) \, d\lambda$$

$$(366) \qquad = \frac{1}{2\sqrt{2}} \oint_{|\lambda|=\ell} i \left(\frac{\lambda - \beta^0(x_-(\alpha))}{(\lambda-\beta^{(1)})(\lambda-\beta^{(2)})(\lambda-\beta^{(3)})} \right)^{1/2} q(\lambda) \; d\lambda \; ,$$

where the integral is taken on a clockwise contour of radius ℓ encircling $(-1, 1)$.

So, expanding the integrand in (366) as $\lambda \to \infty$, we define $c_j, j \leq 3$ by

$$(367) \qquad \left(\frac{\lambda - \beta^0(x_-(\alpha))}{(\lambda-\beta^{(1)})(\lambda-\beta^{(2)})(\lambda-\beta^{(3)})} \right)^{1/2} q(\lambda) =$$
$$c_3\lambda^3 + c_2\lambda^2 + c_1\lambda + c_0 + \frac{c_{-1}}{\lambda} + \; \dots \; ,$$

and we see from (366) and (367) that

$$T_0 \; = \; \frac{\pi c_{-1}}{\sqrt{2}} \; .$$

Differentiating the representation (366), we have

$$(368) \qquad \frac{\partial}{\partial \beta_j} \, T_0 = \frac{1}{2\sqrt{2}} \oint_{|\lambda|=R} i \left(\frac{\lambda - \beta^0(x_-(\alpha))}{(\lambda-\beta^{(1)})(\lambda-\beta^{(2)})(\lambda-\beta^{(3)})} \right)^{1/2} \times$$
$$\frac{1/2}{\lambda - \beta_j} \; q(\lambda) \; d\lambda \; .$$

Expanding the integrand in (368) for large λ, and using (367), we obtain

$$\left(\frac{\lambda - \beta^0(x_-(\alpha))}{(\lambda-\beta^{(1)})(\lambda-\beta^{(2)})(\lambda-\beta^{(3)})} \right)^{1/2} \frac{1/2}{\lambda - \beta_j} \; q(\lambda)$$
$$= \frac{1/2}{\lambda - \beta_j} \left(c_3\lambda^3 + c_2\lambda^2 + c_1\lambda + c_0 + \; \dots \right)$$
$$= \frac{1/2}{\lambda} \left(1 + \frac{\beta_j}{\lambda} + \left(\frac{\beta_j}{\lambda} \right)^2 + \left(\frac{\beta_j}{\lambda} \right)^3 + \; \dots \right) \times$$
$$\left(c_3\lambda^3 + c_2\lambda^2 + c_1\lambda + c_0 + \; \dots \right)$$
$$= d_3\lambda^2 + d_2\lambda + d_1 + \frac{c_0 + c_1\beta_j + c_2\beta_j^2 + c_3\beta_j^3}{2\lambda} + d_{-1}\lambda^{-2} + \dots \; .$$

Therefore,

$$(369) \qquad \frac{\partial}{\partial \beta_j} T_0 \; = \; \frac{\pi}{2\sqrt{2}} \left(c_0 + c_1\beta_j + c_2\beta_j^2 + c_3\beta_j^3 \right) \; .$$

Now we use a lemma, whose proof we temporarily defer.

LEMMA 9.1. *For $-\alpha < \beta^{(1)} < \beta^{(2)} < \beta^{(3)}$ close to β_2,*

$$(370)\quad N = \frac{i}{\sqrt{2}} \int_{\beta^{(1)}}^{\beta^{(2)}} R(\mu)^{1/2} \left(c_0 + c_1\mu + c_2\mu^2 + c_3\mu^3\right) \, d\mu \ ,$$

where $\{c_j\}$, $0 \le j \le 3$ are as in (367).

Important Remark Formula (369) and (370) above, which are for symmetric data, $-\alpha^0(x) = \beta^0(x)$, but not true in general, play a crucial role in proving that, for symmetric data, the maximization problem is solved by a 1 gap ansatz **throughout** $(x_L(t), x_R(t))$. Observe that symmetry enters into the calculation by allowing one to compute the quantity $H\psi_0^* - \theta'(\lambda, t)$ explicitly, through the change of variables $y \leftrightarrow \left(\beta^0(y)\right)^2$ (see (363 - 364)). As long as we have symmetry, the exact form of the initial data $\left(\beta^0(x)\right)^2 = \beta_s^2 + (x - x_s)^{1/3}$, is not essential. We obtain the analog of (369) and (370), as long as the inverse function $\left(\left(\beta^0\right)^2\right)^{-1}$ is analytic in a neighborhood of β_s.

We note, using (369), that if $\frac{\partial}{\partial \beta_j} T_0 = 0$, then the cubic in the integrand of (370) vanishes at $\mu = \beta_j$. We now show that in the case $-\alpha, \beta^{(j)}, 1 \le j \le 3$ are close to β_s, if this occurs, then necessarily $N \neq 0$. We consider $j = 1$ first.

As we are assuming that the cubic vanishes at $\beta^{(1)}$, we can write

$$\begin{aligned} N &= \int_{\beta^{(1)}}^{\beta^{(2)}} R(\mu)^{1/2} \left(c_1(\mu - \beta^{(1)}) + c_2(\mu^2 - \left(\beta^{(1)}\right)^2) + c_3(\mu^3 - \left(\beta^{(1)}\right)^3)\right) d\mu \\ &= \int_{\beta^{(1)}}^{\beta^{(2)}} R(\mu)^{1/2} \, (\mu - \beta^{(1)}) Q(\mu) \, d\mu \ , \end{aligned}$$

where

$$Q(\mu) = c_3(\mu^2 + \mu\beta^{(1)} + \left(\beta^{(1)}\right)^2) \ + \ c_2(\mu + \beta^{(1)}) \ + \ c_1 \ .$$

We will show that Q does not vanish on $(\beta^{(1)}, \beta^{(2)})$.

A simple calculation shows that $Q'(\mu) = 0$ for

$$(371)\quad \mu = -\frac{c_3\beta^{(1)} + c_2}{2c_3} \ .$$

Some tedious calculations show that

$$(372)\quad c_3 = 8 \ ,$$

$$(373)\quad c_2 = 4\left(\beta^{(1)} + \beta^{(2)} + \beta^{(3)} - \beta^0\right) \ ,$$

$$\begin{aligned} (374)\quad c_1 = \ & 4((\beta^0)^2 - 5\beta_s^2) + \\ & +(\beta^{(1)} + \beta^{(2)} + \beta^{(3)} - \beta^0)^2 + \\ & 2\left(\left(\beta^{(1)}\right)^2 + \left(\beta^{(2)}\right)^2 + \left(\beta^{(3)}\right)^2 - (\beta^0)^2\right) \ , \end{aligned}$$

Using the fact that β^0 and the $\beta^{(j)}$'s are close to β_s, it is simple to check that the location of the critical value of Q, (371) is negative, using (372) and (373).

Thus we just need to show that $Q(\beta^{(1)}) > 0$. We have

$$\begin{aligned} Q(\beta^{(1)}) &= 3c_3\left(\beta^{(1)}\right)^2 + 2c_2\beta^{(1)} + c_1 \\ &= 24\left(\beta^{(1)}\right)^2 - 20\beta_s^2 + 8\beta^{(1)}(\beta^{(1)} + \beta^{(2)} + \beta^{(3)} - \beta^0) + \\ &\quad + (\beta^{(1)} + \beta^{(2)} + \beta^{(3)} - \beta^0)^2 + \\ &\quad + 2(\left(\beta^{(1)}\right)^2 + \left(\beta^{(2)}\right)^2 + \left(\beta^{(3)}\right)^2 - (\beta^0)^2) + 4(\beta^0)^2 , \end{aligned}$$

which is positive because of the fact that the β_j's are all close to β_s, which is nonzero.

This implies that Q is positive over the interval $(\beta^{(1)}, \beta^{(2)})$, which shows that $N \neq 0$. This proves the theorem, for the case $j = 1$.

Similarly,

$$\begin{aligned} \frac{\partial}{\partial\beta^{(2)}}T_0 &= 0 \Rightarrow \\ N &= \int_{\beta^{(1)}}^{\beta^{(2)}} R(\mu)^{1/2}(\mu - \beta^{(2)})Q_2(\mu)\, d\mu , \end{aligned}$$

where

$$Q_2(\mu) = c_3(\mu^2 + \mu\beta^{(2)} + \left(\beta^{(2)}\right)^2) + c_2(\mu + \beta^{(2)}) + c_1 .$$

Again, $Q_2'(\mu) = 0$ if

$$\mu = -\frac{c_3\beta^{(2)}}{2c_3} < 0 ,$$

and we have

$$\begin{aligned} Q_2(\beta^{(1)}) &= 8(\left(\beta^{(1)}\right)^2 + \beta^{(1)}\beta^{(2)} + \left(\beta^{(2)}\right)^2) + \\ &\quad +4(\beta^{(1)} + \beta^{(2)} + \beta^{(3)} - \beta^0)(\beta^{(1)} + \beta^{(2)}) + \\ &\quad + 4((\beta^0)^2 - 5\beta_s^2) + (\beta^{(1)} + \beta^{(2)} + \beta^{(3)} - \beta^0)^2 + \\ &\quad + 2(\left(\beta^{(1)}\right)^2 + \left(\beta^{(2)}\right)^2 + \left(\beta^{(3)}\right)^2 - (\beta^0)^2) + 4(\beta^0)^2 \\ &> Q(\beta^{(1)}) > 0 . \end{aligned}$$

Similar arguments show that

$$\begin{aligned} \frac{\partial}{\partial\beta^{(3)}}T_0 &= 0 \Rightarrow \\ N &= \int_{\beta^{(1)}}^{\beta^{(2)}} R(\mu)^{1/2}(\mu - \beta^{(2)})Q_3(\mu)\, d\mu , \end{aligned}$$

for an analogous $Q_3(\mu)$, and similar arguments also show that $Q_3(\mu) > 0$ for $\mu \in (\beta^{(1)}, \beta^{(2)})$.

To complete the proof of Theorem 9.1, we now prove the Lemma. We begin with the formula for $iH\psi(\lambda)$. From (302) and (303), we have for $\lambda \in (\beta^{(1)}, \beta^{(2)})))$,

$$i\ H\psi = \frac{R(\mu)^{1/2}}{\pi i}\left(\int_{\alpha}^{\beta^{(1)}} + \int_{\beta^{(2)}}^{\beta^{(3)}}\right)\frac{i(\theta'(\lambda)-t)}{R(\lambda)^{1/2}}\frac{1}{\lambda-\mu}\,d\lambda\ .$$

Removing the t dependence, we obtain (c.f. (329))

$$\begin{aligned}
i\ H\psi + t &= \frac{R(\mu)^{1/2}}{\pi i}\left(\int_{\alpha}^{\beta^{(1)}} + \int_{\beta^{(2)}}^{\beta^{(3)}}\right)\frac{i\theta'(\lambda)}{R(\lambda)^{1/2}}\frac{1}{\lambda-\mu}\,d\lambda \\
&= \frac{R(\mu)^{1/2}}{\pi i}\int_0^{x_-^0(\alpha)}\left\{\int_{\alpha}^{\alpha^0(y)} + \int_{\beta^0(y)}^{\beta^{(1)}} + \int_{\beta^{(2)}}^{\beta^{(3)}}\right\}\frac{i}{[R(\lambda)R^0(\lambda,y)]^{1/2}}\frac{d\lambda dy}{\lambda-\mu} \\
&+ \frac{R(\mu)^{1/2}}{\pi i}\int_{\beta^0(y)}^{\beta^{(1)}}\int_{x_-^0(\alpha)}^{x_-^0(\lambda)}\frac{i}{[R(\lambda)R^0(\lambda,y)]^{1/2}}\frac{1}{\lambda-\mu}\,dyd\lambda\ + \\
&+ \frac{R(\mu)^{1/2}}{\pi i}\int_{\beta^{(2)}}^{\beta^{(3)}}\int_{x_-^0(\alpha)}^{x_-^0(\lambda)}\frac{i}{[R(\lambda)R^0(\lambda,y)]^{1/2}}\frac{1}{\lambda-\mu}\,dyd\lambda \\
&= \frac{R(\mu)^{1/2}}{\pi i}\int_0^{x_-^0(\alpha)}\pi\frac{1}{2\pi}\oint_{|\lambda-\mu|=\epsilon}\frac{i}{[R(\lambda)R^0(\lambda,y)]^{1/2}}\frac{1}{\lambda-\mu}d\lambda dy \\
&+ \frac{R(\mu)^{1/2}}{\pi}\left(\int_{\beta^0(x_-^0(\alpha))}^{\beta^{(1)}} + \int_{\beta^{(2)}}^{\beta^{(3)}}\right)\frac{\theta'(\lambda) - H\psi_0^*(\lambda\ ;\ x_-^0(\alpha),0)}{[R(\lambda)R^0(\lambda,y)]^{1/2}}\frac{1}{\lambda-\mu}\,d\lambda\ ,
\end{aligned}$$

where the above contour integral is taken counter - clockwise on a circle centered around $\lambda = \mu$, and with (small) radius ϵ, and in the second integral, we have used the formula (363). Evaluating the contour integral explicitly, we have

$$\begin{aligned}
iH\psi + t &= -\frac{R(\mu)^{1/2}}{\pi i}\int_0^{x_-^0(\alpha)}\pi\frac{i}{[R(\mu)R_0(\mu,y)]^{1/2}}\,dy \\
&+ \frac{R(\mu)^{1/2}}{\pi}\left(\int_{\beta^0(x_-^0(\alpha))}^{\beta^{(1)}} + \int_{\beta^{(2)}}^{\beta^{(3)}}\right)\frac{\theta'(\lambda) - H\psi_0^*(\lambda\ ;\ x_-^0(\alpha),0)}{[R(\lambda)R^0(\lambda,y)]^{1/2}}\frac{d\lambda}{\lambda-\mu} \\
&= i\int_0^{x_-^0(\alpha)}\frac{1}{R_0(\mu,y)^{1/2}}\,dy\ + \\
&+ \frac{R(\mu)^{1/2}}{\pi}\left(\int_{\beta^0(x_-^0(\alpha))}^{\beta^{(1)}} + \int_{\beta^{(2)}}^{\beta^{(3)}}\right)\frac{\theta'(\lambda) - H\psi_0^*(\lambda\ ;\ x_-^0(\alpha),0)}{[R(\lambda)R^0(\lambda,y)]^{1/2}}\frac{1}{\lambda-\mu}\,d\lambda.
\end{aligned}$$

Now we observe that

$$i\int_0^{x_-^0(\alpha)} \frac{1}{R_0(\mu,y)^{1/2}}\,dy = i\int_0^{x_-^0(\alpha)} H\psi_{0y}^*(\mu;\; y,0)dy$$
$$= i\; H\psi_0^*(\mu\;;\; x_-^0(\alpha),0)\;.$$

Hence we have

$$iH\psi + t = iH\psi_0^*(\mu\;;\; x_-^0(\alpha),0)\; +$$
$$+\;\frac{R(\mu)^{1/2}}{\pi}\left(\int_{\beta^0(x_-^0(\alpha))}^{\beta^{(1)}} + \int_{\beta^{(2)}}^{\beta^{(3)}}\right)\frac{\theta'(\lambda) - H\psi_0^*(\lambda\;;\; x_-^0(\alpha),0)}{[R(\lambda)R^0(\lambda,y)]^{1/2}}\frac{d\lambda}{\lambda-\mu}\;.$$

Thus,

$$iN = i\int_{\beta^{(1)}}^{\beta^{(2)}} H\psi - \theta' + t d\mu$$
$$= \int_{\beta^{(1)}}^{\beta^{(2)}}\frac{R(\mu)^{1/2}}{\pi}\left(\int_{\beta^0(x_-^0(\alpha))}^{\beta^{(1)}} + \int_{\beta^{(2)}}^{\beta^{(3)}}\right)\frac{\theta'(\lambda) - H\psi_0^*(\lambda\;;\; x_-^0(\alpha),0)}{R(\lambda)^{1/2}}\frac{d\lambda d\mu}{\lambda-\mu}$$
$$+\; i\int_{\beta^{(1)}}^{\beta^{(2)}} H\psi_0^*(\lambda\;;\; x_-^0(\alpha),0) - \theta'(\mu)\; d\mu\;.$$

Inserting the representation (364) of $H\psi_0^*(\lambda\;;\; x_-^0(\alpha),0) - \theta'(\mu)$, we have

$$iN =$$
$$= \int_{\beta^{(1)}}^{\beta^{(2)}}\frac{R(\mu)^{1/2}}{\pi i}\left(\int_{\beta^0(x_-^0(\alpha))}^{\beta^{(1)}} + \int_{\beta^{(2)}}^{\beta^{(3)}}\right)\frac{i}{\sqrt{2}}\cdot$$
$$\cdot\left(\frac{\lambda-\beta^0}{(\lambda-\beta^{(1)})(\lambda-\beta^{(2)})(\lambda-\beta^{(3)})}\right)^{1/2} q(\lambda)\frac{1}{\lambda-\mu}\;d\lambda\;d\mu$$
$$-\; i\int_{\beta^{(1)}}^{\beta^{(2)}}\left(\frac{(\mu-\alpha)(\mu-\beta^0)}{2}\right)^{1/2} q(\mu)\;d\mu$$
$$= \int_{\beta^{(1)}}^{\beta^{(2)}} R(\mu)^{1/2}\frac{i}{\sqrt{2}}\cdot$$
$$\cdot\left[\left(\frac{\mu-\beta^0}{(\mu-\beta^{(1)})(\mu-\beta^{(2)})(\mu-\beta^{(3)})}\right)^{1/2} q(\mu) - (c_3\mu^3 + c_2\mu^2 + c_1\mu + c_0)\right]\;d\mu$$
$$-\; i\int_{\beta^{(1)}}^{\beta^{(2)}}\left(\frac{(\mu-\alpha)(\mu-\beta^0)}{2}\right)^{1/2} q(\mu)\;d\mu$$
$$= \frac{-i}{2}\int_{\beta^{(1)}}^{\beta^{(2)}} R(\mu)^{1/2}(c_3\mu^3 + c_2\mu^2 + c_1\mu + c_0)\;\;d\mu\;\;.$$

This completes the proof of the Lemma, and the Theorem. □
Finally,

THEOREM 9.2. *If* $-\alpha < \beta^{(1)} < \beta^{(2)} < \beta^{(3)}$, *and* $T(\alpha, \beta^{(1)}, \beta^{(2)}, \beta^{(3)}) = 0$, *then*

$$\frac{\partial T_0}{\partial \alpha}(\alpha, \beta^{(1)}, \beta^{(2)}, \beta^{(3)}) < 0 \ .$$

Proof. We begin with the representation (330) from Chapter 8

$$(375) T_0(\alpha, \beta^{(1)}, \beta^{(2)}, \beta^{(3)}) = \left(\int_{\beta^0(x_-^0(\alpha))}^{\beta^{(1)}} + \int_{\beta^{(2)}}^{\beta^{(3)}} \right) \frac{i}{R(\lambda)^{1/2}} \left(\theta'(\lambda) - H\psi_0^*(\lambda \ ; \ x_-^0(\alpha), 0)\right) \, d\lambda \ .$$

Differentiating (375) with repsect to α, we obtain

$$\frac{\partial T_0}{\partial \alpha} = \left(\int_{\beta^0(x_-^0(\alpha))}^{\beta^{(1)}} + \int_{\beta^{(2)}}^{\beta^{(3)}} \right) \frac{i}{R(\lambda)^{1/2}} \left(-\frac{\partial}{\partial \alpha} H\psi_0^*(\lambda \ ; \ x_-^0(\alpha), 0) \right) \, d\lambda$$
$$+ \left(\int_{\beta^0(x_-^0(\alpha))}^{\beta^{(1)}} + \int_{\beta^{(2)}}^{\beta^{(3)}} \right) \frac{i\left(\theta'(\lambda) - H\psi_0^*(\lambda \ ; \ x_-^0(\alpha), 0)\right)}{R(\lambda)^{1/2}} \frac{d\lambda/2}{\lambda - \alpha} ,$$

and, using the identity

$$\frac{\partial}{\partial \alpha} H\psi_0^*(\lambda \ ; \ x_-^0(\alpha), 0) = \frac{\frac{\partial x_-^0(\alpha)}{\partial \alpha}}{\left[(\lambda - \alpha)(\lambda - \beta^0(x_-^0(\alpha)))\right]^{1/2}} ,$$

we have

$$(376) \quad \frac{\partial T_0}{\partial \alpha} = \frac{\pi \frac{\partial x_-^0(\alpha)}{\partial \alpha}}{\left[(\alpha - \beta^0(x_-^0(\alpha)))(\alpha - \beta^{(1)})(\alpha - \beta^{(2)})(\alpha - \beta^{(3)})\right]^{1/2}}$$
$$(377) \quad + \left(\int_{\beta^0(x_-^0(\alpha))}^{\beta^{(1)}} + \int_{\beta^{(2)}}^{\beta^{(3)}} \right) \frac{i\left(\theta'(\lambda) - H\psi_0^*(\lambda \ ; \ x_-^0(\alpha), 0)\right)}{R(\lambda)^{1/2}} \frac{d\lambda/2}{\lambda - \alpha} .$$

Now (376) is clearly negative, and we show that the sum of integrals in (377) is negative as well.

Consider the second integral in (377). We have

$$(378) \quad \int_{\beta^{(2)}}^{\beta^{(3)}} \frac{i}{R(\lambda)^{1/2}} \left(\theta'(\lambda) - H\psi_0^*(\lambda \ ; \ x_-^0(\alpha), 0)\right) \frac{1/2}{\lambda - \alpha} d\lambda <$$
$$\int_{\beta^{(2)}}^{\beta^{(3)}} \frac{i}{R(\lambda)^{1/2}} \left(\theta'(\lambda) - H\psi_0^*(\lambda \ ; \ x_-^0(\alpha), 0)\right) \frac{1/2}{\beta^{(2)} - \alpha} d\lambda ,$$

as $\lambda - \alpha > \beta^{(2)} - \alpha$ over the region of integration. Now the integral on the right hand side of (378) happens to be the second integral in the representation (375). Thus as we are solving $T_0 = 0$, we have

$$\int_{\beta^{(2)}}^{\beta^{(3)}} \frac{i\left(\theta'(\lambda) - H\psi_0^*(\lambda\,;\,x_-^0(\alpha),0)\right)}{R(\lambda)^{1/2}} \frac{d\lambda/2}{\lambda-\alpha} <$$

$$-\frac{1/2}{\beta^{(2)}-\alpha}\int_{\beta^0(x_-^0(\alpha))}^{\beta^{(1)}} \frac{i\left(\theta'(\lambda) - H\psi_0^*(\lambda\,;\,x_-^0(\alpha),0)\right)}{R(\lambda)^{1/2}}\, d\lambda\,,$$

and this in turn implies

$$\left(\int_{\beta^0(x_-^0(\alpha))}^{\beta^{(1)}} + \int_{\beta^{(2)}}^{\beta^{(3)}}\right) \frac{i\left(\theta'(\lambda) - H\psi_0^*(\lambda\,;\,x_-^0(\alpha),0)\right)}{R(\lambda)^{1/2}} \frac{d\lambda/2}{\lambda-\alpha} <$$

$$\int_{\beta^0(x_-^0(\alpha))}^{\beta^{(1)}} \frac{i\left(\theta'(\lambda) - H\psi_0^*(\lambda\,;\,x_-^0(\alpha),0)\right)}{R(\lambda)^{1/2}} \left(\frac{1/2}{\lambda-\alpha} - \frac{1/2}{\beta^{(2)}-\alpha}\right) d\lambda\,,$$

which is clearly negative.

This completes the proof of Theorem 9.2. □

A simple calculation shows that $T_0(\alpha, \beta^{(1)}, \beta^{(2)}, \beta^{(3)}) > 0$ if $\beta^0(x_-(\alpha)) = -\alpha > \beta^{(1)}$. Hence $\beta^0(x_-(\alpha))$ must be less than $\beta^{(1)}$ on a solution of $T_0(\alpha, \beta^{(1)},$
$\beta^{(3)}) = 0$.

Thus Theorems 9.1 and 9.2, imply that $\frac{\partial T_0}{\partial \beta^{(j)}}(\alpha, \beta^{(1)}, \beta^{(2)}, \beta^{(3)}) \neq 0$, $0 \leq j \leq 3$, on a solution of (358 - 361), as desired.

CHAPTER 10

Global Description

In this chapter we will prove that for the class of initial data studied in chapter 8, the first point $x = \bar{x}$ at which the Jacobian $\partial Y/\partial(\alpha, \beta^{(1)}, \beta^{(2)}, \beta^{(3)})$ vanishes is a point at which $\beta^{(1)}$ intersects $\beta^{(2)}$. That is, the new gap closes, and at this point, the pair $(\alpha(\bar{x},t), \beta^{(3)}(\bar{x},t))$ solves the zero gap problem, along with the additional conditions

$$\begin{aligned} \psi_{zero\ gap}(\lambda = \beta^{(1)}(\bar{x},t) = \beta^{(2)}(\bar{x},t), \bar{x}, t) &= 0 \ , \\ \frac{\partial}{\partial\lambda}\psi_{zero\ gap}(\lambda = \beta^{(1)}(\bar{x},t) = \beta^{(2)}(\bar{x},t), \bar{x}, t) &= 0 \ . \end{aligned}$$

We remind the reader that, by the remark at the end of Chapter 8, this immediately implies that the associated function $\psi_{one\ gap}(\lambda, x, t)$ solves the maximization problem for all $x \in (\underline{x}, \bar{x})$, and hence that the continuum limit of the Toda lattice is described **globally** by a small one - gap region, surrounded by two zero gap regions.

We begin by cataloging relevant information.

I1. On a solution to $T_0 = 0$, with $\alpha < 0$, $0 < \beta^{(1)} < \beta^{(2)} < \beta^{(3)}$,

$$\frac{\partial}{\partial\alpha}T_0 < 0 \ .$$

I2. There exists a unique solution $(\alpha, \vec{\beta}, x) = (\alpha, \vec{\beta}, \underline{x})$ to $Y(\alpha, \vec{\beta}) = \pi x \mathbf{e}_3$, with $\beta^{(1)} < \beta^{(2)} = \beta^{(3)}$.

I3. For $\underline{x} \leq x < \underline{x} + \epsilon_1$ (ϵ_1 small), there exists a continuous solution $(\alpha, \vec{\beta})(x,t)$ to $Y = \pi x \mathbf{e}_3$ which extends the solution described in I2, and for $\underline{x} < x < \underline{x} + \epsilon_1$,

$$J = \left.\frac{\partial Y}{\partial(\alpha, \beta^{(1)}, \beta^{(2)}, \beta^{(3)})}\right|_{(\alpha,\vec{\beta})(x,t)} \neq 0 \ . \tag{379}$$

I4. For $\underline{x} < x < \underline{x} + \epsilon_1$, we have

$$\begin{aligned} T_{0,\ \beta^{(1)}}((\alpha,\vec{\beta})(x,t)) &< 0 \\ T_{0,\ \beta^{(2)}}((\alpha,\vec{\beta})(x,t)) &> 0 \\ T_{0,\ \beta^{(3)}}((\alpha,\vec{\beta})(x,t)) &> 0 \\ \alpha_x(x,t), \beta_x^{(2)}(x,t) &< 0 \\ \beta_x^{(1)}(x,t), \beta_x^{(3)}(x,t) &> 0 \ . \end{aligned}$$

I5. There exists $\delta < \epsilon_1$ such that for all $x \in [\underline{x}, \underline{x} + \delta)$, the above solution generates a function ψ which solves the associated maximization problem.

I6. There exists a unique solution $(\alpha, \vec{\beta}, x) = (\alpha, \vec{\beta}, \bar{x})$ to $Y(\alpha, \vec{\beta}) = \pi x \mathbf{e}_3$, with $\beta^{(1)} = \beta^{(2)} < \beta^{(3)}$, and $\beta^{(1)}(\underline{x}, t) < \beta^{(1)}(\bar{x}, t) < \beta^{(3)}(\underline{x}, t)$.

I7. For $\bar{x} - \epsilon_2 < x \le \bar{x}$ (ϵ_2 small), there exists a continuous solution $(\alpha, \vec{\beta})(x,t)$ to $Y = \pi x \mathbf{e}_3$ which extends the solution described in I6, and for $\bar{x} - \epsilon_2 < x < \bar{x}$, we have

$$J = \frac{\partial Y}{\partial(\alpha, \beta^{(1)}, \beta^{(2)}, \beta^{(3)})}\bigg|_{(\alpha,\vec{\beta})(x,t)} \neq 0 \,. \tag{380}$$

I8. For $\bar{x} - \epsilon_2 < x < \bar{x}$, we have

$$\begin{aligned} T_{0,\ \beta^{(1)}}\left((\alpha, \vec{\beta})(x,t)\right) &< 0 \\ T_{0,\ \beta^{(2)}}\left((\alpha, \vec{\beta})(x,t)\right) &> 0 \\ T_{0,\ \beta^{(3)}}\left((\alpha, \vec{\beta})(x,t)\right) &> 0 \\ \alpha_x(x,t), \beta_x^{(2)}(x,t) &< 0 \\ \beta_x^{(1)}(x,t), \beta_x^{(3)}(x,t) &> 0 \,. \end{aligned}$$

I9. There exists $\delta_2 < \epsilon_2$ such that for all $x \in (\bar{x} - \delta_2, \bar{x}]$, the above solution generates a function ψ which solves the associated maximization problem.

LEMMA 10.1. *For $(\alpha, \beta^{(1)}, \beta^{(2)}, \beta^{(3)})$ near $(\alpha_c, \beta_c, \beta_c, \beta_c)$, $\frac{\partial^2}{\partial \beta^{(j)} \beta^{(k)}} T_0 > 0$. In fact, we have*

$$\frac{\partial^2}{\partial \beta^{(j)} \beta^{(k)}}\bigg|_{(\alpha_c, \beta_c, \beta_c, \beta_c)} T_0 = \left\{ \begin{array}{ll} \frac{12\pi}{15(\beta_c - \alpha_c)c^3} & j \neq k \\ \frac{36\pi}{15(\beta_c - \alpha_c)c^3} & j = k \end{array} \right\} . \tag{381}$$

Further, we also have

$$T_{0\beta^{(j)}}\big|_{(\alpha_c, \beta_c, \beta_c, \beta_c)} = 0 \quad for\ j = 1, 2, 3. \tag{382}$$

Proof. We start with the expression (c.f. (330))

$$T_0(\alpha, \vec{\beta}) = \left(\int_{\beta_0(x_-(\alpha))}^{\beta^{(1)}} + \int_{\beta^{(2)}}^{\beta^{(3)}} \right) \frac{i}{R^{1/2}} \left[\theta'(\lambda, 0) - H\psi(\lambda, x_-(\alpha), 0) \right] d\lambda \,,$$

where we remark that the function $T_0(\alpha, \vec{\beta})$ does not depend on the parameter t, and we recall that $\theta'(\lambda, 0) - H\psi(\lambda, x_-(\alpha), 0)$ is given by (c.f. proof of Lemma 4.4)

$$\theta'(\lambda, 0) - H\psi(\lambda, x_-(\alpha), 0) = \int_{x_-(\alpha)}^{x_-(\lambda)} \frac{1}{\left((\lambda - \alpha_0(y))(\lambda - \beta_0(y))\right)^{1/2}} dy \,,$$

where $\alpha_0(y)$ and $\beta_0(y)$ are the initial data.

For y near x_c, let the equation

$$(\lambda - \alpha(y))(\lambda - \beta(y)) = \lambda^2 - \mu$$

define the function $y = f(\mu, \lambda)$. The reader can verify that $\partial\mu/\partial y > 0$ for y near x_c. Then we have

$$\theta'(\lambda, 0) - H\psi(\lambda, x_-(\alpha), 0) = \int_{\mu(\alpha)}^{\lambda^2} \frac{f_\mu(\mu, \lambda)}{(\lambda^2 - \mu)^{1/2}} d\mu \ ,$$

and setting $\lambda^2 - \mu = (\lambda - \alpha)(\lambda - \beta_0(x_-(\alpha)))\frac{1-x}{2}$, we obtain the expression

$$\theta'(\lambda, 0) - H\psi(\lambda, x_-(\alpha), 0) = \left[\frac{(\lambda - \alpha)(\lambda - \beta_0(x_-(\alpha)))}{2}\right]^{1/2} \int_{-1}^{1} \frac{f_\mu(\mu(x, \alpha, \lambda), \lambda)}{(1 - x)^{1/2}} dx$$

Thus

(383)

$$T_0 = \left(\int_{\beta_0(x_-(\alpha))}^{\beta^{(1)}} + \int_{\beta^{(2)}}^{\beta^{(3)}}\right) \frac{i}{\sqrt{2}} \left[\frac{(\lambda - \beta_0(x_-(\alpha)))}{(\lambda - \beta^{(1)})(\lambda - \beta^{(2)})(\lambda - \beta^{(3)})}\right]^{1/2} \cdot \int_{-1}^{1} \frac{f_\mu(\mu(x, \alpha, \lambda), \lambda)}{(1 - x)^{1/2}} dx d\lambda \ .$$

We define μ_c by

$$\lambda^2 - \mu_c = (\lambda - \alpha_c)(\lambda - \beta_c) \ .$$

Using the asymptotic description of the initial curves $\alpha_0(x)$ and $\beta_0(x)$, namely

$$\alpha_0(x) = \alpha_c + c_\alpha(x - x_c) + \dots \ , \quad \beta_0(x) = \beta_c + c(x - x_c)^{1/3} + \dots \ ,$$

we have the following expansion for $f(\mu, \lambda)$, for μ near μ_c:

$$f = x_c + \left(\frac{\mu - \mu_c}{(\lambda - \alpha_c)c}\right)^3 + \dots \ ,$$

from which we obtain

$$\begin{aligned}
(384)\quad f'(\mu, \lambda) &= \frac{3}{[(\lambda - \alpha_c)c]^3} (\mu - \mu_c)^2 + \dots \\
&= \frac{3}{(\lambda - \alpha_c)c^3} \Big[\lambda - \beta_c - (\frac{1 - x}{2})(\lambda - \beta_0) - \\
&\qquad (\frac{1 - x}{2})\frac{\alpha - \alpha_c}{\lambda - \alpha_c}(\lambda - \beta_0)\Big]^2 + \dots \\
&\approx \frac{3}{(\beta_c - \alpha_c)c^3} \left[\lambda - \beta_c - (\frac{1 - x}{2})(\lambda - \beta_0)\right]^2 \ .
\end{aligned}$$

Inserting (384) into (383), we obtain the expression, asymptotically valid for $(\alpha, \vec{\beta})$ near $(\alpha_c, \beta_c(1,1,1))$:

$$T_0 \approx \left(\int_{\beta_0(x_-(\alpha))}^{\beta^{(1)}} + \int_{\beta^{(2)}}^{\beta^{(3)}} \right) \frac{i}{\sqrt{2}} \left[\frac{(\lambda - \beta_0(x_-(\alpha)))}{(\lambda - \beta^{(1)})(\lambda - \beta^{(2)})(\lambda - \beta^{(3)})} \right]^{1/2} \times$$
$$\times \int_{-1}^{1} \frac{3}{(\beta_c - \alpha_c)c^3} \frac{\left[\lambda - \beta_c - (\frac{1-x}{2})(\lambda - \beta_0)\right]^2}{(1-x)^{1/2}} dx d\lambda \,.$$

To obtain the leading behavior of this expression, we note that the integrand is analytic in λ, for $\lambda \notin [\beta_0, \beta^{(1)}] U [\beta^{(2)}, \beta^{(3)}]$, and so we deform the contour of integration,

(385)

$$T_0 \approx \frac{1}{2} \oint_{\ell} \frac{i}{\sqrt{2}} \left[\frac{(\lambda - \beta_0(x_-(\alpha)))}{(\lambda - \beta^{(1)})(\lambda - \beta^{(2)})(\lambda - \beta^{(3)})} \right]^{1/2} \times$$
$$\times \int_{-1}^{1} \frac{3}{(\beta_c - \alpha_c)c^3} \frac{\left[\lambda - \beta_c - (\frac{1-x}{2})(\lambda - \beta_0)\right]^2}{(1-x)^{1/2}} dx d\lambda \,,$$

where the contour ℓ encircles the cuts $[\beta_0, \beta^{(1)}] U [\beta^{(2)}, \beta^{(3)}]$ in the clockwise direction. Differentiation of (385) with respect to any of the $\beta^{(j)}$'s can be done under the integral sign, and we leave the reader to verify (381 - 382). □

LEMMA 10.2. *The quantities $T_m(\alpha, \vec{\beta})$ $(m \geq 0)$ satisfy*

(386) $$2(\beta^{(j)} - \beta^{(k)}) T_{m,\, \beta^{(j)}\beta^{(k)}} = T_{m,\, \beta^{(j)}} - T_{m,\, \beta^{(k)}} \,, \quad j \neq k \,.$$

Proof. From Lemma 8.2, we have

$$T_{m+1,\, \beta^{(j)}} = \beta^{(j)} T_{m,\, \beta^{(j)}} - \frac{1}{2} T_m \,.$$

Thus

(387) $$T_{m+1,\, \beta^{(j)}\beta^{(k)}} = \beta^{(j)} T_{m,\, \beta^{(j)}\beta^{(k)}} - \frac{1}{2} T_{m,\, \beta^{(k)}} \,,$$

and similarly,

(388) $$T_{m+1,\, \beta^{(j)}\beta^{(k)}} = \beta^{(k)} T_{m,\, \beta^{(j)}\beta^{(k)}} - \frac{1}{2} T_{m,\, \beta^{(j)}} \,.$$

Combining (387) and (388) establishes (386). □

LEMMA 10.3. *For $(\alpha, \vec{\beta})$ near $(\alpha_c, \beta_c(1,1,1))$, and $\beta^{(1)} < \beta^{(2)} < \beta^{(3)}$, we have $T_{0\beta^{(1)}} < T_{0\beta^{(2)}} < T_{0\beta^{(3)}}$.*

Proof. This follows easily from equations (386), and the inequalities

$$T_{0\beta^{(j)}\beta^{(k)}} > 0 \,, \quad j,k = 1,2,3 \,,$$
$$\beta^{(1)} < \beta^{(2)} < \beta^{(3)} \,.$$

□

Lemma 10.3, together with the signs of $\{T_{0\beta^{(j)}}(\alpha(x,t),\vec{\beta}(x,t))\}_{j=1}^3$ for x near $\underline{x}$ (see I4 above), demonstrate that for $(\alpha,\vec{\beta})=(\alpha,\vec{\beta})(x,t)$, and x near $\underline{x}$, we have

$$T_{0\beta^{(1)}} < 0 < T_{0\beta^{(2)}} < T_{0\beta^{(3)}} \ . \tag{389}$$

As $T_{0\alpha}<0$ on a solution to $T_0=0$, we may chose $\alpha(\vec{\beta})$ such that $T_0=0$, for $\vec{\beta}$ near $\vec{\beta}(\underline{x}+\epsilon,t)$ (ϵ small). We recall that we must, in part, solve

$$F(\vec{\beta})=T_1(\alpha(\vec{\beta}),\vec{\beta})=\pi t \ . \tag{390}$$

The assumption that t is small, together with (390) and the following lemma, guarantees that $\vec{\beta}$ must remain in a small neighborhood of $\beta_c(1,1,1)$.

LEMMA 10.4. *For $\vec{\beta}$ near $\beta_c(1,1,1)$,*

$$F \geq const. \sum_{j=1}^{3}(\beta^{(j)}-\beta_c)^2 \ . \tag{391}$$

Proof. We will accomplish this by establishing

$$F_{\beta^{(j)}}\big|_{\vec{\beta}=\beta_c(1,1,1)}=0 \ , \quad F_{\beta^{(j)}\beta^{(k)}}\big|_{\vec{\beta}=\beta_c(1,1,1)}= (\beta_c-\alpha)T_{0\beta^{(j)}\beta^{(k)}}\big|_{\vec{\beta}=\beta_c(1,1,1)} \ . \tag{392}$$

Differentiating $F(\beta^{(1)},\beta^{(2)},\beta^{(3)})$ with respect to $\beta^{(j)}$, we have

$$\begin{aligned} F_{\beta^{(j)}} &= T_{1\alpha}\alpha_{\beta^{(j)}}+T_{1\beta^{(j)}} \\ &= \alpha T_{0\alpha}\alpha_{\beta^{(j)}}+\beta^{(j)}T_{0\beta^{(j)}} \\ &= (\beta^{(j)}-\alpha)T_{0\beta^{(j)}} \ , \end{aligned} \tag{393}$$

where we have used the fact that $T_0(\alpha(\vec{\beta}),\vec{\beta})=0$, in the form

$$T_{0\alpha}\alpha_{\beta^{(j)}}+T_{0\beta^{(j)}}=0 \ .$$

Similar calculations yield

$$F_{\beta^{(j)}\beta^{(k)}} = \frac{T_{0\beta^{(k)}}}{T_{0\alpha}}T_{0\beta^{(j)}}+\frac{1}{2}\left(T_{0\beta^{(j)}}-T_{0\alpha}\right)\alpha_{\beta^{(k)}}+ (\beta^{(j)}-\alpha)T_{0\beta^{(j)}\beta^{(k)}} \ , \tag{394}$$

$$F_{\beta^{(j)}\beta^{(j)}} = \left(1+\frac{T_{0\beta^{(j)}}}{T_{0\alpha}}\right)T_{0\beta^{(j)}}+\frac{1}{2}\left(T_{0\beta^{(j)}}-T_{0\alpha}\right)\alpha_{\beta^{(j)}}+ (\beta^{(j)}-\alpha)T_{0\beta^{(j)}\beta^{(j)}} \ . \tag{395}$$

Setting $\vec{\beta}=\beta_c(1,1,1)$ in (393 - 395), and using Lemma 10.1, we establish (392). This, together with (381), imply (391). □

Let

$$\hat{x} = sup \left\{ x : \begin{array}{c} (\alpha, \vec{\beta})(x', t) \text{ is a solution of } Y(\alpha, \vec{\beta}) = \pi x' \mathbf{e}_3 \text{ for} \\ \underline{x} \le x' \le x \text{ which continues the solution for } x \text{ near } \underline{x}, \\ \text{and for which } \partial Y / \partial(\alpha, \beta^{(1)}, \beta^{(2)}, \beta^{(3)}) \neq 0 \text{ for} \\ \underline{x} < x' \le x \end{array} \right\}.$$

Note first that $\hat{x} < \infty$. Indeed, it is clear that

$$T_2 - \beta^{(j)} T_1 = \left(\int_{\alpha}^{\beta^{(1)}} + \int_{\beta^{(2)}}^{\beta^{(3)}} \right) \frac{i\theta'(\lambda, t)}{R^{1/2}} (\lambda - \beta^{(j)}) d\lambda$$

is bounded if $(\alpha, \vec{\beta})$ remains near $(\alpha_c, \beta_c(1,1,1))$. Thus, as we know that $T_1 = \pi t$, we see that $\pi x = T_2$ must remain bounded as well.

We list some properties of the solution.

- 1. All $\beta^{(j)}$'s must remain close to β_c, and α to α_c.
- 2. For $x \in (\underline{x}, \hat{x})$, $\{\beta^{(j)}(x,t)\}_{j=1}^3$ are distinct.
- 3. For $x \in (\underline{x}, \hat{x})$, $T_{0\beta^{(1)}} < 0 < T_{0\beta^{(2)}} < T_{0\beta^{(3)}}$.
- 4. For $x \in (\underline{x}, \hat{x})$, $\beta_x^{(1)} > 0$, $\beta_x^{(2)} < 0$, $\beta_x^{(3)} > 0$, $\alpha_x < 0$.
- 5. $\hat{x} < \infty$.
- 6. $\{\beta^{(j)}\}_{j=1}^3$, α have limits as $x \uparrow \hat{x}$.

It is clear that at $\hat{x}$, the Jacobian must vanish. Thus, from the expression for the Jacobian (see, for example, the proof of Theorem 8.4),

$$\frac{\partial Y(\alpha, \vec{\beta})}{\partial(\alpha, \vec{\beta})} = \tag{396}$$

$$T_{0\alpha} T_{0\beta^{(1)}} T_{0\beta^{(2)}} T_{0\beta^{(3)}} \begin{vmatrix} 1 & 1 & 1 & 1 \\ \alpha & \beta^{(1)} & \beta^{(2)} & \beta^{(3)} \\ \alpha^2 & (\beta^{(1)})^2 & (\beta^{(2)})^2 & (\beta^{(3)})^2 \\ g_0 & g_1 & g_2 & g_3 \end{vmatrix},$$

where g_j are defined in Lemma 8.3, this can only happen if one of the derivatives $T_{0\beta^{(j)}}$, or the determinant in (396), were to vanish. We note from the shape of the curves (see •2 and •4 above) that the determinant can be zero only if $\beta^{(1)} = \beta^{(2)}$.

THEOREM 10.1. *At $\hat{x}$, we have $\beta^{(1)} = \beta^{(2)}$, and hence $\hat{x} = \bar{x}$.*

We will prove this theorem by establishing the following four conclusions. We state them first informally, so that the argument is clear. These statements will be made precise later. By a "turnaround", we refer to an infinite derivative in one of the curves $\beta^{(j)}(x,t)$, with $\{\beta^{(j)}(x,t)\}_{j=1}^3$ distinct.

A. No consecutive turnarounds in any of the curves $\{\beta^{(j)}(x,t)\}_{j=1}^3$ are allowed (Lemma 10.5 and Corollary 10.2 below).

B. To connect from the solution at $\underline{x}$ to the solution at $\bar{x}$, there can be no turnarounds (Lemma 10.6).

C. There exists a second degenerate solution to $Y - \pi x e_3 = 0$, degenerate in the sense that 2 or more of the $\beta^{(j)}$'s coincide, connected to the solution

$(\alpha, \vec{\beta})(\underline{x}, t)$ by a continuous 1-parameter family of solutions to $y - \pi x e_3 = 0$ (Lemma 10.7).

D. There are only 2 degenerate solutions possible, one at $\underline{x}(t)$, and the other at $\bar{x}(t)$ (Lemma 10.8).

LEMMA 10.5. *Suppose $(\alpha, \vec{\beta}, x)(\gamma)$ is a continuous 1-parameter family of solutions to*

$$\begin{aligned}
&Y - \pi x e_3 = 0 \\
&(\alpha, \vec{\beta}, x)(\gamma = 0) = ((\alpha, \vec{\beta})(\underline{x}, t), \underline{x}) \ , \\
&\beta^{(1)}(\gamma) < \beta^{(2)}(\gamma) < \beta^{(3)}(\gamma) \quad for\ \gamma > 0 \ , \\
&(\alpha, \vec{\beta}, x)(\gamma') \neq (\alpha, \vec{\beta}, x)(\gamma'') \quad for\ \gamma' \neq \gamma'' \ ,
\end{aligned}$$

and that at γ_0, $T_{0,\ \beta^{(j)}}((\alpha, \vec{\beta})(\gamma_0)) = 0$.

Then there is no $\gamma_1 > \gamma_0$ such that

$$\left. \frac{\partial(T_0, T_1, T_2, N)}{\partial(\alpha, \vec{\beta})} \right|_{(\alpha, \vec{\beta})(\gamma)} \neq 0 \quad for\ all\ \gamma \in (\gamma_0, \gamma_1) \ , \tag{397}$$

$$\beta^{(j)}(\gamma)\ is\ monotone\ for\ \gamma \in (\gamma_0, \gamma_1) \ , \tag{398}$$

$$T_{0,\beta^{(j)}}((\alpha, \vec{\beta})(\gamma_1)) = 0 \ . \tag{399}$$

Proof. Note from the inequalities contained in I4 above, we have

$$T_{0\beta^{(j)}} = 0 \Rightarrow T_{0\beta^{(k)}} \neq 0 \ , \ k \neq j \ ,$$

and hence we conclude that (c.f. (327))

$$\frac{\partial(T_0, T_1, T_2 - \pi x, N)}{\partial(\alpha, \beta^{(k)}, \beta^{(l)}, x)} = \pi T_{0\alpha} T_{0\beta^{(k)}} T_{0\beta^{(l)}} \begin{vmatrix} 1 & 1 & 1 \\ \alpha & \beta^{(k)} & \beta^{(l)} \\ g_0 & g_k & g_l \end{vmatrix} \neq 0 \ ,$$

where l, k are distinct, and $l, k \neq j$. Thus for $\beta^{(j)}$ near $\beta^{(j)}(\gamma_0)$, we can solve for $(\alpha, \beta^{(k)}, \beta^{(l)}, x)(\beta^{(j)})$. Furthermore, one can calculate that (c.f. proof of

Lemma 8.5)

$$(400)\qquad \frac{d}{d\beta^{(j)}}x(\beta^{(j)}) = -T_{0\beta^{(j)}}\frac{\begin{vmatrix} 1 & 1 & 1 & 1 \\ \alpha & \beta^{(1)} & \beta^{(2)} & \beta^{(3)} \\ (\alpha)^2 & (\beta^{(1)})^2 & (\beta^{(2)})^2 & (\beta^{(3)})^2 \\ g_0 & g_1 & g_2 & g_3 \end{vmatrix}}{\begin{vmatrix} 1 & 1 & 1 \\ \alpha & \beta^{(k)} & \beta^{(l)} \\ g_0 & g_k & g_l \end{vmatrix}},$$

$$(401)\qquad \frac{d^2}{d(\beta^{(j)})^2}\Big|_{\beta^{(j)}=\beta^{(j)}(\gamma_0)}x(\beta^{(j)}) =$$

$$-T_{0\beta^{(j)}\beta^{(j)}}\frac{\begin{vmatrix} 1 & 1 & 1 & 1 \\ \alpha & \beta^{(1)} & \beta^{(2)} & \beta^{(3)} \\ (\alpha)^2 & (\beta^{(1)})^2 & (\beta^{(2)})^2 & (\beta^{(3)})^2 \\ g_0 & g_1 & g_2 & g_3 \end{vmatrix}}{\begin{vmatrix} 1 & 1 & 1 \\ \alpha & \beta^{(k)} & \beta^{(l)} \\ g_0 & g_k & g_l \end{vmatrix}},$$

where in (401), we have used that fact that $\frac{d}{d\beta^{(j)}}x(\beta^{(j)})\Big|_{\beta^{(j)}(\gamma_0)} = 0$.

If there exists $\gamma_1 > \gamma_0$ such that (397 - 399) hold, then we would have

$$(402)\qquad \frac{d^2}{d(\beta^{(j)})^2}\Big|_{\beta^{(j)}=\beta^{(j)}(\gamma_1)}x(\beta^{(j)}) =$$

$$-T_{0\beta^{(j)}\beta^{(j)}}\frac{\begin{vmatrix} 1 & 1 & 1 & 1 \\ \alpha & \beta^{(1)} & \beta^{(2)} & \beta^{(3)} \\ (\alpha)^2 & (\beta^{(1)})^2 & (\beta^{(2)})^2 & (\beta^{(3)})^2 \\ g_0 & g_1 & g_2 & g_3 \end{vmatrix}}{\begin{vmatrix} 1 & 1 & 1 \\ \alpha & \beta^{(k)} & \beta^{(l)} \\ g_0 & g_k & g_l \end{vmatrix}},$$

i.e. $\frac{d^2}{d(\beta^{(j)})^2}x(\beta^{(j)})$ will have the same sign at $\beta^{(j)}(\gamma_0)$ and $\beta^{(j)}(\gamma_1)$. Thus we would have a curve $x(\beta)$ satisfying

$$x_\beta(\beta^{(j)}(\gamma_0)) = x_\beta(\beta^{(j)}(\gamma_1)) = 0 \ ,$$

and for all β in the interior of the interval defined by $\beta^{(j)}(\gamma_0)$ and $\beta^{(j)}(\gamma_1)$,

$$x_\beta(\beta) \neq 0 \ ,$$

and $x_{\beta\beta}(\beta^{(j)}(\gamma_0))$, $x_{\beta\beta}(\beta^{(j)}(\gamma_1))$ are both nonzero and have the same sign.

We leave the reader to verify that such a curve does not exist. □

COROLLARY 10.2. *Under the assumptions of Lemma 10.5, we have that for all* $\gamma > \gamma_0$, $T_{0\beta^{(j)}} \neq 0$.

Proof. We will show that the only way that the quantity $T_{0\beta^{(j)}}$ could vanish for some $\gamma_1 > \gamma_0$ is in such a way that

$$\frac{\partial Y(\alpha,\vec{\beta})}{\partial(\alpha,\vec{\beta})}(\alpha,\vec{\beta})(\gamma) \neq 0 \quad \textit{for all } \gamma \in (\gamma_0,\gamma_1) ,$$

and that the parametrization may be taken such that $\beta^{(j)}(\gamma)$ will be monotone. The result then follows from Lemma 10.5.

From •3 above, we can assume that at γ_0, either $T_{0\beta^{(1)}} = 0$ or $T_{0\beta^{(2)}} = 0$ (either $j = 1$ or $j = 2$). In either case, as in the proof of Lemma 10.5, we can solve for the other quantities $(\alpha, \beta^{(k)}, \beta^{(3)}, x)$, where $k \neq j$, $k \neq 3$, and once again we have (401). Thus the curve $x(\beta^{(j)})$ looks locally quadratic, i.e. the quantity $x_{\beta^{(j)}}$ changes signs, and we conclude that $T_{0\beta^{(j)}}$ changes signs as γ crosses γ_0 (note that the determinant in (400) **does not** change signs).

Suppose that at γ_0, we have that $T_{0\beta^{(1)}} = 0$. Then, for $\gamma_0 < \gamma < \gamma_0 + \delta$ (δ small) and $(\alpha,\vec{\beta}) = (\alpha,\vec{\beta})(\gamma)$, we have

$$0 < T_{0\beta^{(1)}} < T_{0\beta^{(2)}} < T_{0\beta^{(3)}} ,$$

and hence if the Jacobian vanishes again at $\gamma = \gamma_1$, with $\beta^{(1)} < \beta^{(2)} < \beta^{(3)}$ for each $\gamma \in [\gamma_0,\gamma_1]$, then necessarily $T_{0\beta^{(1)}}((\alpha,\vec{\beta})(\gamma_1)) = 0$, and we may clearly take the parametrization such that $\beta^{(1)}(\gamma)$ will be increasing in γ (for $\gamma \in (\gamma_0,\gamma_1)$), as desired. This possibility is thus ruled out by Lemma 10.5.

If, on the other hand,

$$T_{0\beta^{(2)}}((\alpha,\vec{\beta})(\gamma)) = 0 ,$$

then for $\gamma_0 < \gamma < \gamma_0 + \delta$ (δ small) and $(\alpha,\vec{\beta}) = (\alpha,\vec{\beta})(\gamma)$, we have

$$T_{0\beta^{(1)}} < T_{0\beta^{(2)}} < 0 < T_{0\beta^{(3)}} ,$$

and either $T_{0\beta^{(2)}}$ or $T_{0\beta^{(3)}}$ could subsequently vanish for $\gamma_1 > \gamma_0$. We may take the parametrization such that $\beta^{(2)}(\gamma)$ will be decreasing in γ for $\gamma \in (\gamma_0,\gamma_1)$, and so the subsequent vanishing of $T_{0\beta^{(2)}}$ has been ruled out by Lemma 10.5.

If $T_{0\beta^{(3)}}((\alpha,\vec{\beta})(\gamma_1)) = 0$, then we would have, for $\gamma_1 < \gamma < \gamma_1 + \delta$ (δ small) and $(\alpha,\vec{\beta}) = (\alpha,\vec{\beta})(\gamma)$,

$$T_{0\beta^{(1)}} < T_{0\beta^{(2)}} < T_{0\beta^{(3)}} < 0 ,$$

and so $T_{0\beta^{(2)}}$ could not vanish again without $T_{0\beta^{(3)}}$ vanishing first, and following a similar line of reasoning, we see that this occurrence has been ruled out by Lemma 10.5. □

LEMMA 10.6. *Suppose that $(\alpha,\vec{\beta},x)(\gamma)$ is a continuous 1 - parameter family of solutions to*

$$\begin{aligned} Y - \pi x \mathbf{e}_3 &= 0 \,, \\ (\alpha,\vec{\beta},x)(\gamma=0) &= ((\alpha,\vec{\beta})(\underline{x},t),\underline{x}) \,, \\ (\alpha,\vec{\beta},x)(\gamma=1) &= ((\alpha,\vec{\beta})(\bar{x},t),\bar{x}) \,, \\ \beta^{(1)}(\gamma) < \beta^{(2)}(\gamma) < \beta^{(3)}(\gamma) & \qquad \textit{for all } \gamma \in (0,1) \,, \\ (\alpha,\vec{\beta})(\gamma') &\neq (\alpha,\vec{\beta})(\gamma'') \ \textit{ for } \gamma' \neq \gamma'' \,. \end{aligned}$$

Then for all $\gamma \in (0,1)$, we have

$$\frac{\partial Y(\alpha,\vec{\beta})}{\partial(\alpha,\vec{\beta})}((\alpha,\vec{\beta})(\gamma)) \neq 0 \,.$$

Proof. If there is a $\gamma_0 \in (0,1)$ such that $\frac{\partial Y(\alpha,\vec{\beta})}{\partial(\alpha,\vec{\beta})} = 0$, then necessarily one of $\{T_{0\beta^{(k)}}\}_{k=1}^3$, say $T_{0\beta^{(j)}}$, must be zero. However, from the proof of Lemma 10.5, this implies that the curve $\beta^{(j)}(x,t)$ has an infinite derivate at $x(\gamma_0)$, and that all the curves become multi-sheeted.

In order for the solution $(\alpha,\vec{\beta},x)$ to match the known unique solution for γ near 1, (x near $\bar{x}$), the curve $\beta^{(j)}(x,t)$ must exhibit another infinite derivative. Indeed, for $\gamma < \gamma_0$, the slope of the curve $\beta^{(j)}(x,t)$ has the correct sign to match to the solution valid for γ near 1 (see I4, I8 above), while for $\gamma_0 < \gamma$, the slope of $\beta^{(j)}(x,t)$ has changed signs.

However, the only way that $\beta^{(j)}(x)$ can exhibit two distinct locations of infinite derivatives is if there exists $\gamma_1 > \gamma_0$ such that $T_{0\beta^{(j)}}((\alpha,\vec{\beta})(\gamma_1)) = 0$, and $(\alpha,\vec{\beta})(\gamma_1) \neq (\alpha,\vec{\beta})(\gamma_0)$. But this contradicts Corollary 10.2. □

LEMMA 10.7. *There exists a continuous 1-parameter family of solutions $(\alpha,\vec{\beta},x)(\gamma)$ to*

$$\begin{aligned} Y - \pi x \mathbf{e}_3 &= 0 \,, \\ (\alpha,\vec{\beta},x)(\gamma=0) &= ((\alpha,\vec{\beta})(\underline{x},t),\underline{x}) \,, \\ (\alpha,\vec{\beta},x)(\gamma_1) &\neq (\alpha,\vec{\beta},x)(\gamma_2) \ \textit{ for } \gamma_1 \neq \gamma_2 \,, \end{aligned}$$

such that for some $\gamma^* > 0$, two or more of the $\{\beta^{(j)}\}_{j=1}^3$ coincide.

Proof. For x near $\underline{x}$, we have already established the existence of such a 1-parameter family of solutions

$$(\alpha,\vec{\beta},x)(\gamma) = ((\alpha,\vec{\beta})(x,t),x)(\gamma = x) \,.$$

Suppose that we have a solution for all $x \in [\underline{x}, x^*]$, for which $\{\beta^{(j)}\}_{j=1}^3$ are distinct. Then, as in the proof of Lemma 10.5, we may extend this solution. Indeed, either the Jacobian (396) is nonzero, in which case we may extend by increasing x, or one of $\{T_{0\beta^{(k)}}\}_{k=1}^3$, say $T_{\beta^{(j)}}$, is zero. In this case, we have

$$\frac{\partial(T_0, T_1, T_2 - \pi x, N)}{\partial(\alpha, \beta^{(k)}, \beta^{(l)}, x)} \neq 0 \ , \ for\ l \neq j \neq k \neq l \ ,$$

and hence we may solve for $(\alpha, \beta^{(k)}, \beta^{(l)}, x)$ in terms of $\beta^{(j)}$, and we have again extended the solution.

We observe now that for such a continuous 1-parameter family of solutions, the number of times that one of $\{T_{0\beta^{(k)}}\}_{k=1}^3$ can vanish is finite. This is guaranteed by Corollary 10.2 of Lemma 10.5. Thus, as $(\alpha, \vec{\beta})$ is bounded, two or more of the $\beta^{(k)}$'s must cross. Following this procedure, one may conclude easily that $(\alpha, \vec{\beta}, x)(\gamma_1) \neq (\alpha, \vec{\beta}, x)(\gamma_2)$ for $\gamma_1 \neq \gamma_2$. □

LEMMA 10.8. *Suppose $(\alpha, \vec{\beta}, x)(\gamma)$ is a continuous 1-parameter family of solutions to*

$$\begin{aligned} Y - \pi x \mathbf{e}_3 &= 0 \ , \\ (\alpha, \vec{\beta}, x)(\gamma = 0) &= ((\alpha, \vec{\beta})(\underline{x}, t), \underline{x}) \ , \\ (\alpha, \vec{\beta}, x)(\gamma_1) &\neq (\alpha, \vec{\beta}, x)(\gamma_2) \ \ for\ \gamma_1 \neq \gamma_2 \ , \end{aligned}$$

and suppose that $(\alpha, \vec{\beta}, x)(\gamma^*)$ is a degenerate solution, i.e. that two or more of the $\{\beta^{(k)}\}_{k=1}^3$ coincide ($0 < \gamma^*$). Then

$$(\alpha, \vec{\beta}, x)(\gamma^*) = ((\alpha, \vec{\beta})(\bar{x}, t), \bar{x}) \ .$$

Proof. The two solutions $((\alpha, \vec{\beta})(\underline{x}, t), \underline{x})$ and $((\alpha, \vec{\beta})(\bar{x}, t), \bar{x})$ are the unique solutions to $Y - \pi x \mathbf{e}_3 = 0$ that satisfy the degenerate conditions $\beta^{(1)} < \beta^{(2)} = \beta^{(3)}$ and $\beta^{(1)} = \beta^{(2)} < \beta^{(3)}$, respectively.

By continuing the solution valid for x near $\underline{x}$, we find that the only other degenerate solutions allowed are those for which all the $\beta^{(j)}$'s coincide. There are 2 possibilities.

A. For $\gamma \uparrow \gamma^*$, we have

$$\begin{aligned} & 0 < T_{0\beta^{(1)}} < T_{0\beta^{(2)}} < T_{0\beta^{(3)}} \ , \\ & x(\gamma) \downarrow x(\gamma^*) \ , \\ & \beta^{(1)}(\gamma) < \beta^{(2)}(\gamma) < \beta(\gamma^*) < \beta^{(3)}(\gamma) \ , \\ & \alpha(\gamma) < \alpha(\gamma^*), \end{aligned}$$

B. For $\gamma \uparrow \gamma^*$, we have

$$\begin{aligned} & T_{0\beta^{(1)}} < T_{0\beta^{(2)}} < T_{0\beta^{(3)}} < 0 \ , \\ & x(\gamma) \uparrow x(\gamma^*) \ , \\ & \beta^{(1)}(\gamma) < \beta^{(2)}(\gamma) < \beta(\gamma^*) < \beta^{(3)}(\gamma) \ , \\ & \alpha(\gamma^*) < \alpha(\gamma). \end{aligned}$$

Before establishing that neither A nor B can occur, we must establish the following claim.

Claim Let $(\alpha, \vec{\beta}, x)(\gamma)$ be a continuous 1-parameter family of solutions to $Y - \pi x e_3 = 0$, defined for $0 \leq \gamma \leq \gamma^*$, such that $(\alpha, \vec{\beta}, x)(\gamma_1) \neq (\alpha, \vec{\beta}, x)(\gamma_2)$ for $\gamma_1 \neq \gamma_2$, and satisfying

$$\vec{\beta}(\gamma^*) = \beta(\gamma^*)(1,1,1)\ .$$

Then the pair $(\alpha(\gamma^*), \beta(\gamma^*))$ solve the zero gap Riemann - Hilbert problem, and furthermore $(x(\gamma^*), \beta(\gamma^*))$ are the coordinates of a singularity in the multi-sheeted zero - gap curve $\beta(x,t)$.

Proof: Consider, for γ near γ^*,

$$\begin{aligned} F &= T_2 - \beta^{(3)}(\gamma) T_1 \\ &= \left(\int_{\alpha}^{\beta^{(1)}} + \int_{\beta^{(2)}}^{\beta^{(3)}} \right) \frac{i\theta'(\lambda,t)}{R^{1/2}} \lambda(\lambda - \beta^{(3)}) d\lambda\ . \end{aligned}$$

Then, as $\gamma \uparrow \gamma^*$, we have

$$F \to \int_{\alpha(\gamma^*)}^{\beta(\gamma^*)} \frac{i\theta'(\lambda,t)}{\left[(\lambda - \alpha(\gamma^*))(\lambda - \beta(\gamma^*))\right]^{1/2}} \lambda d\lambda\ ,$$

and so $(\alpha(\gamma^*), \beta(\gamma^*))$ solves

$$\int_{\alpha(\gamma^*)}^{\beta(\gamma^*)} \frac{i\theta'(\lambda,0)}{\left[(\lambda - \alpha(\gamma^*))(\lambda - \beta(\gamma^*))\right]^{1/2}} \left(\lambda - \frac{\alpha(\gamma^*) + \beta(\gamma^*)}{2} \right) d\lambda = \pi x\ .$$

Similarly, $G = T_1 - \beta_3(\gamma) T_0$ satisfies

$$G \to \int_{\alpha(\gamma^*)}^{\beta(\gamma^*)} \frac{i\theta'(\lambda,0)}{\left[(\lambda - \alpha(\gamma^*))(\lambda - \beta(\gamma^*))\right]^{1/2}} d\lambda \ - \pi t \ = \ 0\ .$$

Thus the pair $(\alpha(\gamma^*), \beta(\gamma^*))$ solves the zero gap Riemann - Hilbert problem.

To show that $(x(\gamma^*), \beta(\gamma^*))$ are the coordinates of a singularity in the multi-sheeted zero gap curve $\beta(x,t)$, we must establish (see the discussion at the beginning of Chapter 6) $T_\beta(\alpha(\gamma^*), \beta(\gamma^*)) = 0$, where $T(\alpha, \beta)$ is the function given by

$$T(\alpha, \beta) = \int_{\alpha}^{\beta} \frac{i\theta'(\lambda,0)}{\left((\lambda - \alpha)(\lambda - \beta)\right)^{1/2}} d\lambda\ .$$

Recall that we have (c.f. (330))

$$T_0(\alpha, \vec{\beta}) = \int_{\alpha_0(x_-(\beta^{(1)}(\gamma)))}^{\alpha(\gamma)} \frac{1}{R^{1/2}} (-\psi_0) + \int_{\beta^{(2)}(\gamma)}^{\beta^{(3)}(\gamma)} \frac{i}{R^{1/2}} (\theta' - H\psi_0)\, d\lambda\ ,$$

where $\psi_0 = \psi_0(\lambda, x_-(\beta^{(1)}(\gamma)))$ is the maximizer for $t = 0$, i.e. $\psi_0(\lambda, x_-(\beta^{(1)}(\gamma)))$ solves the zero gap Riemann - Hilbert problem for $t = 0$, and

$x_-(\lambda)$ is the inverse function of $\beta_0(y)$. Letting $\gamma \uparrow \gamma^*$, using the fact that for λ near $\beta^{(1)}$,

$$(\theta' - H\psi_0)(\lambda) \approx 2\left(\frac{\lambda - \beta^{(1)}}{\lambda - \alpha_0(x_-(\beta^{(1)}))}\right)^{1/2} \frac{d}{d\beta^{(1)}} x_-(\beta^{(1)}) ,$$

which can be obtained by expanding

$$\theta' - H\psi_0 = \int_{x_-(\beta^{(1)}(\gamma))}^{x_-(\lambda)} \frac{1}{\left((\lambda - \alpha_0(y))(\lambda - \beta_0(y))\right)^{1/2}} dy ,$$

we have

$$\lim_{\gamma\uparrow\gamma^*} T_0 = \int_{\alpha_0(x_-(\beta(\gamma^*)))}^{\alpha} \frac{\psi_0}{\left((\lambda - \alpha(\gamma^*))(\lambda - \beta(\gamma^*))\right)^{1/2}} \frac{d\lambda}{\lambda - \beta(\gamma^*)} + \\ + \frac{2\pi \frac{d}{d\beta} x_-(\beta(\gamma^*))}{\left((\beta(\gamma^*) - \alpha_0)(\beta(\gamma^*) - \alpha(\gamma^*))\right)^{1/2}} = 0 ,$$

which is precisely the equation $T_\beta = 0$ (compare with (245)). This completes the proof of the claim.

To complete the proof of the Lemma, we must show that neither A nor B can occur. Let (x^*, β^*) represent the right-most singularity in the multi-sheeted zero gap curve $\beta(x,t)$, and suppose $(\alpha, \vec{\beta}, x)(\gamma^*) = (\alpha^*, \beta^*(1,1,1), x^*)$. We leave the reader to verify, using (381), that since $\beta^* < \beta_c$, where β_c is the height of the cubic singularity in the initial curve $\beta_0(x)$, $T_{0\beta(k)}\Big|_{(\alpha,\vec{\beta},x)(\gamma^*)} < 0$ for $k = 1,2,3$. Thus this situation corresponds to case B.

We recall

$$R_j(\lambda, y, t) = (\lambda - \alpha_j(y,t))(\lambda - \beta_j(y,t)) \ , \ j = 0,1,2,$$

where $\{\alpha_j(y,t), \beta_j(y,t)\}_{j=0}^2$ represent the multi-sheeted solution to the zero - gap Riemann - Hilbert problem.

We will show that for $\alpha^* < \alpha$, $\beta^{(1)} < \beta^{(2)} < \beta^* < \beta^{(3)}$, $(\alpha, \vec{\beta})$ close to $(\alpha^*, \beta^*(1,1,1))$, and $z \in (\beta^{(1)}, \beta^{(2)})$, we have

$$\text{(403)} \quad H\psi_{1-gap}(z, \alpha, \vec{\beta}) - \theta'(z,t) = \\ R(z)^{1/2}\left[\frac{8/3}{(\beta^* - \alpha^*)\hat{c}^2} + O\left((\alpha, \vec{\beta}) - (\alpha^*, \beta^*(1,1,1))\right)\right] ,$$

and hence $N = \int_{\beta^{(1)}}^{\beta^{(2)}} H\psi - \theta' \, dz \neq 0$. This will rule out case B. As case A is ruled out by a similar argument, we omit it.

For a given $(\alpha, \vec{\beta})$, the associated 1-gap quantity $f = \psi + iH\psi$ is given by

$$f = \frac{R(z)^{1/2}}{\pi i}\left(\int_{\alpha}^{\beta^{(1)}} + \int_{\beta^{(2)}}^{\beta^{(3)}}\right) \frac{i\theta'(\lambda, t)}{R(\lambda)^{1/2}} \frac{d\lambda}{\lambda - z} .$$

Let $z \in (\beta^{(1)}, \beta^{(2)})$. We have, interchanging orders of integration,

$$\frac{1}{\pi i}\left(\int_{\alpha}^{\beta^{(1)}} + \int_{\beta^{(2)}}^{\beta^{(3)}}\right) \frac{i\theta'(\lambda,t)}{R(\lambda)^{1/2}} \frac{d\lambda}{\lambda - z} =$$

$$\frac{1}{\pi i}\left\{\int_{0}^{x_-(\alpha)} \left[\int_{\alpha}^{\alpha_0(y,t)} + \int_{\beta_0(y,t)}^{\beta^{(1)}} + \int_{\beta^{(2)}}^{\beta^{(3)}}\right] + \right.$$

$$+ \int_{x_-(\alpha)}^{x_-(\beta^{(1)})} \left[\int_{\alpha_0(y,t)}^{\alpha} + \int_{\beta_0(y,t)}^{\beta^{(1)}} + \int_{\beta^{(2)}}^{\beta^{(3)}}\right] +$$

$$\left. + \int_{x_-(\beta^{(1)})}^{x_-(z)} \left[\int_{\alpha_0(y,t)}^{\alpha} + \int_{\beta^{(1)}}^{\beta_0(y,t)} + \int_{\beta^{(2)}}^{\beta^{(3)}}\right]\right\} \cdot$$

$$\cdot \frac{i}{(RR_0(\lambda,t))^{1/2}} \frac{d\lambda dy}{\lambda - z} +$$

$$+ \frac{1}{\pi i}\left\{- \int_{\alpha_0}^{\alpha} \int_{x_-(\lambda)}^{x_-(z)} - \int_{\beta^{(1)}}^{z} \int_{x_-(\lambda)}^{x_-(z)} + \int_{\beta^{(2)}}^{\beta^{(3)}} \left(\int_{x_-(z)}^{x^*} + \int_{x^*}^{x_-^1(\lambda,t)}\right)\right\} \cdot$$

$$\cdot \frac{i}{(RR_0(\lambda,y,t))^{1/2}} \frac{dy d\lambda}{\lambda - z}$$

$$= \frac{i\theta'(\lambda,t)}{R(z)^{1/2}} + \frac{1}{\pi i}\left(\int_{\alpha_0}^{\alpha} + \int_{\beta^{(1)}}^{z}\right) \frac{-\psi_0}{R^{1/2}} \frac{d\lambda}{\lambda - z} +$$

$$+ \frac{1}{\pi i} \int_{\beta^{(2)}}^{\beta^{(3)}} \frac{i}{R^{1/2}} \left(\theta' - H\psi_0\right) \frac{d\lambda}{\lambda - z} ,$$

where we have used the representation (328) of the function $\theta'(\lambda, t)$. Thus we have

$$(404) H\psi - \theta'(z,t) =$$

$$= \frac{-R(z)^{1/2}}{\pi}\left(\int_{\alpha_0}^{\alpha} + \int_{\beta^{(1)}}^{z}\right) \frac{-\psi_0}{R^{1/2}} \frac{d\lambda}{\lambda - z} -$$

$$\frac{R(z)^{1/2}}{\pi} \int_{\beta^{(2)}}^{\beta^{(3)}} \frac{i}{R^{1/2}} \left(\theta' - H\psi_0\right) \frac{d\lambda}{\lambda - z} .$$

We now present the following expression for the zero gap quantities in (404), valid for λ near z, or λ near $\alpha_0 \equiv \alpha_0(x_-(z))$:

$$\psi_0 = -i \left[\frac{(\lambda - \alpha_0)(\lambda - z)}{2}\right]^{1/2} \int_{-1}^{1} \frac{f_\mu(\mu(x,z,\lambda),\lambda)}{(1-x)^{1/2}} dx ,$$

$$\theta' - H\psi_0 = \left[\frac{(\lambda - \alpha_0)(\lambda - z)}{2}\right]^{1/2} \int_{-1}^{1} \frac{f_\mu(\mu(x,z,\lambda),\lambda)}{(1-x)^{1/2}} dx ,$$

obtained by the successive variable changes

$$\begin{aligned}(\lambda-\alpha_0(y,t))(\lambda-\beta_0(y,t)) &= \lambda^2-\mu \; (\textit{determines } y=f(\mu,\lambda)) \,, \\ \lambda^2-\mu &= (\lambda-\alpha_0)\,(\lambda-z)\,\frac{1-x}{2} \,.\end{aligned}$$

Then, using the known asymptotic prescription of the zero gap quantities $\alpha_0(y,t)$, $\beta_0(y,t)$, valid for y near x^*,

$$\begin{aligned}\beta_0(y,t) &= \beta^* \pm \hat{c}(x^*-y)^{1/2}+\cdots \,, \\ \alpha_0(y,t) &= \alpha^* + c_\alpha(y-x^*)+\cdots \,,\end{aligned}$$

we obtain, for λ near z,

$$f_\mu \approx \frac{-2}{[\hat{c}(\lambda-\alpha^*)]^2}\left[(\lambda-\alpha^*)(\lambda-\beta^*)-(\lambda-\alpha_0)(\lambda-z)\frac{1-x}{2}\right] \,,$$

and hence

$$\int_{-1}^{1} \frac{f_\mu}{(1-x)^{1/2}}dx \approx \frac{-4\sqrt{2}}{[\hat{c}(\lambda-\alpha^*)]^2}\left[(\lambda-\alpha^*)(\lambda-\beta^*)-\frac{1}{3}(\lambda-\alpha_0)(\lambda-z)\right] \,.$$

Thus we have

$$\begin{aligned}&\frac{1}{\pi i}\int_{\beta(1)}^{z} \frac{-\psi_0}{R^{1/2}}\frac{d\lambda}{\lambda-z}+\frac{1}{\pi i}\int_{\beta(2)}^{\beta(3)} \frac{i}{R^{1/2}}\,(\theta'-H\psi_0)\,\frac{d\lambda}{\lambda-z} \\ &\approx \frac{1}{\pi i}\left(\int_{\beta(1)}^{z}+\int_{\beta(2)}^{\beta(3)}\right)\left[\frac{(\lambda-\alpha_0)(\lambda-z)}{R}\right]^{1/2}\frac{-4i}{[\hat{c}(\lambda-\alpha^*)]^2}\times \\ &\qquad\times\left[(\lambda-\alpha^*)(\lambda-\beta^*)-\frac{1}{3}(\lambda-\alpha_0)(\lambda-z)\right]\frac{d\lambda}{\lambda-z} \\ &\approx \frac{-4}{\hat{c}^2(\beta^*-\alpha^*)}\left[-\frac{2}{3}i\right] \approx \frac{8i}{3\hat{c}^2(\beta^*-\alpha^*)} \,. \qquad (405)\end{aligned}$$

The last step is to obtain the following bound,

$$\left|\frac{1}{\pi i}\int_{\alpha_0}^{\alpha} \frac{-\psi_0}{R^{1/2}}\frac{d\lambda}{\lambda-z}\right| = O\big((\alpha,\vec{\beta})-(\alpha^*,\beta^*(1,1,1))\big) \,. \qquad (406)$$

We have

$$\begin{aligned}&\frac{1}{\pi i}\int_{\alpha_0}^{\alpha} \frac{-\psi_0}{R^{1/2}}\frac{d\lambda}{\lambda-z} \\ &= \frac{1}{\pi i}\int_{\alpha_0}^{\alpha} \frac{i}{\sqrt{2}}\left[\frac{(\lambda-\alpha_0)(\lambda-z)}{R}\right]^{1/2}\int_{-1}^{1}\frac{f_\mu(\mu(x,z,\lambda),\lambda)}{(1-x)^{1/2}}\frac{dx\,d\lambda}{\lambda-z} \,.\end{aligned}$$

Now for $\lambda \in (\alpha_0,\alpha)$, $y \in (x_-(\lambda),x_-(z)) \subset (x_-(\alpha),x_-(z))$, we have that $y=f(\mu,\lambda)$, determined by

$$(\lambda-\alpha_0(y,t))(\lambda-\beta_0(y,t))=\lambda^2-\mu$$

satisfies

$$0 < f_\mu < \frac{1}{\alpha_0'(y,t)\;(\lambda-\beta_0(y,t))} < c_1 \,.$$

Therefore

$$\left|\int_{-1}^{1} \frac{f_\mu}{(1-x)^{1/2}} dx\right| \le c_2 \ ,$$

and hence

$$\left|\frac{1}{\pi i}\int_{\alpha_0}^{\alpha} \frac{i}{\sqrt{2}}\left[\frac{(\lambda-\alpha_0)(\lambda-z)}{R}\right]^{1/2} \int_{-1}^{1} \frac{f_\mu(\mu(x,z,\lambda),\lambda)}{(1-x)^{1/2}} \frac{dx d\lambda}{\lambda - z}\right|$$

$$\le c_3 \left|\frac{1}{\pi i}\int_{\alpha_0}^{\alpha} i\left(\frac{\lambda-\alpha_0}{\lambda-\alpha}\right)^{1/2} \frac{d\lambda}{\left[(\lambda-\beta^{(1)})(\lambda-z)(\lambda-\beta^{(2)})(\lambda-\beta^{(3)})\right]^{1/2}}\right|$$

$$\le c_4 \left|\frac{1}{\pi i}\int_{\alpha_0}^{\alpha} i\left(\frac{\lambda-\alpha_0}{\lambda-\alpha}\right)^{1/2} d\lambda\right| = c_4 \left|\frac{\alpha-\alpha_0}{2}\right| \ .$$

This proves (406). Combining (405), (406), and (404) establishes (403). This completes the proof of Lemma 10.8, and Theorem 10.1.

□

CHAPTER 11

Large time calculations

In this Chapter we consider the general class of data of Chapters 5 - 8, for which $\beta^0(x)$ possesses one (nondegenerate) critical point, a maximum, and $\alpha^0(x)$ is strictly monotonically decreasing, and $\alpha^0(0) = \beta^0(0)$, $\alpha^0(1) = \beta^0(1)$.

We consider the "Right ansatz",

$$H\psi = \theta' \quad , \quad \lambda \in (\alpha \, , \, \beta) \; , \tag{407}$$

$$\psi = 0 \quad , \quad \lambda \in (-\infty \, , \, \alpha) \cup (1, \infty) \; , \tag{408}$$

$$\psi = \phi \quad , \quad \lambda \in (\beta \, , \, 1) \; . \tag{409}$$

In standard fashion, we obtain the equations

$$T(\alpha, \beta) \equiv \int_\alpha^\beta \frac{1}{[(\lambda - \alpha)(\beta - \lambda)]^{1/2}} \theta' d\lambda + \int_\beta^1 \frac{1}{[(\lambda - \alpha)(\lambda - \beta)]^{1/2}} \phi d\lambda = \pi t \; , \tag{410}$$

$$X(\alpha, \beta) \equiv \int_\alpha^\beta \frac{\lambda - m}{[(\lambda - \alpha)(\beta - \lambda)]^{1/2}} \theta' d\lambda + \int_\beta^1 \frac{\lambda - m}{[(\lambda - \alpha)(\lambda - \beta)]^{1/2}} \phi d\lambda = \pi x \; . \tag{411}$$

LEMMA 11.1. *For $-1 < \alpha^0(x_+(\beta)) < \alpha < \beta < 1$, we have $T_\beta < 0$.*

Proof.

We consider $\beta < 0$ first. We have

$$\begin{aligned} T &= \int_\alpha^\beta \int_0^{x_-(\lambda)} \frac{i}{[R(\lambda)R^0(\lambda,y)]^{1/2}}\, dy d\lambda \\ &+ \int_\beta^1 \int_{x_-(\lambda)}^{x_+(\lambda)} \frac{i}{[R(\lambda)R^0(\lambda,y)]^{1/2}}\, dy d\lambda \\ &= \int_0^{x_-(\beta)} \left(\int_\alpha^\beta + \int_{\alpha^0(y)}^{\beta^0(y)} \right) \frac{i}{[R(\lambda)R^0(\lambda,y)]^{1/2}}\, d\lambda dy \\ &+ \int_{x_-(\beta)}^{x_-(\alpha)} \left(\int_\alpha^{\alpha^0(y)} + \int_\beta^{\beta^0(y)} \right) \frac{i}{[R(\lambda)R^0(\lambda,y)]^{1/2}}\, d\lambda dy \\ &+ \int_{x_-(\alpha)}^{x_+(\beta)} \int_\beta^{\beta^0(y)} \frac{i}{[R(\lambda)R^0(\lambda,y)]^{1/2}}\, d\lambda dy\ , \end{aligned}$$

and the first two terms vanish by countour integration, yielding

$$T(\alpha,\beta) = \int_{x_-(\alpha)}^{x_+(\beta)} \int_\beta^{\beta^0(y)} \frac{i}{[R(\lambda)R^0(\lambda,y)]^{1/2}}\, d\lambda dy\ , \tag{412}$$

and, also by contour integration, we have

$$T(\alpha,\beta) = \int_{x_-(\alpha)}^{x_+(\beta)} \int_{\alpha^0(y)}^{\alpha} \frac{-i}{[R(\lambda)R^0(\lambda,y)]^{1/2}}\, d\lambda dy\ . \tag{413}$$

Differentiating (413), we obtain

$$\begin{aligned} T_\beta &= \frac{\pi \partial_\beta x_+(\beta)}{[(\beta-\alpha)(\beta-\alpha^0(x_+(\beta)))]^{1/2}} + \\ &- \int_{x_-(\alpha)}^{x_+(\beta)} \int_{\alpha^0(y)}^{\alpha} \frac{i}{[R(\lambda)R^0(\lambda,y)]^{1/2}} \frac{\frac{1}{2}}{\lambda-\beta}\, d\lambda dy \quad < 0\ . \end{aligned}$$

Similarly, for $\alpha^0(x_+(\beta)) < \alpha < 0 < \beta$, we have the representation for T,

$$T = \int_{x_-(\alpha)}^{x_+(\beta)} \int_\beta^{\beta^0(y)} \frac{i}{[R(\lambda)R^0(\lambda,y)]^{1/2}}\, d\lambda dy\ ,$$

and hence we have again

$$\begin{aligned} T_\beta &= \frac{\pi \partial_\beta x_+(\beta)}{[(\beta-\alpha)(\beta-\alpha^0(x_+(\beta)))]^{1/2}} + \\ &- \int_{x_-(\alpha)}^{x_+(\beta)} \int_{\alpha^0(y)}^{\alpha} \frac{i}{[R(\lambda)R^0(\lambda,y)]^{1/2}} \frac{\frac{1}{2}}{\lambda-\beta}\, d\lambda dy \quad < 0\ . \end{aligned}$$

Similar calculations apply for the case $0 < \alpha < \beta$. □

LEMMA 11.2. *For $\alpha^0(x_+(\beta)) < \alpha < 0$, $\alpha < \beta$, we have $T_\alpha > 0$.*

Proof. This was established in Lemma 5.1. □

A direct calculation using (144), shows $\theta''(\lambda) \to \infty$ as $\lambda \to 1$. Hence, there must exist a $\bar{\lambda} < 1$ such that

$$\lambda > \bar{\lambda} \;\Rightarrow \theta''(\lambda) > 0 \; .$$

We chose β^*, $\bar{\lambda} + \delta < \beta^* < 1$, for some $\delta > 0$. Now we define t^* by

$$t^* \equiv \frac{1}{\pi} \max_{\substack{-1 \le \alpha \le \bar{\lambda} \\ \beta^* \le \beta \le 1}} T(\alpha, \beta) \; < \infty \; , \tag{414}$$

and t^{**} such that

$$t^{**} - \sup_{0<\lambda\le\beta^*} \theta'(\lambda) \; > \; 0 \; , \tag{415}$$

$$\inf_{0<\lambda\le\beta^*} \theta''(\lambda) + \left[t^{**} - \sup_{0<\lambda\le\beta^*} \theta'(\lambda) \right] > 0 \; . \tag{416}$$

Now we take

$$\bar{t} \equiv \max\{t^*, t^{**}\} \; < \; \infty \; . \tag{417}$$

THEOREM 11.1. *(c.f. Theorem 5.2) For $t > \bar{t}$, and all $\beta \in (-1, 1)$, there exists a unique solution $\alpha(\beta, t) < \beta$ to $T(\alpha(\beta, t), \beta) = \pi t$.*

Proof.

We note first that a solution $\alpha(\beta, t)$ to the equation $T = \pi t$ always exists, as the initial data satisfies $T(\alpha^0(x), \beta^0(x)) = 0$, and further if we let $\alpha \uparrow \beta$, then $T(\alpha, \beta) \to \infty$. We will show that for $t > \bar{t}$,

$$T(\alpha, \beta) = \pi t \Rightarrow \; T_\alpha > 0 \;\; .$$

A simple calculation then shows that the solution $\alpha = \alpha(\beta, t)$ of $T(\alpha, \beta)$ is unique for $t > \bar{t}$.

From Lemma 11.2 above, we have that $T_\alpha > 0$ for $\alpha^0(x_+(\beta)) < \alpha < 0$, $\alpha < \beta$, and hence we need only consider $0 < \alpha < \beta$. We begin with $\beta < \beta^*$, and re - write T in the form

$$\begin{aligned} T \;\; &= \;\; \int_{-\pi/2}^{\pi/2} \theta' \left(\frac{\beta - \alpha}{2} \sin z + \frac{\beta + \alpha}{2} \right) \, dz \; + \\ & \quad + \int_\beta^1 \frac{\phi}{R^{1/2}} \, d\lambda \\ &= \;\; I \; + \; II \; . \end{aligned} \tag{418}$$

We compute $\frac{\partial I}{\partial \alpha}$, $\frac{\partial II}{\partial \alpha}$:

$$(419) \quad \frac{\partial}{\partial \alpha} I = \int_{-\frac{\pi}{2}}^{\frac{\pi}{2}} \theta'' \left(\frac{\beta - \alpha}{2} \sin z + \frac{\beta + \alpha}{2} \right) \frac{1 - \sin z}{2} \geq \min_{\lambda \in [\alpha, \beta]} \theta''(\lambda) \pi \ ,$$

$$(420) \quad \frac{\partial}{\partial \alpha} II = \int_\beta^1 \frac{\phi}{R^{1/2}} \frac{1/2}{\lambda - \alpha} \, d\lambda$$

$$> \frac{1/2}{1 - \alpha} \int_\beta^1 \frac{\phi}{R^{1/2}} d\lambda = \frac{1/2}{1 - \alpha} [\pi t - I] \ ,$$

where we have used the fact that α solves $T(\alpha, \beta) = \pi t$. Hence, from (419) and (420), we have

$$T_\alpha > \frac{\pi}{2} \inf_{0 < \lambda \leq \beta^*} \theta''(\lambda) + \frac{1/2}{1 - \alpha} \left[\pi t - \int_{-\pi/2}^{\pi/2} \theta' \left(\frac{\beta - \alpha}{2} \sin z + \frac{\beta + \alpha}{2} \right) dz \right]$$

$$> \frac{\pi}{2} \inf_{0 < \lambda \leq \beta^*} \theta''(\lambda) + \frac{\pi/2}{1 - \alpha} \left[t - \sup_{0 < \lambda \leq \beta^*} \theta'(\lambda) \right] > 0 \ ,$$

where the last inequality follows from (415),(416), and the assumption $t > t^{**}$.

Now we consider $\beta \geq \beta^*$. As we're supposing that α solves $T(\alpha, \beta) = \pi t$, we see from

- $t > \bar{t} \geq t^*$,
- $\beta > \beta^*$,

and the definition (414), that $\alpha > \bar{\lambda}$.

Using (418), we obtain, as above,

$$(421) \qquad T_\alpha = \int_{-\pi/2}^{\pi/2} \frac{1 - \sin z}{2} \theta'' \left(\frac{\beta - \alpha}{2} \sin z + \frac{\beta + \alpha}{2} \right) dz \ + \int_\beta^1 \frac{\phi}{R^{1/2}} \frac{1/2}{\lambda - \alpha} \, d\lambda \ .$$

Now we know that $\frac{\beta-\alpha}{2} \sin z + \frac{\beta+\alpha}{2} > \alpha$. Hence $\theta''(\frac{\beta-\alpha}{2} \sin z + \frac{\beta+\alpha}{2}) > 0$, and so the first term in (421) above is positive. As the second is clearly positive as well, we have

$$T_\alpha > 0 \ ,$$

and thus $T_\alpha > 0$ in the case $0 < \alpha < \beta$. Finally we note that the case $\alpha < \alpha^0(x_+(\beta))$ cannot occur as

$$T(\alpha, \beta) = \int_{x_+(\beta)}^{x_-(\alpha)} \int_\alpha^{\alpha^0(y)} \frac{i}{(R(\lambda) R^0(\lambda, y))^{1/2}} d\lambda dy$$

is negative (c.f. end of Chapter 9). This completes the proof of the Theorem.
□

Inserting the function $\alpha(\beta,t)$ into $X(\alpha,\beta)$, we obtain the familiar function $X(\beta,t)$, which satisfies

$$\partial_\beta X(\beta,t) \quad = \quad (\beta-\alpha)T_\beta \ ,$$

and hence, from Lemma 11.1, we conclude that

$$\partial_\beta X(\beta,t) < 0 \ .$$

Now we define

$$x_\ell(t) \equiv \frac{1}{\pi}X(\alpha(1,t),1) \ < \frac{1}{\pi}X(-1,t) = 1 \ ,$$

and we have a unique solution to the equations (410-411) for $t > \bar{t}$, for $x \in (x_\ell, 1)$. An immediate consequence of the above considerations is that on $(x_\ell, 1)$ we have

$$\begin{pmatrix} \alpha_t & \alpha_x \\ \beta_t & \beta_x \end{pmatrix} = \frac{1}{(\beta-\alpha)T_\alpha T_\beta}\begin{pmatrix} \frac{\beta-\alpha}{2}T_\beta & -T_\beta \\ \frac{\beta-\alpha}{2}T_\alpha & T_\alpha \end{pmatrix}\begin{pmatrix} \pi & 0 \\ 0 & \pi \end{pmatrix} ,$$

and hence

$$\begin{aligned} \alpha_x(x,t) \quad &= \quad -\frac{\pi}{(\beta-\alpha)T_\alpha(\alpha(x,t),\beta(x,t))} \ < \ 0 \ , \\ \beta_x(x,t) \quad &= \quad \frac{\pi}{(\beta-\alpha)T_\beta(\alpha(x,t),\beta(x,t))} \ < \ 0 \ , \end{aligned}$$

i.e. $\alpha(x,t)$ and $\beta(x,t)$ are monotone decreasing on $(x_\ell(t), 1)$.

Using the shape of the curves $\alpha(x,t), \beta(x,t)$, we define $x(t) > x_\ell(t)$ by

(422) $$\beta(x(t),t) \quad = \quad \alpha(x_\ell(t),t) \ .$$

THEOREM 11.2. *For $t > \bar{t}$, $x \in (x(t),1)$, the solution to the Riemann - Hilbert problem (407 -409) with $\alpha(x,t), \beta(x,t)$ from the above considerations solves the maximization problem.*

Proof.

We must show that $\psi^*(\lambda;\ x,t)$ constructed as in (154 - 155) out of $\alpha(x,t)$, $\beta(x,t)$, satisfies

(423) $$\lambda > \beta(x,t) \ \Rightarrow \ \int_{\beta(x,t)}^{\lambda} H\psi^*(\mu;\ x,t) - \theta'(\mu,t)d\mu \ > \ 0 \ ,$$

(424) $$\lambda < \alpha(x,t) \ \Rightarrow \ \int_{\alpha(x,t)}^{\lambda} H\psi^*(\mu;\ x,t) - \theta'(\mu,t)d\mu \ < \ 0 \ ,$$

(425) $$\lambda \in (\alpha(x,t),\beta(x,t)) \Rightarrow 0 \ \le \ \psi^* \ \le \ \phi^*(\lambda;\ x,t) \ .$$

We begin with (423). For a fixed $x \in (x(t),1)$, we use the now familiar relation

$$H\psi_x^*(\lambda;\ x,t) = \frac{1}{R^t(\lambda,x)^{1/2}} \ , \quad \lambda \notin [\alpha(x,t),\beta(x,t)] \ ,$$

where

$$R^t(\lambda, y) = (\lambda - \alpha(x,t))(\lambda - \beta(x,t)) \ ,$$
$$R^t(\lambda, y)^{1/2} > 0 \ , \quad for\ \lambda > \beta(x,t) \ ,$$

to write

$$H\psi^*(\lambda\ ;\ x,t) \ = \ H\psi^*(\lambda\ ;\ x_+(\lambda),t) \ + \ \int_{x_+(\lambda)}^{x} \frac{1}{R^t(\lambda,y)^{1/2}}\, dy \ ,$$

for $\lambda > \beta(x,t)$, where $x_+(\lambda,t)$ is chosen such that $\beta(x_+(\lambda,t),t) = \lambda$. Hence, from the Riemann - Hilbert problem (407 - 409),

$$H\psi^*(\lambda\ ;\ x_+(\lambda),t) \ = \ \theta'(\lambda,t) \ ,$$

and so we conclude

$$H\psi(\lambda\ ;\ x,t) - \theta'(\lambda,t) \ = \ \int_{x_+(\lambda)}^{x} \frac{1}{R^t(\lambda,y)^{1/2}}\, dy \ > \ 0 \ .$$

Similarly, for $\lambda \in (\alpha(x,t),\beta(x,t))$, $\psi^*(\lambda;\ x,t)$ has the two representations

$$(426)\quad \psi^*(\lambda\ ;\ ,x,t) \ = \ \begin{cases} \psi^*(\lambda\ ;\ x_-(\lambda),t) \ + \ \int_{x_-(\lambda)}^{x} \frac{i}{R^t(\lambda,y)^{1/2}}\, dy \\ \psi^*(\lambda\ ;\ x_+(\lambda),t) \ + \ \int_{x_+(\lambda)}^{x} \frac{i}{R^t(\lambda,y)^{1/2}} \end{cases}$$

$$(427)\quad = \ \begin{cases} \int_{x_-(\lambda)}^{x} \frac{i}{R^t(\lambda,y)^{1/2}}\, dy \\ \phi(\lambda) \ - \ \int_{x}^{x_+(\lambda)} \frac{i}{R^t(\lambda,y)^{1/2}} \end{cases} .$$

Note that $x(t)$ was chosen so that for each $x \in (x(t),1)$, we have a well defined intersection between the curve $\alpha(\cdot,t)$ and λ, for all $\lambda \in (\alpha(x,t),\beta(x,t))$.

Now (427) immediately implies (425).

Similar arguments verify (424).

□

THEOREM 11.3. *The function $x(t)$ defined in (422) satisfies*

$$\lim_{t\to\infty} x(t) \ = \ 0 \ .$$

Proof.

We will establish the following limits,

$$(428)\qquad \alpha(1,t) \ \to \ 1 \ ,$$

$$(429)\qquad x_\ell(t) \ = \frac{1}{\pi} X(1,t) \ \to \ 0 \ ,$$

$$(430)\qquad for\ fixed\ \beta \in (-1,1),\ X(\beta,t) \to \Phi(\beta) \ ,$$

as $t \to \infty$. Supposing (428 - 430) have been established, we now show that they imply $x(t) \ \to \ 0$.

Suppose this is not the case. Then there exists a sequence t_n such that $|x(t_n)| > \delta$, for some δ. However, from the definition (422) of $x(t)$, and (428), we see that $\beta(x(t),t) \to 1$.

As $\Phi(1) = 0$, we fix β_c so close to 1 that $\Phi(\beta_c) < \pi\frac{\delta}{4}$. As $\beta(x(t_n), t_n)$ is tending to 1, we will eventually have $\beta_c < \beta(x(t_n), t_n)$. Now we take n so large that t_n is sufficiently large to guarantee

$$|X(\beta_c, t_n) - \Phi(\beta_c)| < \pi\frac{\delta}{4} .$$

Now on this subsequence t_n, we have, by monotonicity, $X(\beta, t) > \pi\delta$ for all $\beta < \beta(x(t_n), t_n)$, and so we have a contradiction, from

$$|X(\beta_c, t)| \le \pi\frac{\delta}{2} .$$

We now prove (428 - 430).

Using the representation for $0 < \alpha < \beta$, obtained through similar calculations to those presented for (413),

$$T = \int_0^{x_-(\alpha)} \int_{\alpha_0(y)}^{\beta_0(y)} \frac{idyd\lambda}{[R(\lambda)R^0(\lambda, y)]^{1/2}} + \int_{x_-(\alpha)}^{x_+(\beta)} \int_{\alpha_0(y)}^{\alpha} \frac{id\lambda dy}{[R(\lambda)R^0(\lambda, y)]^{1/2}} ,$$

we see that for a fixed β, the only way to maintain $T(\alpha, \beta) = \pi t$, as $t \to \infty$, is for $\alpha \uparrow \beta$. In particular $\alpha(1, t) \to 1$ as $t \to \infty$, which proves (428).

We now prove (429). From the representation for $0 < \alpha < \beta$,

$$(431) X(\alpha, \beta) = \int_\alpha^\beta \frac{\lambda - m}{[(\lambda - \alpha)(\beta - \lambda)]^{1/2}} \theta' d\lambda + \int_\beta^1 \frac{\lambda - m}{[(\lambda - \alpha)(\lambda - \beta)]^{1/2}} \phi d\lambda ,$$

we have

$$X(1, t) = X(\alpha(1, t), 1) = \int_\alpha^1 \frac{\lambda - m}{[(\lambda - \alpha)(\beta - \lambda)]^{1/2}} \theta' d\lambda .$$

Using $\alpha(1, t) \to 1$ as $t \to \infty$, and $m = \frac{1+\alpha}{2}$, we see immediately that $\lim_{t\to\infty} X(1, t) = 0$.

Finally, for any $\beta \in (-1, 1)$, we have already shown that

$$T(\alpha(\beta, t), \beta) = \pi t \to \infty$$

implies that $\alpha(\beta, t) \to \beta$. Again evaluating the expression (431) in the limit, we see that

$$X(\alpha(\beta, t), \beta) \to \int_\beta^1 \phi(\lambda) d\lambda = \Phi(\beta) ,$$

where we have used the fact that $\Phi(1) = 0$.

This completes the proof of the Theorem.

□

Remark Note that if we fix $x \in (0, 1)$, we see that the limit

$$T(\alpha(x, t), \beta(x, t)) = \pi t \to \infty$$

implies that $\alpha(x, t) \to \beta(x, t)$ as above. But then

$$\pi x = X(\alpha(x, t), \beta(x, t)) \to \int_{\beta(x,\infty)}^1 \phi(\lambda) d\lambda = \Phi(\beta(x, \infty)) ,$$

and hence we see that

$$\alpha(x,t), \beta(x,t) \rightarrow \Phi^{-1}(\pi x)$$

as $t \rightarrow \infty$.

So we have shown that on the interval $(x(t), 1)$, the zero gap ansatz solves the maximizational problem, and that $x(t) \rightarrow 0$ as t tends to infinity.

Further, we have shown that in the large t limit we have

$$\lim_{t\rightarrow\infty} \beta(x,t) - \alpha(x,t) = 0 \ ,$$

$$\lim_{t\rightarrow\infty} \beta(x,t) = \Phi^{-1}(\pi x) \ ,$$

where we have used that $\Phi(1) = 0$. Thus we have shown that the continuum limit of the Toda lattice acts as a sorter even in the presence of oscillatory shocks.

CHAPTER 12

Appendix I - WKB

1. Introduction

Consider the second order difference equation

$$b_{n-1}y_{n-1} + (a_n - \lambda)y_n + b_n\, y_{n+1} = 0, \qquad b_n > 0 \quad . \tag{432}$$

Which arises from the eigenvalue problem for a tridiagonal matrix. We suppose that the coefficients a_n and b_n are discretizations of two functions $a(x)$ and $b(x)$, defined on $[0,1]$:

$$\begin{aligned} a_n &= a(n\epsilon) \\ b_n &= b(n\epsilon), \\ n &= 1,\ldots,\frac{1}{\epsilon}. \end{aligned}$$

The functions $a(x)$ and $b(x)$ are assumed to be infinitely differentiable, and $b(x)$ is taken to be positive.

The discrete version of the J-WKB approximation is obtained in exactly the same manner as for the continuous problem. We make the guess that the solution to (432) is given by

$$y_n = \exp\left[\frac{1}{\epsilon}s_0(n\epsilon) + s_1(n\epsilon)\right], \tag{433}$$

where $s_0(x)$ and $s_1(x)$ are C^∞ functions defined on $[0,1]$. Now we insert (433)into the difference equation, expand everything in powers of ϵ, and solve the equation to first order in ϵ.

More explicitly, we can express y_{n+1} as

$$\begin{aligned} y_{n+1} &= y_n \exp\left\{\left[\frac{s_0((n+1)\epsilon) - s_0(n\epsilon)}{\epsilon}\right] + s_1((n+1)\epsilon) - s_1(n\epsilon)\right\} \\ &= y_n e^{s_0'(n\epsilon)} \exp\left\{\epsilon\left[\frac{1}{2}s_0''(n\epsilon) + s_1'(n\epsilon)\right] + O(\epsilon^2)\right\} \\ &= y_n e^{s_0'(n\epsilon)}\left[1 + \epsilon\left(\frac{1}{2}s_0''(n\epsilon) + s_1'(n\epsilon)\right) + O(\epsilon^2)\right]. \end{aligned} \tag{434}$$

Similarly for y_{n-1}:

(435) $$y_{n-1} = y_n e^{-s_0'(n\epsilon)}\left[1+\epsilon\left(\frac{1}{2}s_0''(n\epsilon)-s_1'(n\epsilon)\right)+O(\epsilon^2)\right].$$

Inserting (434) and (435) into equation (432), we obtain

$$b_n y_n\left[e^{-s_0'(n\epsilon)}+e^{s_0'(n\epsilon)}\right]+(a_n-\lambda)y_n-(b_n-b_{n-1})y_n e^{-s_0'(n\epsilon)}+$$
$$+\epsilon b_n y_n\left[e^{s_0'(n\epsilon)}\left(\frac{1}{2}s_0''(n\epsilon)+s_1'(n\epsilon)\right)+e^{-s_0'(n\epsilon)}\left(\frac{1}{2}s_0''(n\epsilon)-s_1'(n\epsilon)\right)\right]=$$
$$O(\epsilon^2).$$

Expanding $b_n - b_{n-1}$ and equating terms of the same order, we find, in keeping with ODE terminology, the geometrical and physical optics equations

(436) $$2b_n\cosh s_0' = \lambda - a_n,$$

(437) $$2b_n\left[\left(\frac{1}{2}s_0''\right)\cosh s_0' + (s_1')\sinh s_0'\right] - b'e^{-s_0'} = 0.$$

These are easily solved,

(438) $$s_0(n\epsilon) = \int^{n\epsilon}\cosh^{-1}\left(\frac{\lambda-a}{2b}\right)dx,$$

(439) $$s_1(n\epsilon) = -\frac{1}{2}\log\sinh s_0'(n\epsilon) + \int^{n\epsilon}\frac{b'}{2b}\frac{e^{-s_0'}}{\sinh s_0'}dy.$$

Thus the formal J-WKB approximation is

(440)
$$f_n = \left[\left(\frac{\lambda-a}{2b}\right)^2-1\right]^{-1/4}\exp\left[\pm\frac{1}{\epsilon}\int^{n\epsilon}\cosh^{-1}\left(\frac{\lambda-a}{2b}\right)+\epsilon\frac{b'}{2b}\frac{e^{-s_0'}}{\sinh s_0'}dx\right].$$

Here we make two observations.

1. If we restrict n to be in a region such that $\left|\frac{\lambda-a_n}{2b_n}\right| > 1$, then we find that $\cosh^{-1}\left(\frac{\lambda-a_n}{2b_n}\right)$ is real, and so f_n exhibits exponential behavior. Depending on the sign in the exponent, f_n is either exponentially growing or decaying. This is the case dealt with in [**8**]. There, the authors assume that $\frac{\lambda-a_n}{2b_n}$ lies in the complement of a compact set which contains $[-1,1]$. In particular, this means that $\left|\frac{\lambda-a_n}{2b_n}\right| \geq \delta > 1$ for all n. Under this assumption the above formula (440) is the same as that given in [**8**]

2. If n lies in a region where $\left|\frac{\lambda-a_n}{2b_n}\right| < 1$, then $\cosh^{-1}\left(\frac{\lambda-a_n}{2b_n}\right)$ is purely imaginary, and so the J-WKB approximation will exhibit oscillatory behavior. This region is excluded by the assumptions made in [**8**].

A point of transition between these two regions, that is, a value of n such that $\left|\frac{\lambda-a_n}{2b_n}\right| = 1$, is the discrete analog of a turning point. We are interested in the behavior of solutions to the difference equation (432) for n in both of the above regions, as well as across turning points. This is because typically eigenvectors of Jacobi matrices whose entries come from two discretized functions will exhibit both exponential and oscillatory behavior, the regions of different behavior being connected by a turning point.

Now our task is clear. Formal calculations give us a formula that we believe is an approximate solution to the difference equation (432). We must prove that this is indeed the case. Fortunately, there's an extremely useful theorem in [**8**], which we will use to prove that the formula (440) provide valid asymptotics for the difference equation (432).

We would like to prove that there is a bona fide solution to the difference equation which is close to our approximation. So we make the assumption that the actual solution y_n looks like f_n, where f_n is given in (440),

$$y_n = f_n(1 + \phi_n), \tag{441}$$

then we prove that we can find a ϕ_n with the desired property. That is, we prove that ϕ_n exists (and is small).

2. A Theorem of Geronimo and Smith

Geronimo and Smith [8] present the following variable change:

$$y_n = \prod_{j=j_0}^{n} u_j. \tag{442}$$

In terms of u_j, equation (432) becomes

$$1 + \frac{a_n - \lambda}{b_{n-1}} u_n + \frac{b_n}{b_{n-1}} u_n u_{n+1} = 0. \tag{443}$$

Then the authors suppose they have an approximate solution to this equation , i.e. that

$$1 + \frac{a_n - \lambda}{b_{n-1}} u_n + \frac{b_n}{b_{n-1}} u_n u_{n+1} = \xi_n, \tag{444}$$

where ξ_n is small, and then look for a solution to (432) which is near-by. To do this, they set

$$y_n = \left[\prod_{j=j_0}^{n} u_j\right] (1 + \phi_n). \tag{445}$$

The strategy from here is to find ϕ_n so that y_n solves the difference equation (432). This they accomplish by obtaining a difference equation for ϕ_n (which will involve the error term ξ_n). This equation is then summed. Picard iteration then shows that the summation equation can be solved, and that the solution ϕ_n is bounded by a controllable error term. We state the result.

THEOREM 12.1 (Geronimo and Smith). *Suppose there are solutions f_n and g_n of*

$$1 + \frac{a_n - \lambda}{b_{n-1}} f_n + \frac{b_n}{b_{n-1}} f_n\, f_{n+1} = \xi_n, \tag{446}$$

$$1 + \frac{a_n - \lambda}{b_{n-1}} g_n + \frac{b_n}{b_{n-1}} g_n\, g_{n+1} = \zeta_n, \tag{447}$$

where $g_n \neq 0$, $f_n \neq 0$,

$$\sum_{i=N}^{N_1-1} |\, \xi_i \,| < \infty, \tag{448}$$

$$\sum_{i=N}^{N_1-1} |\, \zeta_i \,| < \infty, \tag{449}$$

and there exists a constant c such that

$$\left| \sum_{k=i}^{n-1} \prod_{j=i}^{k} \left(\frac{b_j}{b_{j-1}} f_j f_{j+1}\right)^{-1} \right| < c, \qquad N \leq i \leq n < N_1, \tag{450}$$

and

$$\left| \sum_{k=n+1}^{i} \prod_{j=k}^{i} \left(\frac{b_j}{b_{j-1}} g_j g_{j+1}\right) \right| < c, \qquad N \leq n < i < N_1. \tag{451}$$

Then there exist solutions $y^{\pm}$ of (432) such that

$$\left| y_n^{+} \left(\prod_{i=N}^{n} f_i\right)^{-1} - 1 \right| \leq \exp\left\{ c \sum_{j=N}^{n} |\xi_j| \right\} - 1, \quad N \leq n < N_1, \tag{452}$$

and

$$\left| y_n^{-} \left(\prod_{i=N}^{n} g_i\right)^{-1} - 1 \right| \leq \exp\left\{ c \sum_{j=N}^{n} |\zeta_j| \right\} - 1, \quad N \leq n < N_1. \tag{453}$$

This theorem turns out to be a useful tool; all it requires is a candidate solution. Given an approximate solution, the task of establishing the existence of a solution to (432) near-by has been reduced to that of proving that the error bound (452) or (453) tends to zero as ϵ tends to zero. Our strategy is to do just that, using the formal WKB approximation (440) as our candidate.

The constant c which appears in (452) and (453) will obviously depend on ϵ, as will the error terms ξ_n and ζ_n. Thus the game is to obtain bounds on c in either (450) or (451), and to bound ξ_n and ζ_n in such a way that

$$\exp\left\{ c(\epsilon) \sum_{j=N}^{n} |\xi_j(\epsilon)| \right\} - 1 \quad \stackrel{\epsilon \to 0}{\longrightarrow} \quad 0 \tag{454}$$

or

$$\exp\left\{ c(\epsilon) \sum_{j=N}^{n} |\zeta_j(\epsilon)| \right\} - 1 \quad \stackrel{\epsilon \to 0}{\longrightarrow} \quad 0. \tag{455}$$

In this way we will establish rigorously the validity of the J-WKB approximation.

3. Exponential Behavior

To show that the formal J-WKB approximation is valid, we must show that

$$\exp\left\{c(\epsilon)\sum_{j=N}^{n}|\zeta_j(\epsilon)|\right\}-1 \quad\longrightarrow\quad 0, \quad as\ \ \epsilon\to 0\ ,$$

from Theorem 12.1. To establish this, we must bound the expression (451), and the error function ζ_n defined by (447).

In terms of the J-WKB approximation, ζ_n can be written (see (447))

$$\zeta_n = \frac{1}{b_{n-1}f_{n-1}}\left[b_{n-1}f_{n-1} \ +(a_n-\lambda)f_n+\ b_nf_{n+1}\right], \tag{456}$$

and the constant $c(\epsilon)$ is obtained by bounding (see (451), with $g_j = h_j = \frac{f_j}{f_{j-1}}$),

$$\sum_{k=n+1}^{i}\prod_{j=k}^{i}\left(\frac{b_j}{b_{j-1}}h_jh_{j+1}\right). \tag{457}$$

The analysis is divided into sections which correspond to the various regions of behavior. In this short section, we establish the result under the assumption that we are bounded away from the oscillatory region. For simplicity we work with the exponentially decreasing formula. The analysis for the exponentially increasing formula is identical.

In following sections, we will consider the exponentially decreasing case for n approaching a turning point, oscillatory behavior in the absence of turning points, oscillatory behavior for n approaching a turning point, and finally the behavior in a neighborhood of a turning point.

For now, we assume

$$\frac{\lambda-a_n}{2b_n}\geq\delta>1, \quad for\ \ n=1,\ldots,\frac{1}{\epsilon}, \tag{458}$$

and we take the J-WKB function

$$f_n = \left[\left(\frac{\lambda-a}{2b}\right)^2-1\right]^{-1/4}\exp\left[-\frac{1}{\epsilon}\int^{n\epsilon}\cosh^{-1}(\frac{\lambda-a}{2b})+\epsilon\frac{b'}{2b}\frac{e^{-s_0'}}{\sinh s_0'}dx\right].$$

LEMMA 12.1. *Under the assumption (458), there exists a solution y_n^- to (432) such that*

$$\left|\frac{y_n^-}{f_n}-1\right| \ \to\ 0 \ \ as\ \ \epsilon\to 0, \ \ for\ n=1,\ldots,\frac{1}{\epsilon}.$$

Proof. We first bound ζ_n, then the expression (457). Using the formula (433), we have

$$
\begin{aligned}
h_n &= \frac{f_n}{f_{n-1}} \\
&= \exp\left(\frac{s_0(n\epsilon) - s_0(n\epsilon - \epsilon)}{\epsilon} + s_1(n\epsilon) - s_1(n\epsilon - \epsilon)\right) \\
&= e^{s_0'}\left[1 + B_n^{(1)}\epsilon + B_n^{(2)}\epsilon^2 + \frac{1}{2}e^{Q_n}\left\{B_n^{(1)}\epsilon + B_n^{(2)}\epsilon^2\right\}^2\right], \qquad (459)
\end{aligned}
$$

where

$$
\begin{aligned}
B_n^{(1)} &= -\frac{1}{2}s_0''(n\epsilon) + s_1'(n\epsilon) \\
B_n^{(2)} &= \frac{1}{3!}s_0'''(z) - \frac{1}{2}s_1''(z_1), \qquad z,\ z_1 \in [n\epsilon - \epsilon,\ n\epsilon], \\
Q_n &\in [0,\ B_n^{(1)}\epsilon + B_n^{(2)}\epsilon^2].
\end{aligned}
$$

Similarly,

$$
\begin{aligned}
h_{n+1} &= \frac{f_{n+1}}{f_n} \\
&= e^{s_0'}\left[1 + B_{n+1}^{(1)}\epsilon + B_{n+1}^{(2)}\epsilon^2 + \frac{1}{2}e^{Q_n}\left\{B_{n+1}^{(1)}\epsilon + B_{n+1}^{(2)}\epsilon^2\right\}^2\right], \qquad (460)
\end{aligned}
$$

where

$$
\begin{aligned}
B_{n+1}^{(1)} &= \frac{1}{2}s_0''(n\epsilon) + s_1'(n\epsilon) \\
B_{n+1}^{(2)} &= \frac{1}{3!}s_0'''(\tilde{z}) + \frac{1}{2}s_1''\tilde{z}_1, \qquad \tilde{z},\ \tilde{z}_1 \in [n\epsilon - \epsilon,\ n\epsilon], \\
Q_{n+1} &\in [0,\ B_{n+1}^{(1)}\epsilon + B_{n+1}^{(2)}\epsilon^2].
\end{aligned}
$$

Now we insert (459) and (460) into equation (456), and obtain

$$
\begin{aligned}
\zeta_n &= 1 + \frac{a_n - \lambda}{b_n}\frac{b_n}{b_{n-1}}u_n + \frac{b_n}{b_{n-1}}u_n u_{n+1} \\
&= 1 + \frac{b_n}{b_{n-1}}u_n\left[\frac{a_n - \lambda}{b_n} + u_{n+1}\right] \\
&= 1 + \left(1 + \epsilon\frac{b'}{b} + \beta\epsilon^2\right)e^{s_0'}\left[1 + B_n^{(1)}\epsilon + B_n^{(2)}\epsilon^2 + \frac{1}{2}e^{Q_n}\left\{B_n^{(1)}\epsilon + B_n^{(2)}\epsilon^2\right\}^2\right] \\
&\quad \times\left[\frac{a_n - \lambda}{b_n} + e^{s_0'}\left(1 + B_{n+1}^{(1)}\epsilon + B_{n+1}^{(2)}\epsilon^2 + \frac{1}{2}e^{Q_{n+1}}\left\{B_{n+1}^{(1)}\epsilon + B_{n+1}^{(2)}\epsilon^2\right\}^2\right)\right].
\end{aligned}
$$

Now just as in the continuous case, if we are bounded away from turning points, then all coefficients are well behaved, i.e. bounded. In addition, the above equation for ζ_n vanishes to second order in ϵ. Indeed, the leading term is

$$1 + e^{s_0'}\left(\frac{a_n - \lambda}{b_n} + e^{s_0'}\right)$$

$$= e^{s_0'}\left(\frac{a_n - \lambda}{b_n} + 2\cosh s_0'\right) \quad = \quad 0.$$

Similarly, the first order term can be reduced to the physical optics equation (437) of Section 1.

Therefore we conclude that

$$\zeta_n \le const.\ \epsilon^2, \quad for \quad n = 1, \ldots, \frac{1}{\epsilon}. \tag{461}$$

Now we wish to bound (457). Using the J-WKB approximation (440), we see

$$\sum_{k=n+1}^{i}\prod_{j=k}^{i}\left(\frac{b_j}{b_{j-1}}h_j h_{j+1}\right) = \sum_{k=n+1}^{i}\left(\frac{b_i}{b_{k-1}}\frac{f_i}{f_{k-1}}\frac{f_{i+1}}{f_k}\right)$$

$$\le \sum_{k=n+1}^{i}\left[\frac{\left\{\left(\frac{\lambda-a_i}{2b_i}\right)^2 - 1\right\}\left\{\left(\frac{\lambda-a_{i+1}}{2b_{i+1}}\right)^2 - 1\right\}}{\left\{\left(\frac{\lambda-a_{k-1}}{2b_{k-1}}\right)^2 - 1\right\}\left\{\left(\frac{\lambda-a_k}{2b_k}\right)^2 - 1\right\}}\right]^{-1/4} \exp\left\{\frac{2}{\epsilon}\int_{k\epsilon}^{i\epsilon} I dx\right\}$$

$$\le \sum_{k=n+1}^{i} Be^{-2m(i-k)} = B\sum_{l=0}^{i-(n+1)}\left(e^{-2m}\right)^l \quad \le \quad c.$$

The constant m which appears above is the minimum value of the integrand, which is bounded away from zero by assumption.

Now we've obtained the necessary bounds to use Theorem 12.1. According to Theorem 12.1, we know that there exists a solution to (432), y_n^-, such that

$$\left|\frac{y_n^-}{f_n} - 1\right| \quad \le \quad \exp\left\{c\sum_{j=N}^{n}|\zeta_j|\right\} - 1.$$

We have shown that the constant c above is independent of ϵ. In addition, we can bound the summation as follows

$$\sum_{j=N}^{n}|\zeta_j| \quad \le \quad B\sum_{j=N}^{n}\epsilon^2 \le B\epsilon.$$

The last inequality is true because we have at most $\frac{1}{\epsilon}$ terms in the summation.

Thus, we have

$$\left|\frac{y_n^-}{f_n} - 1\right| \quad \le \quad e^{c'\epsilon} - 1\ , \quad for \quad n = 1, \ldots, \frac{1}{\epsilon}.$$

The lemma is proven.

This lemma encompasses the case discussed in [**8**].

4. Exponential Behavior: Turning Points

In this section we will allow n to approach a turning point. The point is, (440) blows up at a turning point.

This poses a slight problem, but we know what happens in the continuous case. There, the J-WKB formula is valid in a region of the form

$$\epsilon^{2/3} \ll x .$$

So the independent variable x may approach a turning point, as $\epsilon \to 0$, but it may only approach at rates allowed by the above inequality. The same is true in the discrete case.

LEMMA 12.2. *Suppose that* $\frac{\lambda-a}{2b} \geq 1$, *and there is exactly one linear turning point, at* $x = 0$. *Then there exists a solution* y_n^+ *to (432) satisfying*

$$\left|\frac{y_n^+}{f_n} - 1\right| \to 0 \quad as \quad \epsilon \to 0 \ , \quad for \ \epsilon^{2/3} \ll n\epsilon \leq 1.$$

Proof. Here we follow the same argument as in the preceding section, only we must be more careful, as we no longer have uniform bounds on the coefficients, and the amplitude factor $\left[\left(\frac{\lambda-a}{2b}\right)^2 - 1\right]^{-1/4}$ cannot be uniformly bounded.

As we are dealing with the exponentially growing formula,

$$f_n = \left[\left(\frac{\lambda - a_n}{2b_n}\right)^2 - 1\right]^{-1/4} \exp\left\{\int^{n\epsilon} \cosh^{-1}\left(\frac{\lambda - a}{2b}\right) + \epsilon\frac{b'}{2b}\frac{e^{s_0'}}{\sinh s_0'}\, dx\right\} ,$$

we must use ξ_n and $\sum_{k=i}^{n-1} \prod_{j=i}^{k} \left(\frac{b_j f_j f_{j+1}}{b_{j-1} f_{j-1} f_j}\right)^{-1}$ (c.f. (452)).

We begin by bounding ξ_n. From (446), we have

$$\xi_n = 1 + \frac{a_n - \lambda}{b_{n-1}} g_n + \frac{b_n}{b_{n-1}} g_n g_{n+1} \ ,$$

where

$$\begin{aligned} g_n &= \frac{f_n}{f_{n-1}} \\ &= e^{s_0'}\left[1 + B_n^{(1)}\epsilon + B_n^{(2)}\epsilon^2 + \frac{1}{2}e^{Q_n}\left\{B_n^{(1)}\epsilon + B_n^{(2)}\epsilon^2\right\}^2\right] , \end{aligned}$$

and

$$\begin{aligned} B_n^{(1)} &= -\frac{1}{2}s_0''(n\epsilon) + s_1'(n\epsilon) \\ B_n^{(2)} &= \frac{1}{3!}s_0'''(z) - \frac{1}{2}s_1''(z_1), \qquad z,\ z_1 \in [n\epsilon - \epsilon,\ \ n\epsilon], \\ Q_n &\in [0,\ B_n^{(1)}\epsilon + B_n^{(2)}\epsilon^2]. \end{aligned}$$

As in the previous case, we have

$$g_{n+1} = e^{s_0'}\left[1 + B_{n+1}^{(1)}\epsilon + B_{n+1}^{(2)}\epsilon^2 + \frac{1}{2}e^{Q_{n+1}}\left\{B_{n+1}^{(1)}\epsilon + B_{n+1}^{(2)}\epsilon^2\right\}^2\right] ,$$

where

$$\begin{aligned} B^{(1)}_{n+1} &= \frac{1}{2}s_0''(n\epsilon) + s_1'(n\epsilon) \\ B^{(2)}_{n+1} &= \frac{1}{3!}s_0'''(\tilde{z}) + \frac{1}{2}s_1''(\tilde{z}_1), \qquad \tilde{z},\ \tilde{z}_1 \in [n\epsilon,\ n\epsilon+\epsilon], \\ Q_{n+1} &\in [0,\ B^{(1)}_{n+1}\epsilon + B^{(2)}_{n+1}\epsilon^2]. \end{aligned}$$

As before, we have the equation, this time for ξ_n rather than ζ_n,

$$\begin{aligned} \xi_n = 1 + \left(1 + \epsilon\frac{b'}{b} + \epsilon^2\beta_n\right) e^{s_0'} \times \\ \times \left[1 + B^{(1)}_n\epsilon + B^{(2)}_n\epsilon^2 + \frac{1}{2}e^{Q_n}\left\{B^{(1)}_n\epsilon + B^{(2)}_n\epsilon^2\right\}^2\right] \times \\ \times \left[\frac{a_n-\lambda}{b_n} + e^{s_0'}\left(1 + B^{(1)}_{n+1}\epsilon + B^{(2)}_{n+1}\epsilon^2 + \right.\right. \\ \left.\left. + \frac{1}{2}e^{Q_{n+1}}\left\{B^{(1)}_{n+1}\epsilon + B^{(2)}_{n+1}\epsilon^2\right\}\right)\right]. \end{aligned} \tag{462}$$

As before, this equation for ξ_n vanishes to second order in ϵ. The problem is that the coefficients are not uniformly bounded. Simple calculations show that

$$B^{(1)}_n\ ,\ B^{(1)}_{n+1} \leq \frac{const.}{n\epsilon} \tag{463}$$

$$B^{(2)}_n\ ,\ B^{(2)}_{n+1} \leq \frac{const.}{(n\epsilon)^2}. \tag{464}$$

Since we know (462) vanishes to second order in ϵ, we will analyze those terms in (462) which are of order ϵ^2 or higher, using (463 and 464).

The coefficient of ϵ^2 in (462) is

$$\begin{aligned} c_2 = \beta_n e^{s_0'}\left(\frac{a_n-\lambda}{b_n} + e^{s_0'}\right) + \frac{b'}{b}e^{s_0'}\left[B^{(1)}_n\left(\frac{a_n-\lambda}{b_n} + e^{s_0'}\right) + e^{s_0'}B^{(1)}_{n+1}\right] \\ e^{s_0'}\left[\left(\frac{1}{2}e^{Q_n}\left(B^{(1)}_n\right)^2 + B^{(2)}_n\right)\left(\frac{a_n-\lambda}{b_n} + e^{s_0'}\right) + B^{(1)}_n e^{s_0'} B^{(1)}_{n+1}\right. \\ \left. + e^{s_0'}\left(B^{(2)}_{n+1} + \frac{1}{2}e^{Q_{n+1}}\left(B^{(1)}_{n+1}\right)^2\right)\right]. \end{aligned}$$

We see that

$$|c_2| \leq const.\frac{1}{(n\epsilon)^2}.$$

Analysis of the order ϵ^3 term shows it to be bounded by

$$|c_3| \leq const.\frac{1}{(n\epsilon)^3}.$$

Similarly,

$$|c_j| \leq const. \frac{1}{(n\epsilon)^j}.$$

Thus we have

$$(465) \qquad |\xi_n| \leq const. \left(\frac{\epsilon}{n\epsilon}\right)^2,$$

under the assumption that $\frac{\epsilon}{n\epsilon} < 1$.

Having bounded ξ_n, we would like to control the constant $c(\epsilon)$ appearing in (450), i.e.,

$$\sum_{k=i}^{n-1} \prod_{j=i}^{k} \left(\frac{b_j}{b_{j-1}} f_j f_{j+1}\right)^{-1} \leq c(\epsilon).$$

We know that

$$(466) \qquad \prod_{j=i}^{k} \frac{b_j}{b_{j-1}} \frac{f_j f_{j+1}}{f_{j-1} f_j} = \frac{b_k}{b_{i-1}} \frac{f_k}{f_{i-1}} \frac{f_{k+1}}{f_i} \geq B \left(\frac{f_k}{f_i}\right)^2.$$

In addition, we know from (440) that

$$(467) \quad \frac{f_k}{f_i} = \left[\frac{\left(\frac{\lambda-a_k}{2b_k}\right)^2 - 1}{\left(\frac{\lambda-a_i}{2b_i}\right)^2 - 1}\right]^{-1/4} \exp\left[\frac{1}{\epsilon}\int_{i\epsilon}^{k\epsilon} \cosh^{-1}\left(\frac{\lambda-a}{2b}\right) + \epsilon \frac{b'}{2b} \frac{e^{-s_0'}}{\sinh s_0'} dx\right]$$

$$\geq E \left[\frac{\left(\frac{\lambda-a_k}{2b_k}\right)^2 - 1}{\left(\frac{\lambda-a_i}{2b_i}\right)^2 - 1}\right]^{-1/4} \exp\left[\frac{1}{\epsilon}\int_{i\epsilon}^{k\epsilon} \cosh^{-1}\left(\frac{\lambda-a}{2b}\right) dx\right].$$

Now we calculate

$$\sum_{k=i}^{n-1}\prod_{j=i}^{k}\left(\frac{b_j}{b_{j-1}}f_jf_{j+1}\right)^{-1}$$
$$\leq \sum_{k=i}^{n-1} F\left[\frac{\left(\frac{\lambda-a_k}{2b_k}\right)^2-1}{\left(\frac{\lambda-a_i}{2b_i}\right)^2-1}\right]^{1/2} \exp\left[-\frac{2}{\epsilon}\int_{i\epsilon}^{k\epsilon}\cosh^{-1}\left(\frac{\lambda-a}{2b}\right)\,dx\right]$$
$$\leq F_1\sum_{k=i}^{n-1}\left[\frac{\left(\frac{\lambda-a_k}{2b_k}\right)^2-1}{\left(\frac{\lambda-a_i}{2b_i}\right)^2-1}\right]^{1/2} e^{-\frac{2\theta}{\epsilon}(k-1-i)\epsilon}$$
$$\leq \frac{G}{\sqrt{2i\epsilon}}\sum_{k=i}^{n-1}\sqrt{2k\epsilon}e^{-\frac{2\theta}{\epsilon}(k\epsilon-i\epsilon)}\ ,$$

where F and G are global constants.

We have bounded the exponential term by bounding the integral. Its value is larger than the minimum of the integrand, θ, multiplied by the length of the interval of integration. The terms involving $\sqrt{2i\epsilon}$ and $\sqrt{2k\epsilon}$ have appeared because we know the nature of the root of $\frac{\lambda-a}{2b}$ at the turning point.

We would like to bound the last summation appearing above. We do this by noting that the sum looks a lot like a Riemann sum. In fact, if we knew that the "integrand" was monotone in k, we could certainly bound by the appropriate integral. But, the function that appears is $\sqrt{2x}e^{-\frac{2\theta}{\epsilon}(x-x_i)}$, which is not monotone. We are saved, however, by the fact that this function has only one critical point, occurring at $x_c = \frac{\epsilon}{4\theta}$, which lies to the left of the entire region of integration, if we assert that $i\epsilon \gg \epsilon^{2/3}$.

Indeed, since we know that θ, the minimum value of $\cosh^{-1}\left(\frac{\lambda-a_n}{2b_n}\right)$, occurs at $n = i$, and since we know the behavior of $\frac{\lambda-a}{2b}$ near the turning point, we have

$$\begin{aligned} x_c &= \frac{\epsilon}{4\theta} \leq const\ \frac{\epsilon}{4\sqrt{2i\epsilon}} \\ &\ll c\epsilon^{2/3}\ . \end{aligned}$$

So we are justified in writing

$$\sum_{k=i}^{n-1}\prod_{j=i}^{k}\left(\frac{b_j}{b_{j-1}}f_jf_{j+1}\right)^{-1} \le \frac{G_1}{\sqrt{2i\epsilon}}\frac{1}{\epsilon}\int_{i\epsilon}^{(n-1)\epsilon}\sqrt{2x}e^{-\frac{2\theta}{\epsilon}(x-x_i)}\,dx$$

$$= \frac{G_1}{\sqrt{2i\epsilon}}\frac{e^{2\theta i}}{\epsilon}\int_{2i\theta}^{2(n-1)\theta}\left(\frac{\epsilon u}{\theta}\right)^{1/2}e^{-u}\frac{\epsilon}{2\theta}\,du$$

$$= \frac{G_1}{\sqrt{2i\epsilon}}\frac{e^{2\theta i}}{\epsilon}\frac{1}{2}\left(\frac{\epsilon}{\theta}\right)^{3/2}\int_{2i\theta}^{2(n-1)\theta}\sqrt{u}e^{-u}\,du$$

$$\le \frac{G_1}{\sqrt{2i\epsilon}}\frac{e^{2\theta i}}{2}\frac{\epsilon^{1/2}}{\theta^{3/2}}\int_{2i\theta}^{\infty}\sqrt{u}e^{-u}\,du$$

$$= \frac{G_1}{\sqrt{2i\epsilon}}\frac{e^{2\theta i}}{2}\frac{\epsilon^{1/2}}{\theta^{3/2}}\left[-u^{1/2}e^{-u}|_{2i\theta}^{\infty}+\frac{1}{2}\int_{2i\theta}^{\infty}\frac{e^{-u}}{u^{1/2}}\,du\right]$$

$$\le \frac{G_1}{\sqrt{2i\epsilon}}\frac{e^{2\theta i}}{2}\frac{\epsilon^{1/2}}{\theta^{3/2}}\left[[2i\theta]^{1/2}e^{-2i\theta}+[2i\theta]^{-1/2}e^{-2i\theta}\right]$$

$$= \frac{G_1}{\sqrt{2i\epsilon}}\frac{\epsilon^{1/2}}{\theta^{3/2}}\left[(2i\theta)^{1/2}+(2i\theta)^{-1/2}\right]$$

$$\le \frac{H}{\sqrt{2i\epsilon}}\frac{\epsilon^{1/2}}{(2i\epsilon)^{3/4}}\left[(2i\epsilon)^{1/4}i^{1/2}+(2i\epsilon)^{-1/4}i^{-1/2}\right],$$

where G_1 and H are global constants.

Summarizing, we have

$$|\xi_n| \le const\left(\frac{\epsilon}{n\epsilon}\right)^2,$$

$$\sum_{k=i}^{n-1}\prod_{j=i}^{k}\left(\frac{b_j}{b_{j-1}}f_jf_{j+1}\right)^{-1} \le H\left[\frac{1}{(i\epsilon)^{1/2}}+\frac{\epsilon}{(i\epsilon)^2}\right].$$

Now Theorem 12.1 asserts that there exists a solution y_n^+ to (432), satisfying

$$\left|\frac{y_n^+}{f_n}-1\right| \le \exp\left\{H\left[\frac{1}{(i\epsilon)^{1/2}}+\frac{\epsilon}{(i\epsilon)^2}\right]\sum_{j=i}^{n}|\xi_n|\right\}-1\,,\quad i\le n\le\frac{1}{\epsilon}\,.$$

Clearly,

$$\left[\frac{1}{(i\epsilon)^{1/2}}+\frac{\epsilon}{(i\epsilon)^2}\right]\sum_{j=i}^{n}|\xi_n| \le \left[\frac{1}{(i\epsilon)^{1/2}}+\frac{\epsilon}{(i\epsilon)^2}\right]\frac{\epsilon}{i\epsilon}$$

$$= \frac{\epsilon}{(i\epsilon)^{3/2}}+\frac{\epsilon^2}{(i\epsilon)^3}\,.$$

This certainly tends to zero if we assume that $i\epsilon \gg \epsilon^{2/3}$, $\epsilon \to 0$. Thus, for $i\epsilon \gg \epsilon^{2/3}$, we have established the validity of the exponentially growing

J-WKB formula. The proof for the exponentially decaying formula follows from similar arguments.

Observe that the situation $\frac{\lambda-a}{2b} \le -1$ can be transformed into the case $\frac{\lambda-a}{2b} > 1$ by considering the equation satisfied by $(-1)^j y_j$, rather than y_j. Thus the above analysis applies to the case $\frac{\lambda-a}{2b} < -1$.

In the next section we consider the oscillatory region, $\left|\frac{\lambda-a}{2b}\right| < 1$. First we will work under the assumption that there are no turning points. Then we will allow n to approach a turning point.

Looking ahead, we shall also consider n in a neighborhood of a turning point, and we will be able to patch together solutions on the entire n - axis. That is, we shall construct solutions valid in regions which overlap with the above solutions, and with the forthcoming oscillatory solutions.

For now, we turn to the oscillatory J-WKB formula.

5. Oscillatory Behavior

The oscillatory J-WKB formula can be expressed as

$$f_n = \left[b_n^2 - \left(\frac{\lambda - a_n}{2}\right)^2\right]^{-1/4} \times$$

$$\exp\left\{\pm\frac{i}{\epsilon}\int^{n\epsilon} \cos^{-1}\left(\frac{\lambda - a}{2b}\right) + \epsilon\frac{b'}{2b}\frac{\frac{\lambda-a}{2b}}{\sqrt{1-\left(\frac{\lambda-a}{2b}\right)^2}}\,dx\right\} .$$

We assume for the moment that there are no turning points,

$$\left|\frac{\lambda - a}{2b}\right| \le \delta < 1 .$$

LEMMA 12.3. *Under the above assumption, there exists solutions* $y_n^{\pm}$ *to (432) satisfying*

$$\left|\frac{y_n^{\pm}}{f_n} - 1\right| \to 0 \quad as \quad \epsilon \to 0 , \quad for\ n = 1, \ldots \left[\frac{1}{\epsilon}\right] .$$

Proof. We prove the lemma only for one of the two formulas, the "+" formula. The bound for ξ_n follows exactly as did the bound for ζ_n in section (3). We have

$$|\xi_n| \le const.\epsilon^2 .$$

We must now focus on the "kernel" of (450)

$$K(i,n,\epsilon) = \sum_{k=i}^{n-1}\prod_{j=i}^{k}\left(\frac{b_j}{b_{j-1}}\frac{f_j}{f_{j-1}}\frac{f_{j+1}}{f_j}\right) ,$$

which we can write in the following way:

$$K(i,n,\epsilon) = \sum_{k=i}^{n-1}\gamma_k \exp\frac{i}{\epsilon}\left[\int_{i\epsilon}^{k\epsilon} f(y)dy\right] , \tag{468}$$

where

$$\gamma_k = \frac{b_{i-1}}{b_k}\left(\frac{f_{k+1}}{f_k}\frac{f_i}{f_{i-1}}\right)^{-1}\left(\frac{b_k^2 - \left(\frac{\lambda-a_k}{2}\right)^2}{b_i^2 - \left(\frac{\lambda-a_i}{2}\right)^2}\right)^{1/2} ,$$

$$f(y) = -2\left[\cos^{-1}\left(\frac{\lambda - a}{2b}\right) - \epsilon\frac{b'}{2b}\frac{\left(\frac{\lambda-a}{2b}\right)}{\sqrt{1-\left(\frac{\lambda-a}{2b}\right)^2}}\right] .$$

Standard integration by parts in (468) yields

$$K(i,n,\epsilon) = \sum_{k=i}^{n-2}(-\Delta\gamma_k)\sum_{j=i}^{k-1}\exp\frac{i}{\epsilon}\left[\int_{i\epsilon}^{j\epsilon} f(y)\,dy\right] + $$

$$+ \gamma_{n-1}\sum_{j=i}^{n-1}\exp\frac{i}{\epsilon}\left[\int_{i\epsilon}^{j\epsilon} f(y)dy\right] .$$

We must bound $\Delta\gamma_k$ by something small, and the exponential sum by a constant. Of course, under the assumption that there are no turning points, it is clear that

$$|\Delta\gamma_k| \leq const.\ \epsilon ,$$

so we focus on the exponential sum.

We consider the summation

$$\sum_{j=i}^{k} e^{\frac{1}{\epsilon}\int_{i\epsilon}^{j\epsilon} f(y)dy} . \tag{469}$$

Setting

$$u_j = e^{\frac{i}{\epsilon}\int_{i\epsilon}^{j\epsilon} f(y)dy} ,$$

we can write

$$\frac{u_{j+1}-u_j}{e^{\frac{i}{\epsilon}\int_{j\epsilon}^{(j+1)\epsilon} f(y)dy} - 1} = e^{\frac{i}{\epsilon}\int_{i\epsilon}^{j\epsilon} f(y)dy} .$$

Substituting this into the summation (469), we obtain

$$\sum_{j=i}^{k} e^{\frac{i}{\epsilon}\int_{i\epsilon}^{j\epsilon} f(y)dy} = \sum_{j=i}^{k}\frac{u_{j+1}-u_j}{e^{\frac{i}{\epsilon}\int_{j\epsilon}^{(j+1)\epsilon} f(y)dy} - 1} \tag{470}$$

$$= -\frac{u_i}{e^{\frac{i}{\epsilon}\int_{i\epsilon}^{(i+1)\epsilon} f(y)dy} - 1} + \frac{u_{k+1}}{e^{\frac{i}{\epsilon}\int_{k\epsilon}^{(k+1)\epsilon} f(y)dy} - 1} +$$

$$+ \sum_{j=i+1}^{k-1} u_j \frac{e^{\frac{i}{\epsilon}\int_{(j+1)\epsilon}^{(j+2)\epsilon} f dy} - e^{\frac{i}{\epsilon}\int_{(j)\epsilon}^{(j+1)\epsilon} f dy}}{\left(e^{\frac{i}{\epsilon}\int_{(j)\epsilon}^{(j+1)\epsilon} f dy} - 1\right)\left(e^{\frac{i}{\epsilon}\int_{(j+1)\epsilon}^{(j+2)\epsilon} f dy} - 1\right)} .$$

Since $\left|\frac{\lambda-a}{2b}\right| \leq \delta < 1$, we have $f \geq \delta' > 0$, and so

$$\left|e^{\frac{i}{\epsilon}\int_{(j+1)\epsilon}^{(j+2)\epsilon} f dy} - 1\right| \geq \beta > 0 . \tag{471}$$

In addition, we can express the difference of exponentials as follows:

$$e^{\frac{i}{\epsilon}\int_{(j+1)\epsilon}^{(j+2)\epsilon} f dy} - e^{\frac{i}{\epsilon}\int_{(j)\epsilon}^{(j+1)\epsilon} f dy} = e^{\frac{i}{\epsilon}\int_{(j)\epsilon}^{(j+1)\epsilon} f dy} \times \tag{472}$$

$$\times\left[e^{\frac{i}{\epsilon}\left(\int_{(j+1)\epsilon}^{(j+2)\epsilon} f dy - \int_{(j)\epsilon}^{(j+1)\epsilon} f dy\right)} - 1\right] .$$

Now clearly

$$\left| \int_{(j+1)\epsilon}^{(j+2)\epsilon} f dy - \int_{(j)\epsilon}^{(j+1)\epsilon} f dy \right| \leq c \cdot \epsilon^2 \ . \tag{473}$$

Finally, combining (470) - (473), we have

$$\sum_{j=i}^{k} e^{\int_{i\epsilon}^{j\epsilon} f(y)dy} \leq \frac{2}{\beta} + \frac{1}{\beta^2} \sum_{j=i+1}^{k-1} D\epsilon$$
$$\leq \frac{2}{\beta} + \frac{D}{\beta^2} \leq const. \ .$$

We use this to bound the kernel $K(i, n, \epsilon)$,

$$K(i, n, \epsilon) \leq const \ .$$

Theorem 12.1 says that there is a solution y_n^+ to (432) satisfying

$$\left| \frac{y_n^+}{f_n} - 1 \right| \leq e^{c \sum |\xi_j|} - 1$$
$$\leq e^{const. \epsilon} - 1$$
$$\rightarrow 0 \quad as \quad \epsilon \rightarrow 0 \ , \quad for \ n = \ 1, \ \ldots \ \left[\frac{1}{\epsilon}\right] \ .$$

The Lemma is proven. □

6. Oscillatory Behavior - Turning Points

In this section we allow n to approach a turning point.

LEMMA 12.4. *Suppose that*

$$-1 < \delta \leq \left(\frac{\lambda - a}{2b} \right) \leq 1 \ ,$$

and there is exactly one turning point, at $x = 0$. Then there exists solutions $y_n^\pm$ to (432)

$$\left| \frac{y_n^\pm}{f_n} - 1 \right| \rightarrow 0 \quad as \quad \epsilon \rightarrow 0 \ , \quad for \ \epsilon^{2/3} \ll n\epsilon \leq 1 \ .$$

Proof.
We begin with

$$\zeta_n = \frac{1}{b_{n-1} f_{n-1}} \left[b_{n-1} f_{n-1} + (a_n - \lambda) f_n + b_n f_{n+1} \right] \ .$$

As before, we expand, remembering that ξ_n vanishes to order ϵ, and we must bound the coefficients that appear. We leave the reader to verify, as in (465), the bound

$$|\xi_n| \leq c \left(\frac{\epsilon}{n\epsilon} \right)^2 \ .$$

Next we must bound

$$\sum_{k=n}^{i}\prod_{j=k}^{i}\frac{b_j}{b_{j-1}}\frac{f_j}{f_{j-1}}\frac{f_{j+1}}{f_j}\ .$$

In similar fashion to the previous section, we write

$$\begin{aligned} K(n,i,\epsilon) &= \sum_{k=n}^{i}\prod_{j=k}^{i}\frac{b_j}{b_{j-1}}\frac{f_j}{f_{j-1}}\frac{f_{j+1}}{f_j} \\ &= \sum_{k=n}^{i}\gamma_k e^{\frac{i}{\epsilon}\int_{(k)\epsilon}^{(i)\epsilon} f dy}\ , \end{aligned}$$

where

$$\gamma_k = \frac{b_j}{b_{j-1}}\frac{f_k}{f_{k-1}}\frac{f_{i+1}}{f_i}\left[\frac{b_k^2-\left(\frac{\lambda-a_k}{2}\right)^2}{b_i^2-\left(\frac{\lambda-a_i}{2}\right)^2}\right]^{1/2}\ ,$$

$$f = \cos^{-1}\left(\frac{\lambda-a}{2b}\right) - \epsilon\frac{b'}{2b}\frac{\frac{\lambda-a}{2b}}{\sqrt{1-\left(\frac{\lambda-a}{2b}\right)^2}}\ .$$

Again, we have

$$\begin{aligned} K(n,i,\epsilon) &= \sum_{k=n}^{i}\gamma_k \exp\frac{i}{\epsilon}\left[\int_{i\epsilon}^{k\epsilon} f(y)dy\right] = e^{\frac{i}{\epsilon}\int_{n\epsilon}^{i\epsilon} f dy}\sum_{k=n}^{i}\gamma_k e^{-\frac{i}{\epsilon}\int_{n\epsilon}^{k\epsilon} f dy} \\ &= e^{-\frac{i}{\epsilon}\int_{n\epsilon}^{k\epsilon} f dy}\left[\sum_{k=n}^{i-1}-(\Delta\gamma)_k\sum_{j=n}^{k-1}e^{-\frac{i}{\epsilon}\int_{n\epsilon}^{j\epsilon} f dy}+\gamma_i\sum_{j=n}^{i}e^{-\frac{i}{\epsilon}\int_{n\epsilon}^{j\epsilon} f dy}\right]\ . \end{aligned}$$

Thus our task is to bound $\Delta\gamma_k$, and the exponential sum. Some algebra shows

$$\begin{aligned} \gamma_k &= \frac{b_i}{b_k}\left\{\frac{\left[b_k^2-\left(\frac{\lambda-a_k}{2}\right)^2\right]\left[b_{k-1}^2-\left(\frac{\lambda-a_{k-1}}{2}\right)^2\right]}{\left[b_i^2-\left(\frac{\lambda-a_i}{2}\right)^2\right]\left[b_{i+1}^2-\left(\frac{\lambda-a_{i+1}}{2}\right)^2\right]}\right\}^{1/4} e^{\frac{i}{\epsilon}\int_{(k-1)\epsilon}^{k\epsilon} f dy} e^{\frac{i}{\epsilon}\int_{i\epsilon}^{(i+1)\epsilon} f dy} \\ &= \tilde{\tilde{\gamma}}_i\tilde{\gamma}_k\ , \end{aligned}$$

where

$$\tilde{\gamma}_k = \sqrt{\frac{b_k}{b_{k-1}}}\left\{\left[1-\left(\frac{\lambda-a_k}{2b_k}\right)^2\right]\left[1-\left(\frac{\lambda-a_{k-1}}{2b_{k-1}}\right)^2\right]\right\}^{1/4} e^{\frac{i}{\epsilon}\int_{(k-1)\epsilon}^{k\epsilon} f dy}\ ,$$

$$\tilde{\tilde{\gamma}}_i = \sqrt{\frac{b_i}{b_{i-1}}}\left\{\left[1-\left(\frac{\lambda-a_i}{2b_i}\right)^2\right]\left[1-\left(\frac{\lambda-a_{i+1}}{2b_{i+1}}\right)^2\right]\right\}^{-1/4} e^{\frac{i}{\epsilon}\int_{i\epsilon}^{(i+1)\epsilon} f dy}$$

We must only obtain a bound for $(\Delta\gamma)_k$, because $(\Delta\gamma)_k = \tilde{\tilde{\gamma}}_i(\Delta\tilde{\gamma})_k$. We difference $\tilde{\gamma_k}$, dropping $\sqrt{\frac{b_k}{b_{k-1}}}$, as it is smooth, and obtain

$$|\tilde{\gamma_{k+1}} - \tilde{\gamma_k}| \leq \left| \left[1 - \left(\frac{\lambda - a_k}{2b_k}\right)^2\right]^{-1/4} e^{-\frac{i}{\epsilon}\int_{(k-1)\epsilon}^{k\epsilon} f dy} . \right.$$
$$\cdot \left\{ \left[1 - \left(\frac{\lambda - a_{k+1}}{2b_{k+1}}\right)^2\right]^{-1/4} e^{\frac{i}{\epsilon}\int_{(k)\epsilon}^{(k+1)\epsilon} f dy - \int_{(k-1)\epsilon}^{k\epsilon} f dy} \right.$$
$$\left. \left. - \left[1 - \left(\frac{\lambda - a_{k-1}}{2b_{k-1}}\right)^2\right]^{-1/4} \right\} \right|,$$

and hence

$$\begin{aligned}
|\tilde{\gamma_{k+1}} - \tilde{\gamma_k}| &\leq c(2k\epsilon)^{1/4} \left| \left\{ (2(k+1)\epsilon)^{1/4} e^{i\epsilon(2k\epsilon)^{-1/2}} - (2(k-1)\epsilon)^{1/4} \right\} \right| \\
&\leq c_1(2k\epsilon)^{1/2} \left| \left\{ (1 + \frac{\epsilon}{k\epsilon})^{1/4} e^{i\epsilon(2k\epsilon)^{-1/2}} - (2(k-1)\epsilon)^{1/4} \right\} \right| \\
&\leq c_2(2k\epsilon)^{1/2} \left| \left\{ (1 + \frac{\epsilon}{4k\epsilon})(1 + i\frac{\epsilon}{(2k\epsilon)^{1/2}}) - (1 - \frac{\epsilon}{4k\epsilon}) \right\} \right| \\
&\leq c_3(2k\epsilon)^{1/2} \frac{\epsilon}{k\epsilon} \left| \left\{ 1 + i(k\epsilon)^{1/2} \right\} \right| \leq c\frac{\epsilon}{(k\epsilon)^{1/2}} .
\end{aligned}$$

This implies that we can bound

$$|(\Delta\gamma)_k| \leq c\frac{\epsilon}{(k\epsilon)^{1/2}} \frac{1}{(i\epsilon)^{1/2}} . \tag{474}$$

Now we focus on the exponential sum. We set

$$u_j = e^{-\frac{i}{\epsilon}\int_{n\epsilon}^{j\epsilon} f(y)dy} ,$$

then

$$\frac{u_j - u_{j-1}}{1 - e^{\frac{i}{\epsilon}\int_{(j-1)\epsilon}^{j\epsilon} f(y)dy}} = e^{-\frac{i}{\epsilon}\int_{n\epsilon}^{j\epsilon} f(y)dy} .$$

Thus

$$\sum_{j=n}^{k} e^{-\frac{i}{\epsilon}\int_{n\epsilon}^{j\epsilon} f(y)dy} = \sum_{j=n}^{k} \frac{u_j - u_{j-1}}{1 - e^{\frac{i}{\epsilon}\int_{(j-1)\epsilon}^{j\epsilon} f(y)dy}}$$
$$= -\frac{u_{n-1}}{1 - e^{\frac{i}{\epsilon}\int_{(n-1)\epsilon}^{n\epsilon} f(y)dy}} + \frac{u_k}{1 - e^{\frac{i}{\epsilon}\int_{(k-1)\epsilon}^{k\epsilon} f(y)dy}} +$$
$$+ \sum_{j=n}^{k-1} u_j \frac{e^{\frac{i}{\epsilon}\int_{(j)\epsilon}^{(j+1)\epsilon} f dy} - e^{\frac{i}{\epsilon}\int_{(j-1)\epsilon}^{(j)\epsilon} f dy}}{\left(1 - e^{\frac{i}{\epsilon}\int_{(j-1)\epsilon}^{(j)\epsilon} f dy}\right)\left(1 - e^{\frac{i}{\epsilon}\int_{(j)\epsilon}^{(j+1)\epsilon} f dy}\right)} .$$

Now some calculations using the nature of the WKB formulae (440) near the turning point show (c.f. (469 - 473) above)

$$\left|1 - e^{\frac{i}{\epsilon}\int_{(j-1)\epsilon}^{(j)\epsilon} f dy}\right| \geq c(j\epsilon)^{1/2} ,$$

$$e^{\frac{i}{\epsilon}\int_{(j)\epsilon}^{(j+1)\epsilon} f dy} - e^{\frac{i}{\epsilon}\int_{(j-1)\epsilon}^{(j)\epsilon} f dy} \leq c_5 \frac{\epsilon}{(j\epsilon)^{1/2}} .$$

So we have

$$\text{(475)} \qquad \sum_{j=n}^{k} e^{-\frac{i}{\epsilon}\int_{n\epsilon}^{j\epsilon} f(y)dy} \leq c_1(n\epsilon)^{-1/2} + \sum_{j=n}^{k-1} c_2 \frac{\epsilon}{(j\epsilon)^{3/2}}$$

$$\leq c_1(n\epsilon)^{-1/2} .$$

Using the two bounds (474), (475), we get

$$|K(n,i,\epsilon)| \leq \sum_{k=n}^{i-1} |(\Delta\gamma)_k| \left|\sum_{j=n}^{k-1} e^{-\frac{i}{\epsilon}\int_{n\epsilon}^{j\epsilon} f(y)dy}\right| + |\gamma_i| \left|\sum_{j=n}^{i} e^{-\frac{i}{\epsilon}\int_{n\epsilon}^{j\epsilon} f(y)dy}\right|$$

$$\leq D \sum_{k=n}^{i-1} \frac{\epsilon}{(k\epsilon)^{1/2}} \frac{1}{(i\epsilon)^{1/2}} \frac{1}{(n\epsilon)^{1/2}} + D_2 |\gamma_i| \frac{1}{(n\epsilon)^{1/2}}$$

$$\leq D_3 \frac{1}{(i\epsilon)^{1/2}} + D_4 \frac{1}{(n\epsilon)^{1/2}}$$

$$\leq const. \frac{1}{(n\epsilon)^{1/2}} .$$

From Theorem 12.1, we know that there exists a solution y_n^+ to (432) satisfying

$$\left|\frac{y_n^+}{f_n} - 1\right| \leq e^{const.\frac{1}{(n\epsilon)^{1/2}}\left[\frac{\epsilon}{n\epsilon}\right]} - 1$$

$$\leq e^{const.\frac{\epsilon}{(n\epsilon)^{3/2}}} - 1 ,$$

and so we see that if $n\epsilon \gg \epsilon^{2/3}$, the error bounds tend to zero, and the Lemma is proven. □

We have now established the existence of solutions to the difference equation (432) which are asymptotic to the WKB formula (440). These asymptotics are valid for all n satisfying

$$|n\epsilon| \gg \epsilon^{2/3} .$$

The task of matching across a turning point lies ahead. We intend to construct a solution with asymptotics valid in a small neighborhood of the turning point. The small neighborhood will be large enough to have an overlap with the region of validity of the above asymptotics, and comparing asymptotic forms in this overlap region will give us the matching conditions.

Th ability to do this is granted by the form of the difference equation (432) in the vicinity of a turning point. We suppose there us one turning point, occurring at the origin. In addition, we suppose for simplicity that the "potential", $\frac{\lambda-a}{2b}$ has slope 1 at $x = 0$:

$$\frac{\lambda - a}{2b} \approx 1 + x + \alpha x_2/2 .$$

Then the difference equations (432) can be written as

$$y_{n-1} + \frac{a_n - \lambda}{b_n}\frac{b_n}{b_{n-1}} y_n + \frac{b_n}{b_{n-1}} y_{n+1} = 0 ,$$

which looks, to first order, like the simpler difference equation

(476) $$y_{n-1} - 2(1 + n\epsilon) y_n + y_{n+1} = 0 ,$$

for $n\epsilon$ in a vicinity of the origin.

It is this latter equation which we use to get the asymptotics. We thank A. Fedotov, (see also [**3**]) for pointing out to us that (476) is just precisely the 3 term recurrence relation for Bessel's equation, and hence can be analyzed explicitly.

Indeed, set

$$\nu = \frac{1 + n\epsilon}{\epsilon} ,$$

$$z = \frac{1}{1 + n\epsilon} .$$

Then, taking y_n to solve (476), we define G_ν by

$$G_\nu = y_n ,$$

and we obtain an equation for G_ν,

$$G_{\nu-1} + G_{\nu+1} = \frac{2\nu}{\nu z} G_\nu ,$$

which is the recurrence relation satisfied by any of the parabolic cylinder functions, (indexed by ν and evaluated at νz), the Bessel function $J_\nu(\nu z)$, and the Hankel functions $H_\nu^j(\nu z)$, $j = 1, 2$.

We shall prove that there is a solution to (432) asymptotic to the Hankel functions for n in a small neighborhood of the origin,and that there is a solution asymptotic to the Bessel function for n positive and near the origin. The actual region will overlap with the previous asymptotic regions.

7. Turning points - special functions

LEMMA 12.5. *Suppose that there is exactly one turning point, occurring at the origin, and that locally we have*

$$\frac{\lambda - a}{2b} \approx 1 + n\epsilon + \alpha(n\epsilon)^2 \ .$$

Then there exists y_n^1, y_n^2, y_n^- solving (432), satisfying

$$\left| \frac{y_n^1}{H^1_{\nu_n}(\nu_n z_n)} - 1 \right| \to 0 \quad as \quad \epsilon \to 0 \ , \quad for \ |n\epsilon| \ll \epsilon^{1/2} \ ,$$

$$\left| \frac{y_n^2}{H^2_{\nu_n}(\nu_n z_n)} - 1 \right| \to 0 \quad as \quad \epsilon \to 0 \ , \quad for \ |n\epsilon| \ll \epsilon^{1/2} \ ,$$

$$\left| \frac{y_n^-}{J_{\nu_n}(\nu_n z_n)} - 1 \right| \to 0 \quad as \quad \epsilon \to 0 \ , \quad for 0 < \ n\epsilon \ll \epsilon^{1/2} \ .$$

Proof. First, we shall consider the solution y_n^-, and then the other two solutions. We follow the same argument as previously, commencing with the error function (either ξ_n or ζ_n), then bounding the kernel.

Throughout, we are aided by the asymptotic representations of $J_\nu(\nu z)$ and $H^j_\nu(\nu z)$,

$$J_\nu(\nu z) = \frac{1}{\nu^{1/3}} \left(\frac{4\zeta}{1 - z^2} \right)^{1/4} \left\{ Ai(\nu^{2/3}\zeta) + \delta_J(\nu, \zeta) \right\} \ ,$$

$$H^1_\nu(\nu z) = \frac{1}{\nu^{1/3}} \left(\frac{4\zeta}{1 - z^2} \right)^{1/4} \left\{ Ai(\nu^{2/3}\zeta) - iBi(\nu^{2/3}\zeta) + \delta^1_H(\nu, \zeta) \right\} \ ,$$

$$H^2_\nu(\nu z) = \frac{1}{\nu^{1/3}} \left(\frac{4\zeta}{1 - z^2} \right)^{1/4} \left\{ Ai(\nu^{2/3}\zeta) + iBi(\nu^{2/3}\zeta) + \delta^2_H(\nu, \zeta) \right\} \ ,$$

where

$$\frac{2}{3}\zeta = \int_z^1 \frac{(1 - t^2)^{1/2}}{t} dt \ ,$$

and the error terms δ_J, δ^j_H satisfy

$$\delta_J << Ai(\nu^{2/3}\zeta) \quad as \quad \nu \to \infty \ ,$$

$$\left|\delta^1_H\right| \ll \left| Ai(\nu^{2/3}\zeta) + Bi(\nu^{2/3}\zeta) \right| \quad as \quad \nu \to \infty \ ,$$

$$\left|\delta^2_H\right| \ll \left| Ai(\nu^{2/3}\zeta) - Bi(\nu^{2/3}\zeta) \right| \quad as \quad \nu \to \infty \ .$$

A good Reference for these facts is the book by Olver [**17**]. We take

$$g_n = J_{\nu_n}(\nu_n z_n) \ ,$$

then,

$$\begin{aligned} g_{n-1}\zeta_n &= g_{n-1} + \frac{a_n - \lambda}{b_n}\frac{b_n}{b_{n-1}}g_n + \frac{b_n}{b_{n-1}}g_{n+1} \\ &= g_{n-1} + \frac{b_n}{b_{n-1}}\left[\frac{a_n - \lambda}{b_n}g_n + g_{n+1}\right] \\ &= g_{n-1} + \frac{b_n}{b_{n-1}}\left[-2(1+n\epsilon)g_n + g_{n+1} + \alpha_n(n\epsilon)^2 g_n\right] \\ &= \left(1 - \frac{b_n}{b_{n-1}}\right) g_{n-1} + \tilde{\alpha}_n(n\epsilon)^2 g_n \ . \end{aligned}$$

So we have

$$|\zeta_n| \le \beta\epsilon + \tilde{\alpha}(n\epsilon)^2 \frac{g_n}{g_{n-1}} \ . \tag{477}$$

To be sure that ζ_n remains this small, we must bound the ratio $\frac{g_n}{g_{n-1}}$. To do this, we use the asymptotic prescription of $J_\nu(\nu z)$. We have

$$\begin{aligned} \frac{g_n}{g_{n-1}} &= \frac{J_{\nu_n}(\nu_n z_n)}{J_{\nu_{n-1}}(\nu_{n-1} z_{n-1})} \\ &= \left(\frac{\nu_{n-1}}{\nu_n}\right)^{1/3}\left(\frac{1 - z_{n-1}^2}{1 - z_n}\right)^{1/4}\left(\frac{\zeta_n}{\zeta_{n-1}}\right)^{1/4}\frac{Ai(\nu_n^{2/3}\zeta_n) + \epsilon_J(\nu_n, \zeta_n)}{Ai(\nu_{n-1}^{2/3}\zeta_{n-1}) + \epsilon_J(\nu_{n-1}, \zeta_{n-1})} \ . \end{aligned} \tag{478}$$

Now we bound the first two terms in the above product,

$$\frac{\nu_{n-1}}{\nu_n} = \frac{1 + n\epsilon - \epsilon}{1 + n\epsilon} \le 1 \ , \tag{479}$$

$$\begin{aligned} \frac{1 - z_{n-1}^2}{1 - z_n} &= \frac{2(n-1)\epsilon + ((n-1)\epsilon)^2}{2(n)\epsilon + (n\epsilon)^2} \cdot \frac{(1+n\epsilon)^2}{(1 + (n-1)\epsilon)^2} \\ &\le \left(\frac{1 + n\epsilon}{1 + (n-1)\epsilon}\right)^2 \le (1 + 2\epsilon)^2 \le 1.1 \ . \end{aligned} \tag{480}$$

Now to bound $\frac{\zeta_n}{\zeta_{n-1}}$, we study

$$\begin{aligned} \frac{2}{3}\zeta &= \int_{z_n}^{1} \frac{(1-t^2)^{1/2}}{t} dt = \int_{\frac{1}{1+n\epsilon}}^{1} \frac{(1-t^2)^{1/2}}{t} \\ &= \epsilon \int_0^{n-1} (2\ell\epsilon)^{1/2} \frac{\left(1 + \frac{\ell\epsilon}{2}\right)^{1/2}}{(1 + \ell\epsilon)^2} d\ell \\ &\epsilon \int_0^{n-1} (2\ell\epsilon)^{1/2} \frac{\left(1 + \frac{\ell\epsilon}{2}\right)^{1/2}}{(1 + \ell\epsilon)^2} d\ell + \epsilon \int_{n-1}^{n} (2\ell\epsilon)^{1/2} \frac{\left(1 + \frac{\ell\epsilon}{2}\right)^{1/2}}{(1 + \ell\epsilon)^2} d\ell \\ &= \frac{2}{3}\zeta_{n-1}^{3/2} + \epsilon \int_{n-1}^{n} (2\ell\epsilon)^{1/2} \frac{\left(1 + \frac{\ell\epsilon}{2}\right)^{1/2}}{(1 + \ell\epsilon)^2} d\ell \ . \end{aligned}$$

Thus

$$\left(\frac{\zeta_n}{\zeta_{n-1}}\right)^{3/2} = 1 + \frac{\epsilon \int_{n-1}^{n} (2\ell\epsilon)^{1/2} \frac{\left(1+\frac{\ell\epsilon}{2}\right)^{1/2}}{(1+\ell\epsilon)^2} d\ell}{\epsilon \int_0^{n-1} (2\ell\epsilon)^{1/2} \frac{\left(1+\frac{\ell\epsilon}{2}\right)^{1/2}}{(1+\ell\epsilon)^2} d\ell} \tag{481}$$

$$\leq 1 + \frac{(2n\epsilon)^{1/2} \frac{(1+\frac{n\epsilon}{2})^{1/2}}{(1+(n-1)\epsilon)^2}}{\int_{n-2}^{n-1} (2\ell\epsilon)^{1/2} \frac{\left(1+\frac{\ell\epsilon}{2}\right)^{1/2}}{(1+\ell\epsilon)^2} d\ell}$$

$$\leq 1 + \frac{(2n\epsilon)^{1/2} \frac{(1+\frac{n\epsilon}{2})^{1/2}}{(1+(n-1)\epsilon)^2}}{(2(n-1)\epsilon)^{1/2} \frac{(1+\frac{(n-2)\epsilon}{2})^{1/2}}{(1+(n-1)\epsilon)^2}}$$

$$\leq 1 + \left(\frac{n}{n-2}\right)^{1/2} \left(\frac{1+\frac{n\epsilon}{2}}{1+\frac{(n-2)\epsilon}{2}}\right)^{1/2} \leq const. \quad , for\ n > 2 \ .$$

To complete the bound on ζ_n in (477), we must bound the ratio of Airy functions.

$$\left|\frac{Ai(\nu_n^{2/3}\zeta_n) + \epsilon_J(\nu_n, \zeta_n)}{Ai(\nu_{n-1}^{2/3}\zeta_{n-1}) + \epsilon_J(\nu_{n-1}, \zeta_{n-1})}\right| \tag{482}$$

$$\leq \frac{Ai(\nu_n^{2/3}\zeta_n)}{Ai(\nu_{n-1}^{2/3}\zeta_{n-1})} \frac{1 + \left|\frac{\epsilon_J(\nu,\zeta)}{Ai(\nu_n^{2/3}\zeta_n)}\right|}{\left|\frac{\epsilon_J(\nu,\zeta)}{Ai(\nu_{n-1}^{2/3}\zeta_{n-1})}\right|}$$

$$\leq const. \frac{Ai(\nu_n^{2/3}\zeta_n)}{Ai(\nu_{n-1}^{2/3}\zeta_{n-1})} \leq const.$$

The last inequality is because, for positive argument, the Airy function $Ai(x)$ is decreasing. Combining (478 - 482) with (477), we have

$$\zeta_n \leq \beta\epsilon + \alpha(n\epsilon)^2 \ .$$

We now must focus on the kernel. We have

$$K(n,i,\epsilon) = \sum_{k=n}^{i} \prod_{j=k}^{i} \frac{b_j}{b_{j-1}} \frac{g_j}{g_{j-1}} \frac{g_{j+1}}{g_j}$$

$$= \sum_{k=n}^{i} \prod_{j=k}^{i} \frac{b_i}{b_{k-1}} \frac{g_i}{g_{k-1}} \frac{g_{i+1}}{g_k} \leq \beta \sum_{k=n}^{i} \frac{g_{i+1}}{g_i} \frac{g_k}{g_{k-1}} \left(\frac{g_i}{g_k}\right)^2$$

$$\leq const. \sum_{k=n}^{i} \left(\frac{g_i}{g_k}\right)^2 \ .$$

So to bound the Kernel, we must bound the ratio $\frac{g_i}{g_k}$. $(k \leq i)$. We have

$$\frac{g_i}{g_k} = \frac{J_{\nu_i}(\nu_i z_i)}{J_{\nu_k}(\nu_k z_k)} = \left(\frac{\nu_k}{\nu_i}\right)^{1/3} \left(\frac{1-z_k^2}{1-z_i^2}\right) \left(\frac{\zeta_i}{\zeta_k}\right)^{1/4} \frac{Ai(\nu_i^{2/3}\zeta_i) + \epsilon_J(\nu_i, \zeta_i)}{Ai(\nu_k^{2/3}\zeta_k) + \epsilon_J(\nu_k, \zeta_k)} ,$$

and analysis similar to the previous calculations shows

$$\frac{\nu_k}{\nu_i} \leq 1 ,$$

$$\frac{1-z_k^2}{1-z_i^2} \leq c\frac{k}{i} ,$$

$$\left(\frac{\zeta_i}{\zeta_k}\right)^{3/2} \leq c\frac{i}{k} .$$

Thus,

$$\frac{g_i}{g_k} \leq const \frac{Ai(\nu_i^{2/3}\zeta_i) + \epsilon_J(\nu_i, \zeta_i)}{Ai(\nu_k^{2/3}\zeta_k) + \epsilon_J(\nu_k, \zeta_k)}$$

$$\leq c\frac{Ai(\nu_i^{2/3}\zeta_i)}{Ai(\nu_k^{2/3}\zeta_k)} \frac{1 + \epsilon_J/Ai(\nu_i^{2/3}\zeta_i)}{1 + \epsilon_J(\nu_k^{2/3}\zeta_k)} \leq\leq c\frac{Ai(\nu_i^{2/3}\zeta_i)}{Ai(\nu_k^{2/3}\zeta_k)}$$

$$\leq const.$$

The last inequality is due to the fact that $Ai(x)$ is decreasing in x. So we have

$$K(n, i, \epsilon) \leq const \sum_{k=n}^{i} 1 \leq const.\epsilon^{-1/2} ,$$

where the final inequality is because if we're in the situation $0 \leq n\epsilon \ll \epsilon^{1/2}$, then the number of terms in the sum is certainly less than $\epsilon^{-1/2}$.

Summarizing, we have

$$\sum_{n=0}^{\epsilon^{-1/2}} |\zeta_n| \leq \sum_{n=0}^{N} \beta\epsilon + \alpha(n\epsilon)^2 \leq \beta N\epsilon + \frac{\alpha(n\epsilon)^3}{\epsilon} ,$$

and

$$K(n, i, \epsilon) \leq const\epsilon^{-1/2} .$$

From Theorem 1, y^- exists, and satisfies

$$\left|\frac{y_n^-}{g_n} - 1\right| \leq e^{c\epsilon^{-1/2}\left(\beta N\epsilon + \frac{\alpha}{\epsilon}(N\epsilon)^3\right)} - 1$$

$$= e^{c_1 \frac{N\epsilon}{\epsilon^{1/2}} + c_2 \left(\frac{N\epsilon}{\epsilon^{1/2}}\right)^3} - 1 .$$

So, if $0 \leq N\epsilon \ll \epsilon^{1/2}$, the error term tends to zero, and so we have established the first part of the Lemma.

We turn to the existence of the solution y_n^1. For this case, we bound the error term ξ_n and the kernel

$$K(i,n,\epsilon)=\sum_{k=i}^{n-1}\prod_{j=i}^{k}\frac{b_j}{b_{j-1}}\frac{g_j}{g_{j-1}}\frac{g_{j+1}}{g_j}\ ,$$

where now

$$g_n = H_{\nu_n}(\nu_n z_n)\ .$$

Now, as in the derivation of (477), we have the bound

$$\xi_n \le \beta\epsilon + \tilde{\alpha}(n\epsilon)^2\frac{g_n}{g_{n-1}}\ .$$

So we must bound

$$\begin{aligned}\left|\frac{g_n}{g_{n-1}}\right| &= \left|\frac{H_{\nu_n}(\nu_n z_n)}{H_{\nu_n}(\nu_n z_n)}\right|\\ &= \left|\left(\frac{\nu_{n-1}}{\nu_n}\right)^{1/3}\left(\frac{1-z_{n-1}^2}{1-z_n^2}\right)^{1/4}\left(\frac{\zeta_n}{\zeta_{n-1}}\right)^{1/4}\times\right.\\ &\qquad\left.\times\frac{Ai(\nu_n^{2/3}\zeta_n)+iBi(\nu_n^{2/3}\zeta_n)+\epsilon_+H^1(\nu_n,\zeta_n)}{Ai(\nu_{n-1}^{2/3}\zeta_{n-1})+iBi(\nu_{n-1}^{2/3}\zeta_{n-1})+\epsilon_H^1(\nu_{n-1},\zeta_{n-1})}\right|\\ &= \omega_n\left|\frac{Ai(\nu_n^{2/3}\zeta_n)+iBi(\nu_n^{2/3}\zeta_n)}{Ai(\nu_{n-1}^{2/3}\zeta_{n-1})+iBi(\nu_{n-1}^{2/3}\zeta_{n-1})}\right|\left|\frac{1+\frac{\epsilon_H^1(\nu_n,\zeta_n)}{\left|Ai(\nu_n^{2/3}\zeta_n)+iBi(\nu_n^{2/3}\zeta_n)\right|}}{1+\frac{\epsilon_H^1(\nu_{n-1},\zeta_{n-1})}{\left|Ai(\nu_{n-1}^{2/3}\zeta_{n-1})+iBi(\nu_{n-1}^{2/3}\zeta_{n-1})\right|}}\right|\ ,\end{aligned}$$

where

$$\omega_n=\left(\frac{\nu_{n-1}}{\nu_n}\right)^{1/3}\left(\frac{1-z_{n-1}^2}{1-z_n^2}\right)^{1/4}\left(\frac{\zeta_n}{\zeta_{n-1}}\right)^{1/4}.$$

As before, ω_n is bounded by a constant, independent of ϵ. So, we have

$$\frac{g_n}{g_{n-1}}\le const.\left|\frac{Ai(\nu_n^{2/3}\zeta_n)+iBi(\nu_n^{2/3}\zeta_n)}{Ai(\nu_{n-1}^{2/3}\zeta_{n-1})+iBi(\nu_{n-1}^{2/3}\zeta_{n-1})}\right|.\tag{483}$$

to bound this, we must work separately for $n>0$, and $n<0$. We take $n>0$. Then

$$\begin{aligned}\left|Ai(\nu_n^{2/3}\zeta_n)+iBi(\nu_n^{2/3}\zeta_n)\right| &\le 2\max\left\{Ai(\nu_n^{2/3}\zeta_n),Bi(\nu_n^{2/3}\zeta_n)\right\}\\ &= 2Bi(\nu_n^{2/3}\zeta_n)\ ,\end{aligned}$$

and

$$\left|Ai(\nu_{n-1}^{2/3}\zeta_{n-1})+iBi(\nu_{n-1}^{2/3}\zeta_{n-1})\right|\ge Bi(\nu_{n-1}^{2/3}\zeta_{n-1})\ ,$$

and so

$$\left|\frac{Ai(\nu_n^{2/3}\zeta_n)+iBi(\nu_n^{2/3}\zeta_n)}{Ai(\nu_{n-1}^{2/3}\zeta_{n-1})+iBi(\nu_{n-1}^{2/3}\zeta_{n-1})}\right|\le 2\frac{Bi(\nu_n^{2/3}\zeta_n)}{Bi(\nu_{n-1}^{2/3}\zeta_{n-1})}\ .\tag{484}$$

To bound the right hand side of (484), we will establish the following bound:

$$\left|\nu_n^{2/3}\zeta_n - \nu_{n-1}^{2/3}\zeta_{n-1}\right| \le c\epsilon^{1/3} . \tag{485}$$

This is true because we can write

$$\zeta_n = 2^{1/3}(n\epsilon) + O(n\epsilon)^2 ,$$

and

$$\nu_n^{2/3} = \epsilon^{-2/3}(1 + \frac{2}{3}(n\epsilon) + O(n\epsilon)^2) .$$

Using these two formulae, we get

$$\nu_n^{2/3}\zeta_n = \frac{2^{1/3}(n\epsilon)}{\epsilon^{2/3}}\left(1 - \frac{1}{30}n\epsilon + O(n\epsilon)^2\right) ,$$

implying

$$\nu_n^{2/3}\zeta_n - \nu_{n-1}^{2/3}\zeta_{n-1} = 2^{1/3}\epsilon^{1/3}\left[1 - \frac{1}{30}(n\epsilon + (n-1)\epsilon) + \frac{O(n\epsilon)^3}{\epsilon}\right] \le c\epsilon^{1/3} .$$

This last inequality is true because n satisfies $n\epsilon \ll \epsilon^{1/2}$. Now we use (485) as follows. We write

$$\begin{aligned}\frac{Bi(\nu_n^{2/3}\zeta_n)}{Bi(\nu_{n-1}^{2/3}\zeta_{n-1})} &= 1 + \frac{Bi(\nu_n^{2/3}\zeta_n) - Bi(\nu_{n-1}^{2/3}\zeta_{n-1})}{Bi(\nu_{n-1}^{2/3}\zeta_{n-1})} \\ &= 1 + \gamma_n ,\end{aligned}$$

and we bound γ_n, as follows,

$$|\gamma_n| = \left|\frac{(\nu_n^{2/3}\zeta_n - \nu_{n-1}^{2/3}\zeta_{n-1}) \cdot Bi'(r)}{Bi(\nu_{n-1}^{2/3}\zeta_{n-1})}\right| , \quad r \in [\nu_{n-1}^{2/3}\zeta_{n-1}, \nu_n^{2/3}\zeta_n] ,$$

and hence

$$\begin{aligned}|\gamma_n| &\le \frac{c\epsilon^{1/3}\left[Bi'(0) + \int_0^r xBi(x)dx\right]}{Bi(\nu_{n-1}^{2/3}\zeta_{n-1})} \\ &\le \frac{c\epsilon^{1/3}\left[Bi'(0) + (\nu_n^{2/3}\zeta_n)^2 Bi(\nu_n^{2/3}\zeta_n)\right]}{Bi(\nu_{n-1}^{2/3}\zeta_{n-1})} \\ &\le c\frac{\epsilon^{1/3}Bi'(0)}{Bi(\nu_{n-1}^{2/3}\zeta_{n-1})} + c_1\epsilon^{1/3}(n\epsilon^{1/3})^2\frac{Bi(\nu_n^{2/3}\zeta_n)}{Bi(\nu_{n-1}^{2/3}\zeta_{n-1})} .\end{aligned}$$

So we have the following inequality,

$$\frac{Bi(\nu_n^{2/3}\zeta_n)}{Bi(\nu_{n-1}^{2/3}\zeta_{n-1})} \le 1 + c\frac{\epsilon^{1/3}Bi'(0)}{Bi(\nu_{n-1}^{2/3}\zeta_{n-1})} + c_1\epsilon^{1/3}(n\epsilon^{1/3})^2\frac{Bi(\nu_n^{2/3}\zeta_n)}{Bi(\nu_{n-1}^{2/3}\zeta_{n-1})} ,$$

which can be rewritten

$$\frac{Bi(\nu_n^{2/3}\zeta_n)}{Bi(\nu_{n-1}^{2/3}\zeta_{n-1})}\left(1 - c_1(n^2\epsilon)\right) \le 1 + c_0\epsilon^{1/3} . \tag{486}$$

Now since $n^2\epsilon \ll 1$ as $\epsilon \to 0$, we have

$$\frac{Bi(\nu_n^{2/3}\zeta_n)}{Bi(\nu_{n-1}^{2/3}\zeta_{n-1})} \leq const. \quad , \quad for \quad n > 0 \ .$$

Combining (483 - 486), we have $\frac{g_n}{g_{n-1}} < c$, for $n > 0$.

We now obtain a similar bound for $n < 0$. This is done with the auxiliary function used in [**17**],

$$M(x) = (Ai(x)^2 + Bi(x)^2)^{1/2} \ .$$

For we can write

$$\left| \frac{Ai(\nu_n^{2/3}\zeta_n) + iBi(\nu_n^{2/3}\zeta_n) + \epsilon_H^1(\nu_n, \zeta_n)}{Ai(\nu_{n-1}^{2/3}\zeta_{n-1}) + iBi(\nu_{n-1}^{2/3}\zeta_{n-1}) + \epsilon_H^1(\nu_{n-1}, \zeta_{n-1})} \right| \leq \frac{M(\nu_n^{2/3}\zeta_n)}{M(\nu_{n-1}^{2/3}\zeta_{n-1})} \left| \frac{1 + \frac{\epsilon_H^1(\nu_n,\zeta_n)}{M(\nu_n^{2/3}\zeta_n)}}{1 + \frac{\epsilon_H^1(\nu_{n-1},\zeta_{n-1})}{M(\nu_{n-1}^{2/3}\zeta_{n-1})}} \right|$$

$$\leq c\frac{M(\nu_n^{2/3}\zeta_n)}{M(\nu_{n-1}^{2/3}\zeta_{n-1})}$$

$$= 1 + \frac{M(\nu_n^{2/3}\zeta_n) - M(\nu_{n-1}^{2/3}\zeta_{n-1})}{M(\nu_{n-1}^{2/3}\zeta_{n-1})}$$

$$= 1 + \frac{\Delta(\nu_n^{2/3}\zeta_n)M'(r)}{M(\nu_{n-1}^{2/3}\zeta_{n-1})} \quad , \quad r \in [\nu_{n-1}^{2/3}\zeta_{n-1}, \nu_n^{2/3}\zeta_n] \ ,$$

and so

$$\left| \frac{Ai(\nu_n^{2/3}\zeta_n) + iBi(\nu_n^{2/3}\zeta_n) + \epsilon_H^1(\nu_n, \zeta_n)}{Ai(\nu_{n-1}^{2/3}\zeta_{n-1}) + iBi(\nu_{n-1}^{2/3}\zeta_{n-1}) + \epsilon_H^1(\nu_{n-1}, \zeta_{n-1})} \right| \leq 1 + c\epsilon^{1/3}\frac{M'(r)}{M(\nu_{n-1}^{2/3}\zeta_{n-1})} \ .$$

Now M' is bounded, so

$$c\epsilon^{1/3}\frac{M'(r)}{M(\nu_{n-1}^{2/3}\zeta_{n-1})} \leq const.\frac{\epsilon^{1/3}}{M(\nu_{n-1}^{2/3}\zeta_{n-1})} \ .$$

In Olver [**17**], the asymptotic prescription

$$M \sim (-x)^{-1/4} \quad , as \ x \to -\infty$$

is proven. From this, we obtain the bound

$$M(\nu_{n-1}^{2/3}\zeta_{n-1}) \geq c\epsilon^{1/8} \ ,$$

for $|n\epsilon| \ll \epsilon^{1/2}$. Thus

$$\epsilon^{1/3}\frac{M'(r)}{M(\nu_{n-1}^{2/3}\zeta_{n-1})} \leq const.\epsilon^{5/24} ,$$

and so

$$\left|\frac{Ai(\nu_n^{2/3}\zeta_n) + iBi(\nu_n^{2/3}\zeta_n) + \epsilon_H^1(\nu_n, \zeta_n)}{Ai(\nu_{n-1}^{2/3}\zeta_{n-1}) + iBi(\nu_{n-1}^{2/3}\zeta_{n-1}) + \epsilon_H^1(\nu_{n-1}, \zeta_{n-1})}\right| \leq const .$$

Thus we now have

$$\frac{g_n}{g_{n-1}} \leq const.$$

for all n satisfying $|n\epsilon| \ll \epsilon^{1/2}$. And from this we conclude that

$$|\xi_n| \leq \beta\epsilon + \tilde{\alpha}(n\epsilon)^2 , \quad |n\epsilon| \ll \epsilon^{1/2} ,$$

which implies

$$\sum_{n=\ell}^{N} \xi_n \leq \beta(N-\ell)\epsilon + \tilde{\alpha}\left[(n\epsilon)^3 - (\ell\epsilon)^3\right] .$$

Now we must bound the kernel

$$K(i,n,\epsilon) = \sum_{k=i}^{n-1} \frac{b_{i-1}}{b_k}\frac{g_{i-1}}{g_k}\frac{g_i}{g_{k+1}} \leq B\sum_{k=i}^{n-1}\left(\frac{g_{i-1}}{g_{k+1}}\right)^2 .$$

Now we have

$$\left|\frac{g_i}{g_k}\right| = \left|\left(\frac{\nu_k}{\nu_i}\right)^{1/3}\left(\frac{1-z_k^2}{1-z_i^2}\right)^{1/4}\left(\frac{\zeta_i}{\zeta_k}\right)^{1/4} \times \right.$$

$$\left. \times \left\{\frac{Ai(\nu_i^{2/3}\zeta_i) + iBi(\nu_i^{2/3}\zeta_i) + \epsilon_H^1(\nu_i, \zeta_i)}{Ai(\nu_k^{2/3}\zeta_k) + iBi(\nu_k^{2/3}\zeta_k) + \epsilon_H^1(\nu_k, \zeta_k)}\right\}\right|$$

$$\leq \omega_k^i \left|\frac{Ai(\nu_i^{2/3}\zeta_i) + iBi(\nu_i^{2/3}\zeta_i)}{Ai(\nu_k^{2/3}\zeta_k) + iBi(\nu_k^{2/3}\zeta_k)}\right| \left|\frac{1 + \frac{\epsilon_H^1(\nu_i,\zeta_i)}{Ai(\nu_i^{2/3}\zeta_i)+iBi(\nu_i^{2/3}\zeta_i)}}{1 + \frac{\epsilon_H^1(\nu_k,\zeta_k)}{Ai(\nu_k^{2/3}\zeta_k)+iBi(\nu_k^{2/3}\zeta_k)}}\right|$$

$$\leq c\left|\frac{Ai(\nu_i^{2/3}\zeta_i) + iBi(\nu_i^{2/3}\zeta_i)}{Ai(\nu_k^{2/3}\zeta_k) + iBi(\nu_k^{2/3}\zeta_k)}\right| .$$

For $i, k > 0$, we bound as before,

$$\left|\frac{Ai(\nu_i^{2/3}\zeta_i) + iBi(\nu_i^{2/3}\zeta_i)}{Ai(\nu_k^{2/3}\zeta_k) + iBi(\nu_k^{2/3}\zeta_k)}\right| \leq 2\frac{Bi(\nu_i^{2/3}\zeta_i)}{Bi(\nu_k^{2/3}\zeta_k)} \leq 2 .$$

If $i < 0$, but $k > 0$, a similar argument will work. Suppose both i and k are negative. Then

$$\left|\frac{Ai(\nu_i^{2/3}\zeta_i) + iBi(\nu_i^{2/3}\zeta_i)}{Ai(\nu_k^{2/3}\zeta_k) + iBi(\nu_k^{2/3}\zeta_k)}\right| = \left|\frac{M(\nu_i^{2/3}\zeta_i)}{M(\nu_k^{2/3}\zeta_k)}\right| \ , \quad (i < k < 0) \ .$$

In [**17**], Olver proves that M is increasing on $(-\infty, \hat{x})$, where $\hat{x}$ is the first negative root of $Ai(x) - Bi(x)$. So we have

$$\frac{M(\nu_i^{2/3}\zeta_i)}{M(\nu_k^{2/3}\zeta_k)} \leq 1 \ , \ for \ \ \nu_i^{2/3}\zeta_i, \ \nu_k^{2/3}\zeta_k \leq \hat{x} \ .$$

Now for $\hat{x} < x < 0$, we have

$$c < M(x) < d$$

for two positive constants c and d, and so

$$\frac{M(\nu_i^{2/3}\zeta_i)}{M(\nu_k^{2/3}\zeta_k)} \leq \frac{d}{c}$$

if $\nu_i^{2/3}\zeta_i, \ \nu_k^{2/3}\zeta_k \geq \hat{x}$. If $\nu_i^{2/3}\zeta_i \leq \hat{x}$, but $\nu_k^{2/3}\zeta_k \geq \hat{x}$, then

$$\frac{M(\nu_i^{2/3}\zeta_i)}{M(\nu_k^{2/3}\zeta_k)} \leq \frac{M(\hat{x})}{c} \leq const. \ .$$

This finally implies that

$$\left|\frac{g_i}{g_k}\right| \leq const. \ ,$$

which implies

$$K(i,n,\epsilon) \leq B \sum_{k=i}^{n} \left(\frac{g_i}{g_k}\right)^2 \leq \sum_{k=i}^{n} const. \leq \frac{c}{\epsilon^{1/2}} \ .$$

The two inequalities we have obtained are

$$K(i,n,\epsilon) \leq \frac{c}{\epsilon^{1/2}} \ ,$$

and

$$\sum_{n=\ell}^{N} \xi_n \leq \beta(N-\ell)\epsilon + \tilde{\alpha}\left[(n\epsilon)^3 - (\ell\epsilon)^3\right] \ .$$

The error term from Theorem 1 is thus

$$\exp\left[\frac{c}{\epsilon^{1/2}}\left(\beta(N-\ell)\epsilon + \frac{\alpha}{\epsilon}\left[(n\epsilon)^3 - (\ell\epsilon)^3\right]\right)\right] - 1 \ .$$

Now we suppose that $|N|, |\ell| \leq N_1$, and so we can bound the above by

$$\exp\left[2\beta\frac{N_1\epsilon}{\epsilon^{1/2}} + \frac{2\alpha(N_1\epsilon)^3}{\epsilon^{3/2}}\right] - 1 \ .$$

We see that under the assumption that $|N_1\epsilon| \ll \epsilon^{1/2}$, the error term tends to zero, and so Theorem 1 asserts the existence of the solution y_n^1 to (432), which is asymptotic to the Hankel function $H_{\nu_n}(\nu_n z_n)$, for $|n\epsilon| \ll \epsilon^{1/2}$, $\epsilon \to 0$. The other Hankel - type solution y_n^2 is handled in identical fashion. We shall omit the proof.

Thus the proof is complete. □

The previous work can be summarized with the following statement of the result.

THEOREM 12.2. *Suppose a_n and b_n arise from the discretization of two functions,*

$$a_n = a(n\epsilon) \quad ,$$
$$b_n = b(n\epsilon) \quad ,$$

for $n = 1, \ldots, \frac{1}{\epsilon}$, and that there is one turning point at x_0, $0 < x_0 < 1$. Then we have 7 solutions of the difference equation (432), each with asymptotics valid in one of four regions.

- *I* $\quad x < x_0,\ |x - x_0| \gg \epsilon^{2/3}$,
- *II* $\quad x > x_0,\ |x - x_0| \gg \epsilon^{2/3}$,
- *III* $\quad |x - x_0| \ll \epsilon^{1/2}$,
- *IV* $\quad x > x_0,\ |x - x_0| \ll \epsilon^{1/2}$.

The solutions are $y_{osc}^{\pm}$, $y_{exp}^{\pm}$, y_H^j $(j = 1, 2)$, and y_J.

- *In region I, the solutions $y_{osc}^{\pm}(n)$ have asymptotics given by the oscillatory J-WKB formulae.*
- *In region II, the solutions $y_{exp}^{\pm}(n)$ have asymptotics given by the exponential J-WKB formulae.*
- *In region III, the solutions $y_H^j(n)$ have asymptotics given by the Hankel functions H^j.*
- *In region IV, the singe solution $y_J(n)$ has asymptotics given by the Bessel -type functions $J_{\nu_n}(\nu_n z_n)$.*

We are prepared to use this Theorem to match solutions across a turning point. The result of this analysis is that the standard WKB analysis for ODE systems applies to the difference equation (432), with coefficients arising from the discretization of two smooth functions.

8. Matching and Eigenvalues

Our first goal is to match the solution $y^-_{exp}(n)$ from the "exterior region" $\epsilon^{2/3} \ll n\epsilon$, into the "interior region" $0 \le n\epsilon \ll \epsilon^{1/2}$. In the exterior region, we can express y^-_{exp} as

$$y^-_{exp}(n) \approx \left[\left(\frac{\lambda - a}{2b}\right)^2 - 1\right]^{-1/4} \exp\left[-\frac{1}{\epsilon}\int^{n\epsilon} \cosh^{-1}(\frac{\lambda - a}{2b}) + \epsilon\frac{b'}{2b}\frac{e^{-s'_0}}{\sinh s'_0}dx\right]$$

$$\approx (2n\epsilon)^{-1/4}\exp\left[-\frac{1}{\epsilon}\frac{2\sqrt{2}}{3}(n\epsilon)^{3/2}\right],$$

and these are valid, in particular, for $\epsilon^{2/3} \ll n\epsilon \ll \epsilon^{1/2}$. In the same region, $\epsilon^{2/3} \ll n\epsilon \ll \epsilon^{1/2}$, the solution $y_J(n)$ has the form

$$y_J(n) \approx J_{\nu_n}(\nu_n z_n) = \frac{1}{\nu_n}\left(\frac{4\zeta_n}{1 - z_n^2}\right)^{1/4}\left\{Ai(\nu_n^{2/3}z_n) + \epsilon_J\right\}$$

$$\approx \frac{1}{\nu_n^{1/3}}\left(\frac{4\zeta_n}{1 - z_n^2}\right)^{1/4}\frac{\exp\left[-\frac{2}{3}\nu_n z_n^{3/2}\right]}{2\pi^{1/2}\nu_n^{1/6}\zeta_n^{1/4}}$$

$$\approx \frac{1}{\nu_n^{1/2}}\frac{4}{(1 - z_n^2)^{1/4}}\frac{1}{2\pi^{1/2}}\exp\left[-\frac{1}{\epsilon}\frac{2\sqrt{2}}{3}(n\epsilon)^{3/2}\right]$$

$$\approx \frac{\epsilon^{1/2}}{\sqrt{2\pi}}(2n\epsilon)^{1/2}\exp\left[-\frac{1}{\epsilon}\frac{2\sqrt{2}}{3}(n\epsilon)^{3/2}\right].$$

Thus we have

$$y^-_{exp}(n) = \frac{\sqrt{2\pi}}{\epsilon^{1/2}}y_J(n)\,, \tag{487}$$

and so we know the asymptotic prescription of y^-_{exp} for all values of n to the right of the turning point. We would like to obtain the asymptotic prescription of $y^-_{exp}(n)$ for negative values of n. To accomplish this, we match the solution $y_J(n)$, whose asymptotics are now known for all $n > 0$, through the turning point.

For $|n\epsilon| \ll \epsilon^{1/2}$, we have the following asymptotics for the solutions $y^1_H(n)$, $y^2_H(n)$,

$$y^1_H(n) \approx H^1_{\nu_n}(\nu_n z_n) \approx \frac{1}{\nu_n^{1/3}}\left(\frac{4\zeta_n}{1 - z_n^2}\right)^{1/4}\left\{Ai(\nu_n^{2/3}z_n) - iBi(\nu_n^{2/3}z_n)\right\},$$

$$y^2_H(n) \approx H^2_{\nu_n}(\nu_n z_n) \approx \frac{1}{\nu_n^{1/3}}\left(\frac{4\zeta_n}{1 - z_n^2}\right)^{1/4}\left\{Ai(\nu_n^{2/3}z_n) + iBi(\nu_n^{2/3}z_n)\right\}.$$

Clearly we must have, for $n < 0$, $|n\epsilon| \ll \epsilon^{1/2}$,

$$y_J(n) = \frac{1}{2}\left[y^1_H(n) + y^1_H(n)\right]. \tag{488}$$

In order to match into the exterior region

$$\epsilon^{2/3} \ll |n\epsilon|, \quad n < 0 ,$$

we must match the Hankel type solution $y_H^j(n)$, whose asymptotics are known for $|n\epsilon| \ll \epsilon^{1/2}$, into the exterior region. From (12.2), we know that for $n < 0$, $\epsilon^{2/3} \ll |n\epsilon| \ll \epsilon^{1/2}$,

$$\begin{aligned} y_H^1(n) &\approx \frac{1}{\nu_n}\left(\frac{4\zeta_n}{1-z_n^2}\right)^{1/4} \frac{1}{\pi^{1/2}\nu_n^{1/6}(-\zeta_n)^{-1/4}} \cdot \\ &\cdot \left[\cos\left(\frac{1}{\epsilon}\frac{2\sqrt{2}}{3}(n\epsilon)^{3/2} - \frac{\pi}{4}\right) + i\sin\left(\frac{1}{\epsilon}\frac{2\sqrt{2}}{3}(n\epsilon)^{3/2} - \frac{\pi}{4}\right)\right] \\ &\approx \frac{2\epsilon^{1/2}}{\pi^{1/2}}(-2n\epsilon)^{-1/4} \exp i\left(\frac{1}{\epsilon}\frac{2\sqrt{2}}{3}(n\epsilon)^{3/2} - \frac{\pi}{4}\right) , \end{aligned}$$

and

$$\begin{aligned} y_H^2(n) &\approx \frac{1}{\nu_n}\left(\frac{4\zeta_n}{1-z_n^2}\right)^{1/4} \frac{1}{\pi^{1/2}\nu_n^{1/6}(-\zeta_n)^{-1/4}} \cdot \\ &\cdot \left[\cos\left(\frac{1}{\epsilon}\frac{2\sqrt{2}}{3}(n\epsilon)^{3/2} - \frac{\pi}{4}\right) + i\sin\left(\frac{1}{\epsilon}\frac{2\sqrt{2}}{3}(n\epsilon)^{3/2} - \frac{\pi}{4}\right)\right] \\ &\approx \frac{2\epsilon^{1/2}}{\pi^{1/2}}(-2n\epsilon)^{-1/4} \exp -i\left(\frac{1}{\epsilon}\frac{2\sqrt{2}}{3}(n\epsilon)^{3/2} - \frac{\pi}{4}\right) . \end{aligned}$$

We also know that in this region we have

$$y_{osc}^{\pm} = b_n^{-1/2}(2n\epsilon)^{-1/4} \exp \pm i\left[\frac{1}{\epsilon}\frac{2\sqrt{2}}{3}(n\epsilon)^{3/2}\right] ,$$

and so we can write

$$\begin{aligned} y_H^1(n) &= b_n^{1/2} e^{-\frac{\pi}{4}} \sqrt{\frac{2\epsilon}{\pi}}\, y_{osc}^+ , \\ y_H^2(n) &= b_n^{1/2} e^{-\frac{\pi}{4}} \sqrt{\frac{2\epsilon}{\pi}}\, y_{osc}^- . \end{aligned}$$

This extends the asymptotic knowledge of y_H^j, and we can use this to extend the asymptotic knowledge of $y_J(n)$. We have

$$y_J(n) = \sqrt{\frac{2\epsilon}{\pi}\frac{b_n^{1/2}}{2}}\left[y_{osc}^+(n)e^{-i\frac{\pi}{4}} + y_{osc}^-(n)e^{i\frac{\pi}{4}}\right]$$

$$\approx \sqrt{\frac{2\epsilon}{\pi}\frac{b_n^{1/2}}{2}}\left[1-\left(\frac{\lambda-a_n}{2b_n}\right)^2\right]^{-1/4} \times$$

$$\times \cos\left[\frac{1}{\epsilon}\int_{n\epsilon}^{0}\cos^{-1}\left(\frac{\lambda-a}{2b}\right)\epsilon\frac{b'}{2b}\frac{\left(\frac{\lambda-a}{2b}\right)}{\sqrt{1-\left(\frac{\lambda-a}{2b}\right)^2}} - \frac{\pi}{4}\right],$$

for $n<0$, $\epsilon^{2/3} \ll |n\epsilon|$.

We can now match the exponentially decaying solution, defined for $n\epsilon \gg \epsilon^{2/3}$ all the way through the turning point, and we have

$$y_{exp}^-(n) \approx \left[1-\left(\frac{\lambda-a_n}{2b_n}\right)^2\right]^{-1/4} \times$$

$$\times \exp\left[-\frac{1}{\epsilon}\int_{n\epsilon}^{0}\cosh^{-1}\left(\frac{\lambda-a}{2b}\right) - \epsilon\frac{b'}{2b}\frac{e^{-s_0'}}{\sqrt{1-\left(\frac{\lambda-a}{2b}\right)^2}}\,dy\right],$$

for $\epsilon^{2/3} \ll n\epsilon$,

$$y_{exp}^-(n) = \sqrt{\frac{2\pi}{\epsilon}}\,y_J(n),$$

for $0 < n\epsilon \ll \epsilon^{1/2}$,

$$y_{exp}^-(n) = \sqrt{\frac{2\pi}{\epsilon}}\frac{1}{2}\left[y_H^1(n) + y_H^2(n)\right],$$

for $|n\epsilon| \ll \epsilon^{1/2}$,

$$y_{exp}^-(n) \approx 2\left[1-\left(\frac{\lambda-a_n}{2b_n}\right)^2\right]^{-1/4} \cdot \tag{489}$$

$$\cdot \cos\left[\frac{1}{\epsilon}\int_{n\epsilon}^{0}\cos^{-1}\left(\frac{\lambda-a}{2b}\right) - \epsilon\frac{b'}{2b}\frac{\left(\frac{\lambda-a}{2b}\right)}{\sqrt{1-\left(\frac{\lambda-a}{2b}\right)^2}}dy - \frac{\pi}{4}\right],$$

for $n<0$, $\epsilon^{2/3} \ll |n\epsilon|$. Similarly, we can write

$$y_{exp}^+(n) \approx \left[1-\left(\frac{\lambda-a_n}{2b_n}\right)^2\right]^{-1/4} \times$$

$$\times \exp\left[\frac{1}{\epsilon}\int_{n\epsilon}^{0}\cosh^{-1}\left(\frac{\lambda-a}{2b}\right) - \epsilon\frac{b'}{2b}\frac{e^{-s_0'}}{\sqrt{1-\left(\frac{\lambda-a}{2b}\right)^2}}\,dy\right],$$

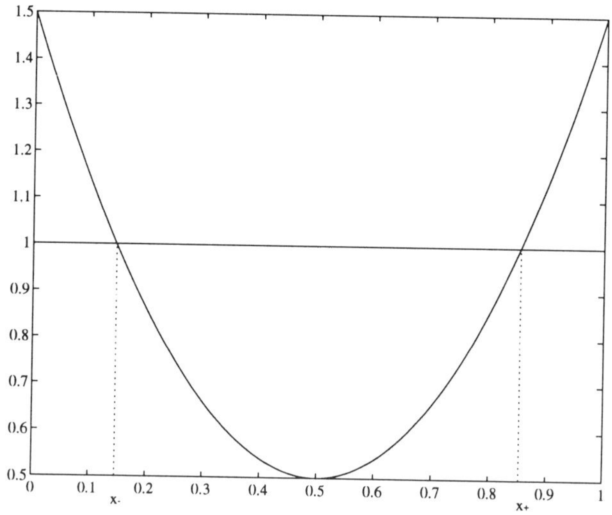

FIGURE 1. Plot of the quantity $\frac{\lambda-a(x)}{2b(x)}$, and the values of x_- and x_+.

for $\epsilon^{2/3} \ll n\epsilon$,

$$y^+_{exp}(n) = \sqrt{\frac{2\pi}{\epsilon}\frac{i}{2}}\left[y^1_H(n) - y^2_H(n)\right] ,$$

for $|n\epsilon| \ll \epsilon^{1/2}$, and

$$y^+_{exp}(n) \approx 2\left[1 - (\frac{\lambda - a_n}{2b_n})^2\right]^{-1/4} \times$$

$$\times \sin\left[\frac{1}{\epsilon}\int_{n\epsilon}^{0} \cos^{-1}\left(\frac{\lambda - a}{2b}\right) - \epsilon\frac{b'}{2b}\frac{\left(\frac{\lambda-a}{2b}\right)}{\sqrt{1-\left(\frac{\lambda-a}{2b}\right)^2}}dy - \frac{\pi}{4}\right] ,$$

for $n < 0$, $\epsilon^{2/3} \ll |n\epsilon|$.

The above asymptotic prescriptions were obtained under the assumption that $\frac{\lambda-a}{2b}$ has slope 1 at the turning point. However, the matching problems, as well as the existence proofs, are exactly the same, for any nonzero slope. The above formulas are valid matching formulas across any linear turning point x_0. One need only replace the value $n = 0$ with $n_0\epsilon = x_0$.

We will now consider the two turning point problem, and obtain as asymptotic eigenvalue condition.

We suppose $\frac{\lambda-a}{2b}$ is as shown in Figure (1), and we will let y^L_n be the exponentially decaying solution for n to the left of the left turning point, and y^R_n the exponentially decaying solution for n to the right of the turning point.

From (489), we have the following representations for n in the oscillatory region $[x_-(\lambda), x_+(\lambda)]$,

$$y_n^L \approx 2\left[1-(\frac{\lambda-a_n}{2b_n})^2\right]^{-1/4} \times$$
$$\times \cos\left[\frac{1}{\epsilon}\int_{x_-}^{n\epsilon} \cos^{-1}\left(\frac{\lambda-a}{2b}\right) - \epsilon\frac{b'}{2b}\frac{\left(\frac{\lambda-a}{2b}\right)}{\sqrt{1-\left(\frac{\lambda-a}{2b}\right)^2}}dy - \frac{\pi}{4}\right],$$

for $\epsilon^{2/3} \ll |n\epsilon - x_-|$, $n\epsilon > x_-$, and

$$y_n^R \approx 2\left[1-(\frac{\lambda-a_n}{2b_n})^2\right]^{-1/4} \times$$
$$\times \cos\left[\frac{1}{\epsilon}\int_{n\epsilon}^{x_+} \cos^{-1}\left(\frac{\lambda-a}{2b}\right) - \epsilon\frac{b'}{2b}\frac{\left(\frac{\lambda-a}{2b}\right)}{\sqrt{1-\left(\frac{\lambda-a}{2b}\right)^2}}dy - \frac{\pi}{4}\right],$$

for $\epsilon^{2/3} \ll |n\epsilon - x_+|$, $n\epsilon < x_+$.

Now we ask when could we have a solution which is exponentially decreasing in both exterior regions? This can happen if the two interior functions are different representations of the same function, give or take a minus sign. This will be the case if

$$\frac{1}{\epsilon}\int_{x_-}^{x_+} \cos^{-1}\left(\frac{\lambda-a}{2b}\right) - \epsilon\frac{b'}{2b}\frac{\left(\frac{\lambda-a}{2b}\right)}{\sqrt{1-\left(\frac{\lambda-a}{2b}\right)^2}}dy - \frac{\pi}{2} = k\pi\ , \quad k = 0, 1, \cdots$$

that is, if

$$(490) \int_{x_-}^{x_+} \cos^{-1}\left(\frac{\lambda-a}{2b}\right) - \epsilon\frac{b'}{2b}\frac{\left(\frac{\lambda-a}{2b}\right)}{\sqrt{1-\left(\frac{\lambda-a}{2b}\right)^2}}dy = \left(k+\frac{1}{2}\right) = \pi\epsilon\ ,$$

for $n = 0, 1, \cdots$. Finally we have obtained a condition which must be asymptotically correct for there to exist a solution to (432) which exhibits exponential decay in both exterior regions. Observe that this must be the case of the solution is to (432) is to be an eigenvector of the tridiagonal matrix arising from the discretization of the functions $a(x)$ and $b(x)$.

A second auxiliary function will be necessary in the analysis of the zero dispersion limit of the Toda lattice is the asymptotic norming constant. If y_n is an eigenfunction of the tridiagonal matrix with entries obtained by the discretization of two functions as described above, and satisfying

$$\sum_{n=1}^{\frac{1}{\epsilon}} y_n^2 = 1\ ,$$

then one can determine that asymptotically y_1 must satisfy

$$y_1 \approx e^{-\frac{1}{\epsilon}\int_0^{x_-(\lambda)} \cosh^{-1}\left(\frac{\lambda-a}{2b}\right)dy}\ ,$$

and so we define the asymptotic norming constant by

$$\theta(\lambda) = \int_0^{x_-(\lambda)} \cosh^{-1}\left(\frac{\lambda - a}{2b}\right) dy \,, \tag{491}$$

and we have

$$y_1 \approx e^{-\frac{\theta(\lambda)}{\epsilon}} \,.$$

Now we present a class of data for which exactly two turning points exist, for λ in an interval, and none exist for λ outside the interval.

Suppose $\alpha(x)$ and $\beta(x)$ are monotone decreasing functions on $(0,1)$, and that $\beta(0) = \alpha(0) = 1$, $\beta(1) = \alpha(1) = -1$. Then

$$a(x) = \frac{\beta(x) + \alpha(x)}{2} \,,$$

$$b(x) = \frac{\beta(x) - \alpha(x)}{2}$$

possess precisely two turning points, where $\left|\frac{\lambda - a}{2b}\right| = 1$, for each value of $\lambda \in (-1,1)$.

Indeed, the equation $\left|\frac{\lambda-a}{2b}\right| = 1$ is equivalent to

$$\lambda = \alpha(x)$$

or

$$\lambda = \beta(x) \,.$$

For each $\lambda \in (-1,1)$, there is precisely one intersection of the curve $\alpha(x)$ with λ, and also precisely one intersection of $\beta(x)$ with λ. Furthermore, for $\lambda \in \mathbf{R}\backslash[-1,1]$, there are no intersections , and hence no turning points.

We leave the reader to verify that the same phenomenon is true if we take data for which $\beta(x) = a(x) + 2b(x)$ has one nondegenerate critical point, and $\alpha(x) = a(x) - 2b(x)$ is strictly monotone decreasing, and $\alpha(0) = \beta(0) = 0$, $\alpha(1) = \beta(1) = -1$. Note that in this case, as $\beta(x)$ is increasing, for x near 0, that for each $\lambda < 0$, the lower turning point $x_-(\lambda)$ used in the definition (491) is an intersection with the curve $\alpha(x)$, while for $\lambda > 0$, the turning point is defined by an intersection with $\beta(x)$.

Note that the choice of the value of $\alpha(0) = \beta(0)$ and $\alpha(1) = \beta(1)$ are not restricted, we chose 0 and -1 for simplicity.

There is a second class of eigenfunctions, which can be observed numerically. These exhibit, in contrast to the above "bound state" eigenfunctions, a type of "potential barrier" behavior. Two turning points exists, and in the interior region the potential $\frac{\lambda-a}{2b}$ is greater than 1 in magnitude. For simplicity, suppose that $b(x) = equiv\frac{1}{2}$, and that $a(x)$ is symmetric about $x = \frac{1}{2}$, and possesses one minimum at $x = \frac{1}{2}$.

The occurrence of these eigenfunctions is always in pairs, and two eigenfunctions in a pair are exponentially close to each other.

To obtain a spectral density for these eigenvalues, use the known nature of eigenvectors of tridiagonal matrices. They solve the associated difference equation, and in addition, they vanish at endpoints. So we take a solution which vanishes at one endpoint, patch it through to the other endpoint, and obtain a condition for the solution to vanish at the other endpoint. This procedure will give us a nonlinear equation for λ, similar to (490), which we interpret as an approximate condition for λ to be an eigenvalue of the original matrix.

The best we will be able to do is obtain one equation for each pair of eigenvectors.

Suppose y_n is the solution which vanishes at $n = 0$. From the analysis presented, we have the following representation, asymptotically,

$$y_n \approx \left[1 - (\frac{\lambda - a_n}{2b_n})^2\right]^{-1/4} \times$$
$$\times \sin\left[\frac{1}{\epsilon}\int_0^{n\epsilon} \cos^{-1}\left(\frac{\lambda - a}{2b}\right) - \epsilon\frac{b'}{2b}\frac{\left(\frac{\lambda-a}{2b}\right)}{\sqrt{1 - \left(\frac{\lambda-a}{2b}\right)^2}}dy - \frac{\pi}{4}\right].$$

Patching across the turning point $x_-(\lambda)$, we obtain a representation for y_n, and by patching this solution across the next turning point, $x_+(\lambda)$, we obtain the representation

$$y_n \approx \left[1 - (\frac{\lambda - a_n}{2b_n})^2\right]^{-1/4} \left[\cdot\right.$$
$$Ae^{\frac{1}{\epsilon}\int_{x_-(\lambda)}^{x_+(\lambda)} \cosh^{-1}\left(\frac{\lambda-a}{2b}\right)dy} \times$$
$$\times \cos\left(\frac{1}{\epsilon}\int_{x_+(\lambda)}^{n\epsilon} \cos^{-1}\left(\frac{\lambda - a}{2b}\right) - \epsilon\frac{b'}{2b}\frac{\left(\frac{\lambda-a}{2b}\right)}{\sqrt{1 - \left(\frac{\lambda-a}{2b}\right)^2}}dy - \frac{\pi}{4}\right) +$$
$$Be^{-\frac{1}{\epsilon}\int_{x_-(\lambda)}^{x_+(\lambda)} \cosh^{-1}\left(\frac{\lambda-a}{2b}\right)dy} \times$$
$$\left.\times \sin\left(\frac{1}{\epsilon}\int_{x_+(\lambda)}^{n\epsilon} \cos^{-1}\left(\frac{\lambda - a}{2b}\right) - \epsilon\frac{b'}{2b}\frac{\left(\frac{\lambda-a}{2b}\right)}{\sqrt{1 - \left(\frac{\lambda-a}{2b}\right)^2}}dy - \frac{\pi}{4}\right)\right],$$

where

$$A = -\left(\sin \frac{1}{\epsilon} \int_0^{x-} \cos^{-1}\left(\frac{\lambda - a}{2b}\right) - \epsilon \frac{b'}{2b} \frac{\left(\frac{\lambda-a}{2b}\right)}{\sqrt{1 - \left(\frac{\lambda-a}{2b}\right)^2}} dy \right. \tag{492}$$

$$\left. + \cos \frac{1}{\epsilon} \int_0^{x-} \cos^{-1}\left(\frac{\lambda - a}{2b}\right) - \epsilon \frac{b'}{2b} \frac{\left(\frac{\lambda-a}{2b}\right)}{\sqrt{1 - \left(\frac{\lambda-a}{2b}\right)^2}} dy \right) \frac{1}{\sqrt{2}}, \tag{493}$$

$$B = \left(\sin \frac{1}{\epsilon} \int_0^{x-} \cos^{-1}\left(\frac{\lambda - a}{2b}\right) - \epsilon \frac{b'}{2b} \frac{\left(\frac{\lambda-a}{2b}\right)}{\sqrt{1 - \left(\frac{\lambda-a}{2b}\right)^2}} dy \right.$$

$$\left. - \cos \frac{1}{\epsilon} \int_0^{x-} \cos^{-1}\left(\frac{\lambda - a}{2b}\right) - \epsilon \frac{b'}{2b} \frac{\left(\frac{\lambda-a}{2b}\right)}{\sqrt{1 - \left(\frac{\lambda-a}{2b}\right)^2}} dy \right) \frac{1}{\sqrt{2}}. \tag{494}$$

Thus, at $n = \frac{1}{\epsilon}$, we have

$$y_{\frac{1}{\epsilon}} \approx \left[1 - \left(\frac{\lambda - a_{\frac{1}{\epsilon}}}{2b_{\frac{1}{\epsilon}}}\right)^2\right]^{-1/4} \left[A e^{\frac{1}{\epsilon} \int_{x_-(\lambda)}^{x_+(\lambda)} \cosh^{-1}\left(\frac{\lambda-a}{2b}\right) dy} \times \right.$$

$$\times \cos\left(\frac{1}{\epsilon} \int_{x_+(\lambda)}^{1} \cos^{-1}\left(\frac{\lambda - a}{2b}\right) - \epsilon \frac{b'}{2b} \frac{\left(\frac{\lambda-a}{2b}\right)}{\sqrt{1 - \left(\frac{\lambda-a}{2b}\right)^2}} dy - \frac{\pi}{4}\right) +$$

$$+ B e^{-\frac{1}{\epsilon} \int_{x_-(\lambda)}^{x_+(\lambda)} \cosh^{-1}\left(\frac{\lambda-a}{2b}\right) dy} \times$$

$$\left. \times \sin\left(\frac{1}{\epsilon} \int_{n\epsilon}^{1} \cos^{-1}\left(\frac{\lambda - a}{2b}\right) - \epsilon \frac{b'}{2b} \frac{\left(\frac{\lambda-a}{2b}\right)}{\sqrt{1 - \left(\frac{\lambda-a}{2b}\right)^2}} dy - \frac{\pi}{4}\right)\right].$$

If we expect λ to be an eigenvalue, the term

$$\left[1 - \left(\frac{\lambda - a_n}{2b_n}\right)^2\right]^{-1/4} \left[A e^{\frac{1}{\epsilon} \int_{x_-(\lambda)}^{x_+(\lambda)} \cosh^{-1}\left(\frac{\lambda-a}{2b}\right) dy} \times \right.$$

$$\left. \times \cos\left(\frac{1}{\epsilon} \int_{x_+(\lambda)}^{1} \cos^{-1}\left(\frac{\lambda - a}{2b}\right) - \epsilon \frac{b'}{2b} \frac{\left(\frac{\lambda-a}{2b}\right)}{\sqrt{1 - \left(\frac{\lambda-a}{2b}\right)^2}} dy - \frac{\pi}{4}\right)\right]$$

Should not appear, and we obtain the condition

$$A \cos\left(\frac{1}{\epsilon} \int_{x_+(\lambda)}^{1} \cos^{-1}\left(\frac{\lambda - a}{2b}\right) - \epsilon \frac{b'}{2b} \frac{\left(\frac{\lambda-a}{2b}\right)}{\sqrt{1 - \left(\frac{\lambda-a}{2b}\right)^2}} dy - \frac{\pi}{4}\right) = 0. \tag{495}$$

Now as we have supposed that the functions $a(x)$ and $b(x)$ are symmetric about $x = \frac{1}{2}$, we conclude that the integrals over the oscillatory regions

$(0, x_-(\lambda))$, and $(x_+(\lambda), 1)$ are interchangeable, and we obtain the simplification

$$y_{\frac{1}{\epsilon}} \approx \left[1 - \left(\frac{\lambda - a_{\frac{1}{\epsilon}}}{2b_{\frac{1}{\epsilon}}}\right)^2\right]^{-1/4} \left[- A^2 e^{\frac{1}{\epsilon}\int_{x_-(\lambda)}^{x_+(\lambda)} \cosh^{-1}\left(\frac{\lambda-a}{2b}\right)dy} + \right.$$

$$\left. + B^2 e^{-\frac{1}{\epsilon}\int_{x_-(\lambda)}^{x_+(\lambda)} \cosh^{-1}\left(\frac{\lambda-a}{2b}\right)dy}\right] .$$

The asymptotic eigenvalue condition becomes

$$(496)\quad y_{\frac{1}{\epsilon}} \approx -A^2 e^{\frac{1}{\epsilon}\int_{x_-(\lambda)}^{x_+(\lambda)} \cosh^{-1}\left(\frac{\lambda-a}{2b}\right)dy} + B^2 e^{-\frac{1}{\epsilon}\int_{x_-(\lambda)}^{x_+(\lambda)} \cosh^{-1}\left(\frac{\lambda-a}{2b}\right)dy} = 0 .$$

Now, if we suppose that $A = 0$, then we see from the above asymptotic representation that $y_{\frac{1}{\epsilon}} > 0$. If we perturb λ a little bit, then A^2 will no longer vanish, and the exponentially growing term will cause the first term in (496) to dominate, and so (asymptotically) we have $y_{\frac{1}{\epsilon}} < 0$. This is true independent of which way we perturb λ, and so we surmise that there will be two eigenvalues near each root of the coefficient

$$A = \cos\left(\frac{1}{\epsilon}\int_{x_+(\lambda)}^{1} \cos^{-1}\left(\frac{\lambda - a}{2b}\right) - \epsilon\frac{b'}{2b}\frac{\left(\frac{\lambda-a}{2b}\right)}{\sqrt{1-\left(\frac{\lambda-a}{2b}\right)^2}}dy - \frac{\pi}{4}\right) .$$

Thus a spectral density function for these exponentially close pairs of "barrier type" eigenfunctions is,

$$(497)\quad \int_{x_+(\lambda)}^{1} \cos^{-1}\left(\frac{\lambda - a}{2b}\right) - \epsilon\frac{b'}{2b}\frac{\left(\frac{\lambda-a}{2b}\right)}{\sqrt{1-\left(\frac{\lambda-a}{2b}\right)^2}}dy = \left(k + \frac{3}{4}\right)\pi\epsilon .$$

Finally we remark that the phenomenon of oscillations for x in a neighborhood of $x = 0$, $x = 1$, arises only if $b(x)$ does not vanish at $x = 0$. Indeed, if $b(0) = b(1) = 0$, then clearly $\left|\frac{\lambda-a}{2}\right| \to \infty$ as $x \to 0$, for $\lambda \neq 0$, and so no oscillations could exist for $x \to 0$. For the case of the continuum limit of the Toda lattice, we will take data for which $b(0) = b(1) = 0$.

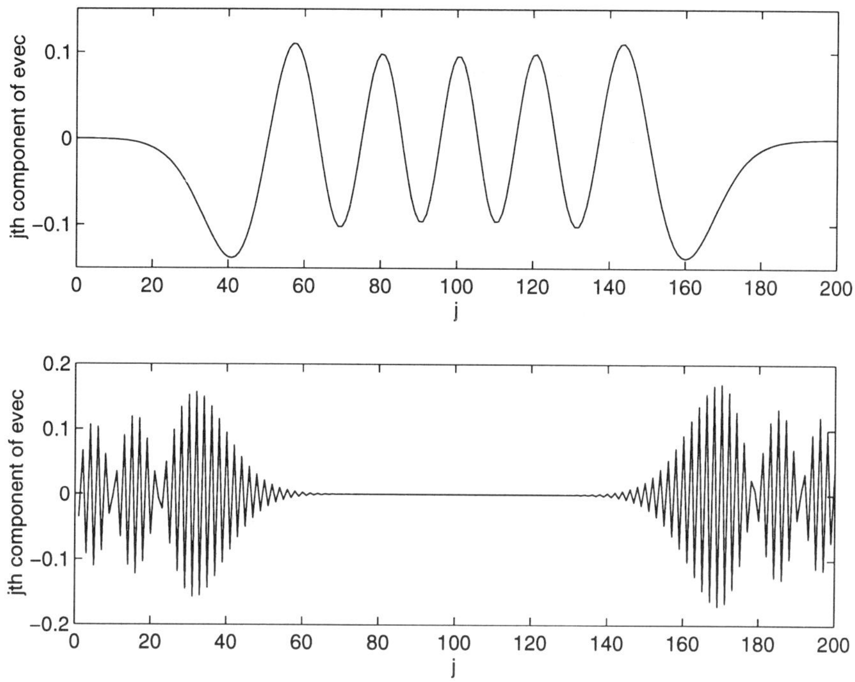

FIGURE 2. Eigenvectors of the matrix $L(0)$, obtained from the functions $a^0(x) = \frac{1}{10}\sin \pi x$, $b(x) = \frac{1}{2}$. The top figure shows an eigenvector which exhibits interior oscillations, surrounded by regions of exopnential decay. The bottom figure shows an eigenvector which exhibits exponential behavior, with oscillations in the wings. $L(0)$ is a 200×200 matrix.

1.09753712306951	1.09753290000000
1.09262826246980	1.09262400000000
1.08775329044608	1.08774900000000
1.08291241556924	1.08290810000000
1.07810585558226	1.07810150000000
1.07333383790758	1.07332950000000
1.06859660019411	1.06859220000000
1.06389439090704	1.06389000000000
1.05922746995580	1.05922300000000
1.05459610929349	1.05459160000000
1.05000059301636	1.04999610000000
1.04544121427641	1.04543660000000
1.04091825609947	1.04091370000000
1.03643190397386	1.03642750000000
1.03198191634714	1.03197850000000
1.02756659360069	1.02756690000000
1.02318017432573	1.02319320000000
1.01880783845678	1.01885780000000

FIGURE 3. For the same matrix $L(0)$, the 18 largest eigenvalues, matched with the predictions obtained by solving the asymptotic eigenvalue condition. The left column contains the exact eigenvalues, computed using Matlab, and the right contains the numerical solution of the approximate eigenvalue condition.

CHAPTER 13

Appendix II

Claim 1.

$$\int_{\beta_0(x,t)}^{\lambda} H\psi_0^*(\mu;\ x,t) - \theta'(\mu,t)d\mu = -\int_x^{x_2(t)} \cosh^{-1}\left(\frac{\lambda - \frac{\beta_0+\alpha_0}{2}}{\frac{\beta_0-\alpha_0}{2}}\right) dy$$

$$+\int_{x_-^1(\lambda)}^{x_2(t)} \cosh^{-1}\left(\frac{\lambda - \frac{\beta_1+\alpha_1}{2}}{\frac{\beta_1-\alpha_1}{2}}\right) dy ,$$

for all $\lambda \in (\beta_0(x_2(t),t) , \beta_1(x_1(t),t))$.

Proof

Familiar computations yield the following representations:

$$H\psi_0^*(\lambda\ ;\ x,t) - \theta'(\lambda,t) =$$

$$-\int_x^{x_-^0(\lambda,t)} \frac{1}{R_0(\lambda,y)^{1/2}}dy , \quad \lambda \in (\beta_0(x,t) , \beta_0(x_2(t),t))$$

$$(498) = -\int_x^{x_2(t)} \frac{1}{R_0(\lambda,y)^{1/2}}dy$$

$$(499) + \int_{x_-^1(\lambda)}^{x_2(t)} \frac{1}{R_1(\lambda,y)^{1/2}}dy , \quad \lambda \in (\beta_0(x_2(t),t) , \beta_1(x_1(t),t)) .$$

Straightforward algebra and integration by parts shows

$$\int_{\beta_0(x,t)}^{\lambda} H\psi_0^*(\mu;\ x,t) - \theta'(\mu,t)d\mu = -\int_x^{x_2(t)}\int_{\beta_0(y,t)}^{\lambda} \frac{1}{R_0(\lambda,y)^{1/2}}d\lambda dy$$

$$+\int_{x_-^1(\lambda)}^{x_2(t)}\int_{\beta_1(y,t)}^{\lambda} \frac{1}{R_1(\lambda,y)^{1/2}}d\lambda dy$$

$$= -\int_x^{x_2(t)} \cosh^{-1}\left(\frac{\lambda - \frac{\beta_0+\alpha_0}{2}}{\frac{\beta_0-\alpha_0}{2}}\right) dy + \int_{x_-^1(\lambda)}^{x_2(t)} \cosh^{-1}\left(\frac{\lambda - \frac{\beta_1+\alpha_1}{2}}{\frac{\beta_1-\alpha_1}{2}}\right) dy .$$

Claim 2 For t small,

$$\lambda \in (\beta_1(x_1(t),t) , \beta_2(x_2(t),t)) \Rightarrow \frac{d}{d\lambda}\left(H\psi_0^*(\lambda;\ x,t) - \theta'(\lambda,t)\right) < 0 .$$

Proof

For λ in the allowed region, we have, as in (498) above, the identity

$$
\begin{aligned}
H\psi_0^*(\lambda;\, x,t) - \theta'(\lambda,t) \quad &= \quad -\int_x^{x_2(t)} \frac{1}{R_0(\lambda,y)^{1/2}}dy + \int_{x_1(t)}^{x_2(t)} \frac{1}{R_1(\lambda,y)^{1/2}}dy \\
(500) \qquad\qquad &\qquad - \int_{x_1(t)}^{x_-^2(\lambda)} \frac{1}{R_2(\lambda,y)^{1/2}}dy \ .
\end{aligned}
$$

We label the integrals I, II, and III, in the order in which they appear. Differentiating the first two, we obtain

$$
\frac{\partial}{\partial\lambda}(I+II) = \int_x^{x_2(t)} \frac{\lambda - \frac{\beta_0+\alpha_0}{2}}{R_0^{3/2}} - \frac{\lambda - \frac{\beta_1+\alpha_1}{2}}{R_1^{3/2}}dy
$$

$$
- \int_{x_1(t)}^{x} \frac{\lambda - \frac{\beta_1+\alpha_1}{2}}{R_1^{3/2}}dy \ .
$$

Now the last term is negative, and familiar calculations (see, for example, the proof of Lemma 7.1 in Chapter 7), using the general shape of the curves, shows that the integrand in the first term is negative as well.

To prove that $\frac{d}{d\lambda}III \ < \ 0$, we set $z = \lambda - \beta_2(y)$ in III and obtain

$$
(501) \ III = -\int_0^{\lambda-\beta_2(x_1(t),t)} \frac{1}{z^{1/2}} \frac{1}{(\lambda - \alpha_2 \circ y(\lambda - z))^{1/2}} \frac{1}{\beta_2' \circ y(\lambda - z)}dz \ .
$$

Using the familiar relationship between β_x and T_β,

$$
\beta_x \quad = \quad \frac{\pi}{(\beta-\alpha)T_\beta} \ ,
$$

we can rewrite 501 as

$$
III = -\int_0^{\lambda-\beta_2(x_1(t),t)} \frac{1}{z^{1/2}} \frac{(\beta-\alpha(\beta))T_\beta}{(\lambda-\alpha(\beta))^{1/2}} \ dz \ ,
$$

where $\alpha(\beta)$ denotes the function $\alpha(\beta,t)$ which solves $T(\alpha,\beta) \ = \ \pi t$, and $\beta = \beta(z)$ is understood to be given by $\beta(z) = \lambda - z$.

As T_β vanishes at $x_1(t)$ by definition, we observe that differentiating III with respect to λ only involves differentiation under the integral sign. We have

$$
\begin{aligned}
(502) \qquad \frac{\partial}{\partial\lambda}III = -\int_0^{\lambda-\beta_2(x_1(t),t)} &\frac{1}{z^{1/2}} \frac{1-\alpha_\beta}{(\lambda-\alpha(\beta))^{1/2}}T_\beta + \\
-\frac{1}{z^{1/2}} \frac{\beta-\alpha}{(\lambda-\alpha(\beta))^{3/2}} \frac{1-\alpha_\beta}{2}T_\beta &+ \frac{1}{z^{1/2}} \frac{\beta-\alpha(\beta)}{(\lambda-\alpha(\beta))^{1/2}}T_{\beta\beta}\, dz.
\end{aligned}
$$

Now we use the fact that at $t \ = \ 0$, we have, in a vicinity of β_s, $T_\beta \sim (\beta-\beta_s)^2$, $T_{\beta\beta} \sim (\beta-\beta_s)$, and hence

$$
T_\beta << T_{\beta\beta} \ ,
$$

and we see that for t small, the integrand in (502) has the same sign as the term

$$-\frac{\beta-\alpha}{[z(\lambda-\alpha_2\circ y(\lambda-z))]^{1/2}}T_{\beta\beta}\ ,$$

which is negative.

This completes the proof of the claim.

Bibliography

[1] R. W. Brockett. Dynamical Systems that Sort Lists, Solve linear Programming Problems and Diagonalize Symmetric Matrices. *Proceedings of the IEEE Conference on Decision and Control*, 1988.

[2] R. W. Brockett and A. M. Bloch. Sorting With the Dispersionless Limit of the Toda Lattice. *preprint*, 1990.

[3] V. Buslaev and A. Fedotov. Complex WKB Method for Harper's Equation. *Preprint*, 1992/93.

[4] O. Costin. *Preprint*, 1994.

[5] P. Deift, S. Venakides, and X. Zhou. private communication.

[6] H. Flaschka. On the Toda Lattice II. *Prog. Theor. Phys.*, 51(3):703–716, 1974.

[7] H. Flaschka. The Toda Lattice. I. Existence of Integrals. *Phys. Rev.*, B9:1924–1925, 1974.

[8] J. Geronimo and D. Smith. WKB Analysis of Second Order Difference Equations and Applications. *Journal of Approximation Theory*, 69(3):269–301, 1992.

[9] J. Goodman and P. Lax. On Dispersive Difference Schemes I. *Comm. Pure. Appl. Math.*, 41:591–613, 1988.

[10] T. Hou and P. Lax. Dispersive Approximations in Fluid Dynamics. *Comm. Pure. Appl. Math.*, 44:1–40, 1991.

[11] Y. Kodama. Solutions of the dispersionless Toda equation. *Physics Letters A*, 147(8,9):477–482, 1990.

[12] P. Koosis. *Introduction to H^p Spaces.* LMS Lecture Notes 40, 1980.

[13] P. D. Lax and C. D. Levermore. The Small Dispersion Limit of the Korteweg -de Vries Equation I, II, III. *Comm. Pure Appl. Math.*, 36:253–290, 571–593, 809–829, 1983.

[14] S. V. Manakov. Complete Integrability and Stochastization of Discrete Dynamical Systems. *Zh. Exp. Teor. Fiz.*, 67:543–555, 1974.

[15] K. D. T.-R. McLaughlin. A Continuum Limit of the Toda Lattice. *Ph. D. Dissertation, New York University*, 1994.

[16] J. Moser. *Finitely Many Mass Points on the Line Under the Influence of an Exponential Potential - An Integrable System.* Springer, New York, 1975.

[17] F. Olver. *Introduction to Asymptotics and Special Functions.* Academic Press, New York, 1974.

[18] G. Szego. *Orthogonal Polynomials.* AMS Colloquium Publications, v. 23, 1939.

[19] F. Tian. Oscillations of the Zero Dispersion Limit of the Korteweg de Vries Equation , (preprint). *Comm. Pure Appl. Math.*, 1993.

[20] S. Tsarev. On Poisson Brackets and One - Dimensional Systems of Hydrodynamic Type. *Sov. Math. Doklady*, 31:488–491, 1985.

[21] S. Venakides. The Zero Dispersion Limit of the Korteweg - de Vries Equation With Periodic Initial Data. *AMS Trans.*, 301:189–225, 1987.

[22] S. Venakides, P. Deift, and R. Oba. The Toda Shock Problem. *Comm. Pure Appl. Math.*, 44:1171–1242, 1991.

[23] G. Whitham. *Linear and Nonlinear Waves.* J. Wiley, 1974.

Editorial Information

To be published in the *Memoirs*, a paper must be correct, new, nontrivial, and significant. Further, it must be well written and of interest to a substantial number of mathematicians. Piecemeal results, such as an inconclusive step toward an unproved major theorem or a minor variation on a known result, are in general not acceptable for publication. *Transactions* Editors shall solicit and encourage publication of worthy papers. Papers appearing in *Memoirs* are generally longer than those appearing in *Transactions* with which it shares an editorial committee.

As of September 30, 1997, the backlog for this journal was approximately 8 volumes. This estimate is the result of dividing the number of manuscripts for this journal in the Providence office that have not yet gone to the printer on the above date by the average number of monographs per volume over the previous twelve months, reduced by the number of issues published in four months (the time necessary for preparing an issue for the printer). (There are 6 volumes per year, each containing at least 4 numbers.)

A Copyright Transfer Agreement is required before a paper will be published in this journal. By submitting a paper to this journal, authors certify that the manuscript has not been submitted to nor is it under consideration for publication by another journal, conference proceedings, or similar publication.

Information for Authors and Editors

Memoirs are printed by photo-offset from camera copy fully prepared by the author. This means that the finished book will look exactly like the copy submitted.

The paper must contain a *descriptive title* and an *abstract* that summarizes the article in language suitable for workers in the general field (algebra, analysis, etc.). The *descriptive title* should be short, but informative; useless or vague phrases such as "some remarks about" or "concerning" should be avoided. The *abstract* should be at least one complete sentence, and at most 300 words. Included with the footnotes to the paper, there should be the 1991 *Mathematics Subject Classification* representing the primary and secondary subjects of the article. This may be followed by a list of *key words and phrases* describing the subject matter of the article and taken from it. A list of the numbers may be found in the annual index of *Mathematical Reviews*, published with the December issue starting in 1990, as well as from the electronic service e-MATH [**telnet e-MATH.ams.org** (or **telnet 130.44.1.100**). Login and password are **e-math**]. For journal abbreviations used in bibliographies, see the list of serials in the latest *Mathematical Reviews* annual index. When the manuscript is submitted, authors should supply the editor with electronic addresses if available. These will be printed after the postal address at the end of each article.

Electronically prepared papers. The AMS encourages submission of electronically prepared papers in AMS-TEX or AMS-LATEX. The Society has prepared author packages for each AMS publication. Author packages include instructions for preparing electronic papers, the *AMS Author Handbook*, samples, and a style file that generates the particular design specifications of that publication series for both AMS-TEX and AMS-LATEX.

Authors with FTP access may retrieve an author package from the Society's Internet node `e-MATH.ams.org` (130.44.1.100). For those without FTP access, the author package can be obtained free of charge by sending e-mail to `pub@ams.org` (Internet) or from the Publication Division, American Mathematical Society, P.O. Box 6248, Providence, RI 02940-6248. When requesting an author package, please specify AMS-TEX or AMS-LATEX, Macintosh or IBM (3.5) format, and the publication in which your paper will appear. Please be sure to include your complete mailing address.

Submission of electronic files. At the time of submission, the source file(s) should be sent to the Providence office (this includes any TEX source file, any graphics files, and the DVI or PostScript file).

Before sending the source file, be sure you have proofread your paper carefully. The files you send must be the EXACT files used to generate the proof copy that was accepted for publication. For all publications, authors are required to send a printed copy of their paper, which exactly matches the copy approved for publication, along with any graphics that will appear in the paper.

TEX files may be submitted by email, FTP, or on diskette. The DVI file(s) and PostScript files should be submitted only by FTP or on diskette unless they are encoded properly to submit through e-mail. (DVI files are binary and PostScript files tend to be very large.)

Files sent by electronic mail should be addressed to the Internet address `pub-submit@ams.org`. The subject line of the message should include the publication code to identify it as a Memoir. TEX source files, DVI files, and PostScript files can be transferred over the Internet by FTP to the Internet node `e-math.ams.org` (130.44.1.100).

Electronic graphics. Figures may be submitted to the AMS in an electronic format. The AMS recommends that graphics created electronically be saved in Encapsulated PostScript (EPS) format. This includes graphics originated via a graphics application as well as scanned photographs or other computer-generated images.

If the graphics package used does not support EPS output, the graphics file should be saved in one of the standard graphics formats—such as TIFF, PICT, GIF, etc.—rather than in an application-dependent format. Graphics files submitted in an application-dependent format are not likely to be used. No matter what method was used to produce the graphic, it is necessary to provide a paper copy to the AMS.

Authors using graphics packages for the creation of electronic art should also avoid the use of any lines thinner than 0.5 points in width. Many graphics packages allow the user to specify a "hairline" for a very thin line. Hairlines often look acceptable when proofed on a typical laser printer. However, when produced on a high-resolution laser imagesetter, hairlines become nearly invisible and will be lost entirely in the final printing process.

Screens should be set to values between 15% and 85%. Screens which fall outside of this range are too light or too dark to print correctly.

Any inquiries concerning a paper that has been accepted for publication should be sent directly to the Editorial Department, American Mathematical Society, P. O. Box 6248, Providence, RI 02940-6248.

Editors

This journal is designed particularly for long research papers (and groups of cognate papers) in pure and applied mathematics. Papers intended for publication in the *Memoirs* should be addressed to one of the following editors:

Ordinary differential equations, partial differential equations, and applied mathematics to JOHN MALLET-PARET, Division of Applied Mathematics, Brown University, Providence, RI 02912-9000; electronic mail: jmp@cfm.brown.edu.

Harmonic analysis, representation theory, and Lie theory to ROBERT J. STANTON, Department of Mathematics, The Ohio State University, 231 West 18th Avenue, Columbus, OH 43210-1174; electronic mail: stanton@math.ohio-state.edu.

Ergodic theory, dynamical systems, and abstract analysis to DANIEL J. RUDOLPH, Department of Mathematics, University of Maryland, College Park, MD 20742; e-mail: djr@math.umd.edu.

Real and harmonic analysis and geometric partial differential equations to WILLIAM BECKNER, Department of Mathematics, University of Texas at Austin, Austin, TX 78712; e-mail: beckner@math.utexas.edu.

Algebra and algebraic geometry to EFIM ZELMANOV, Department of Mathematics, Yale University, 10 Hillhouse Avenue, New Haven, CT 06520-8283; e-mail: zelmanov@math.yale.edu

Algebraic topology and cohomology of groups to STEWART PRIDDY, Department of Mathematics, Northwestern University, 2033 Sheridan Road, Evanston, IL 60208-2730; e-mail: s_priddy@math.nwu.edu.

Global analysis and differential geometry to CHUU-LIAN TERNG, Department of Mathematics, Northeastern University, Huntington Avenue, Boston, MA 02115-5096; e-mail: terng@neu.edu.

Probability and statistics to RODRIGO BAÑUELOS, Department of Mathematics, Purdue University, West Lafayette, IN 47907-1968; e-mail: banuelos@math.purdue.edu.

Combinatorics and Lie theory to PHILIP J. HANLON, Department of Mathematics, University of Michigan, Ann Arbor, MI 48109-1003; e-mail: hanlon@math.lsa.umich.edu.

Logic and universal algebra to THEODORE SLAMAN, Department of Mathematics, University of California at Berkeley, Berkeley, CA 94720; e-mail: slaman@math.berkeley.edu.

Number theory and arithmetic algebraic geometry to ALICE SILVERBERG, Department of Mathematics, Harvard University, 1 Oxford St.–Science Center, Cambridge, MA 02138; e-mail: silver@math.ohio-state.edu.

Complex analysis and complex geometry to DANIEL M. BURNS, Department of Mathematics, University of Michigan, Ann Arbor, MI 48109-1003; e-mail: dburns@umich.edu.

Algebraic geometry and commutative algebra to LAWRENCE EIN, Department of Mathematics, University of Illinois, 851 S. Morgan (M/C 249), Chicago, IL 60607-7045; email: lawrence.man.ein@uic.edu.

Geometric topology to JOHN LUECKE, Department of Mathematics, University of Texas at Austin, Austin, TX 78712; e-mail: luecke@math.utexas.edu.

All other communications to the editors should be addressed to the Managing Editor, PETER SHALEN, Department of Mathematics, University of Illinois, 851 S. Morgan (M/C 249), Chicago, IL 60607-7045; e-mail: shalen@math.uic.edu.

Selected Titles in This Series

(Continued from the front of this publication)

595 **Eldar Straume,** Compact connected Lie transformation groups on spheres with low cohomogeneity, II, 1997

594 **Solomon Friedberg and Hervé Jacquet,** The fundamental lemma for the Shalika subgroup of $GL(4)$, 1996

593 **Ajit Iqbal Singh,** Completely positive hypergroup actions, 1996

592 **P. Kirk and E. Klassen,** Analytic deformations of the spectrum of a family of Dirac operators on an odd-dimensional manifold with boundary, 1996

591 **Edward Cline, Brian Parshall, and Leonard Scott,** Stratifying endomorphism algebras, 1996

590 **Chris Jantzen,** Degenerate principal series for symplectic and odd-orthogonal groups, 1996

589 **James Damon,** Higher multiplicities and almost free divisors and complete intersections, 1996

588 **Dihua Jiang,** Degree 16 Standard L-function of $GSp(2) \times GSp(2)$, 1996

587 **Stéphane Jaffard and Yves Meyer,** Wavelet methods for pointwise regularity and local oscillations of functions, 1996

586 **Siegfried Echterhoff,** Crossed products with continuous trace, 1996

585 **Gilles Pisier,** The operator Hilbert space OH, complex interpolation and tensor norms, 1996

584 **Wayne W. Barrett, Charles R. Johnson, and Raphael Loewy,** The real positive definite completion problem: Cycle completability, 1996

583 **Jin Nakagawa,** Orders of a quartic field, 1996

582 **Darryl McCollough and Andy Miller,** Symmetric automorphisms of free products, 1996

581 **Martin U. Schmidt,** Integrable systems and Riemann surfaces of infinite genus, 1996

580 **Martin W. Liebeck and Gary M. Seitz,** Reductive subgroups of exceptional algebraic groups, 1996

579 **Samuel Kaplan,** Lebesgue theory in the bidual of $C(X)$, 1996

578 **Ale Jan Homburg,** Global aspects of homoclinic bifurcations of vector fields, 1996

577 **Freddy Dumortier and Robert Roussarie,** Canard cycles and center manifolds, 1996

576 **Grahame Bennett,** Factorizing the classical inequalities, 1996

575 **Dieter Heppel, Idun Reiten, and Sverre O. Smalø,** Tilting in Abelian categories and quasitilted algebras, 1996

574 **Michael Field,** Symmetry breaking for compact Lie groups, 1996

573 **Wayne Aitken,** An arithmetic Riemann-Roch theorem for singular arithmetic surfaces, 1996

572 **Ole H. Hald and Joyce R. McLaughlin,** Inverse nodal problems: Finding the potential from nodal lines, 1996

571 **Henry L. Kurland,** Intersection pairings on Conley indices, 1996

570 **Bernold Fiedler and Jürgen Scheurle,** Discretization of homoclinic orbits, rapid forcing and "invisible" chaos, 1996

569 **Eldar Straume,** Compact connected Lie transformation groups on spheres with low cohomogeneity, I, 1996

568 **Raúl E. Curto and Lawrence A. Fialkow,** Solution of the truncated complex moment problem for flat data, 1996

567 **Ran Levi,** On finite groups and homotopy theory, 1995

566 **Neil Robertson, Paul Seymour, and Robin Thomas,** Excluding infinite clique minors, 1995

565 **Huaxin Lin and N. Christopher Phillips,** Classification of direct limits of even Cuntz-circle algebras, 1995

564 **Wensheng Liu and Héctor J. Sussmann,** Shortest paths for sub-Riemannian metrics on rank-two distributions, 1995

(See the AMS catalog for earlier titles)